Medical Big Data and Internet of Medical Things

Advances, Challenges and Applications

Edited by
Aboul Ella Hassanien
Nilanjan Dey
Surekha Borra

CRC Press is an imprint of the
Taylor & Francis Group, an **Informa** business

CRC Press
Taylor & Francis Group
6000 Broken Sound Parkway NW, Suite 300
Boca Raton, FL 33487-2742

First issued in paperback 2023

Publisher's Note
The publisher has gone to great lengths to ensure the quality of this reprint but points out that some imperfections in the original copies may be apparent.

Library of Congress Cataloging-in-Publication Data

Names: Hassanien, Aboul Ella, editor. | Dey, Nilanjan, 1984- editor. | Borra, Surekha, editor.
Title: Medical big data and internet of medical things : advances, challenges and applications / editors, Aboul Ella Hassanien, Nilanjan Dey and Surekha Borra.
Description: Boca Raton : Taylor & Francis, [2019] | Includes bibliographical references.
Identifiers: LCCN 2018023410 (print) | LCCN 2018024427 (ebook) | ISBN 9781351030380 (eBook General) | ISBN 9781351030373 (Adobe Pdf) | ISBN 9781351030366 (ePUB) | ISBN 9781351030359 (Mobipocket) | ISBN 9781138492479 (hardback : alk. paper)
Subjects: | MESH: Medical Informatics--methods | Data Mining--methods | Internet | Databases as Topic
Classification: LCC R855.3 (ebook) | LCC R855.3 (print) | NLM W 26.5 | DDC 610.285--dc23
LC record available at https://lccn.loc.gov/2018023410

Visit the Taylor & Francis Web site at
http://www.taylorandfrancis.com

and the CRC Press Web site at
http://www.crcpress.com

ISBN 13: 978-1-032-65329-7 (pbk)
ISBN 13: 978-1-138-49247-9 (hbk)
ISBN 13: 978-1-351-03038-0 (ebk)

DOI: 10.1201/9781351030380

Medical Big Data and Internet of Medical Things

Advances, Challenges and Applications

Contents

Preface

The revolution in digital information, its management and retrieval systems has led to, among other things, rapid advancements in wearable devices as well as the design and development of several effective prediction systems and mobile health Apps. Big data and the Internet of Things (IoT) plays a vital role in prediction systems used in biological and medical applications, particularly for resolving issues related to disease biology at different scales. Modelling and integrating medical big data with the IoT helps in building effective prediction systems for automatic recommendations of diagnosis and treatment. Machine-learning techniques, related algorithms, methods, tools, standards and infrastructures are used for modelling the prediction systems and management of the data. In life sciences and medicine, the main concerns are the provision of privacy, security and the establishment of standards. Also important are scalability, availability, user friendliness and the continuous improvement of the tools and technologies. The ability to mine, process, analyse, characterize, classify and cluster a variety and wide volume of medical data is a challenging task. There is a great demand for the design and development of methods dealing with capturing medical data from imaging systems and IoT sensors. Most importantly, this data then needs de-noising, enhancement, segmentation, feature extraction, feature selection, classification, integration, storage, searching, indexing, transferring, sharing and visualizing.

This book addresses recent advances in mining, learning and analysis of big volumes of medical images resulting from both high rate real-time systems and off-line systems. It comes in a variety of data formats, data types and has complex and large sets of features that have many and complex dimensions. The book includes advances, challenges, applications and security issues of medical big data and the IoT in healthcare analytics.

Chapter 1 explains how the data mining technology is more convenient in integrating medical data for a variety of applications such as disease diagnosis, prevention and hospital administration. Further, it discusses the practicality of big data analytics, and methodological and technical issues such as data quality, inconsistency and instability. Also addressed are analytical and legal issues, and the integration of big data analytics with clinical practice and clinical utility. The chapter also aims at exploring the potential of Hadoop in managing healthcare data, along with possible future work.

Chapter 2 presents a theoretical study of different aspects used in the field of healthcare like big data, IoT, soft computing, machine learning and augmented reality. Two other techniques are also discussed: organs on chip and personalized drugs used for healthcare. Case studies on the next generation medical chair, electronic nose and pill cam are also presented.

Chapter 3 discusses implantable electronics, integration of bio-interfaces, and wearable sensors and devices. Different types of fast and rapidly evolving sensors (and their proper placement) are also discussed. A practical body area network (BAN) providing real-time monitoring is presented along with its challenges including isolation of noise and associated filter design scope.

Chapter 4 considers web systems for biomedical signal analysis, and describes their software architecture design aspects. It examines applications, and focuses on challenges involving data privacy, frontend workflow, frontend and backend interactions, database design, integration of data analysis and reporting libraries, programming language issues and so on. The chapter also emphasizes the scope for research on architectural considerations.

Chapter 5 addresses the classification improvement of imbalanced medical big datasets using an enhanced *adjacent extreme mix neighbours over sampling technique* (AEMNOST). The results are verified using the UCI repository datasets in real-time applications using benchmarking parameters such as geometric mean and area under the curve values; the experimental outcomes clearly demonstrate the superiority of the AEMNOST.

Chapter 6 presents a framework for removing misclassified instances to improve classification performance of biomedical big data. The framework has four main stages: preparation, feature selection, instance reduction and classification stages, which employ fuzzy-rough nearest neighbour to remove mislabelled instances. Experimental results proved the effectiveness of the proposed technique on biomedical big data, enhancing the classification accuracy up to 89.24%.

Chapter 7 presents a fuzzy C-mean and density-based spatial clustering technique for IoT data processing. The architecture uses KNN to clean and replace noisy and missing data, and the SVD is used to reduce data to save time. The mutual information is implemented to uncover relationships between the data and detect semantic clustering to achieve a high accuracy and to speed up the running time. The performance of FCM-DBSCAN with its varied approaches for data reduction is analysed, where it was found that the FCM-DBSCAN with SVD led to high accuracy and speed of retrieval when applied to MapReduce and Spark.

Mining gene expression data produced from microarray experiments is an important study in bioinformatics. Chapter 8 discusses the use of an optimization algorithm – K-means clustering hybridized with differential evolution and ant colony optimization, to cluster breast cancer gene expression data. Further, the MapReduce programming model was applied to parallelize the computationally intensive tasks and exhibited good scalability and accuracy.

In Chapter 9, an evolutionary MapReduce approach has been described to analyse data after the successful management of noisy, irrelevant data. The chapter also provides a comparison between pre-processing with and without the MapReduce-based approach. Comparisons are made for various sizes of datasets. For clarification, the accuracy, specificity and sensitivity of the algorithms are also measured.

Chapter 10 presents the advances, challenges and different applications of IoT and robotics in the field of healthcare, which includes sensors, master tool manipulators (MTMs) and patient side manipulators (PSM).

Chapter 11 presents a new remote healthcare monitoring system to collect wearable sensor data linked to an IoMT-based cloud server for storage and processing. A prototype of a remote patient monitoring architecture is proposed and is built to demonstrate its performance and advantages. Herein, the concept of IoMT and the customized network architecture is detailed. Finally, the chapter presents the interactions, advantages, limitations, challenges and future perspectives of remote healthcare monitoring systems.

Chapter 12 presents a comparative analysis of classical cryptography versus quantum cryptography for the Web of Medical Things (WoMT); it presents an authentication (Handshake) protocol using elliptic curve cryptography (ECC) to reduce the computational/communicational cost as well as to improve the attack resistance capability. The security analysis of the model confirms a guaranteed security in WoMT environments.

This book is useful to researchers, practitioners, manufacturers, professionals and engineers in the field of biomedical systems engineering and may be referred to students for advanced material.

We would like to express our gratitude to the authors for their contributions. Our gratitude is also extended to the reviewers for their diligence in reviewing the chapters. Special thanks to our publisher, Taylor & Francis Group/CRC Press.

As editors, we hope this book will stimulate further research in developing algorithms and optimization approaches related to medical big data and the IoT.

Aboul Ella Hassanien
Faculty of Computers & Information
Giza, Egypt

Nilanjan Dey
Techno India College of Technology
Kolkata, India

Surekha Borra
K.S. Institute of Technology
Bangalore, Karnataka, India

Editors

Aboul Ella Hassanien is the Founder and Chair of the Egyptian Scientific Research Group and a Professor of Information Technology Department at the Faculty of Computer and Information, Cairo University. Professor Hassanien has 700+ scientific research papers published in prestigious international journals and more than 40+ books covering such diverse topics as data mining, medical images, intelligent systems, social networks and smart environment.

Nilanjan Dey was born in Kolkata, India, in 1984. He earned his B.Tech. in Information Technology from West Bengal University of Technology in 2005, M.Tech. in Information Technology in 2011 from the same university and his Ph.D. in digital image processing in 2015 from Jadavpur University, India. In 2011, he was appointed as an Asst. Professor in the Department of Information Technology at JIS College of Engineering, Kalyani, India followed by Bengal College of Engineering College, Durgapur, India in 2014. He is now employed as an Asst. Professor in Department of Information Technology, Techno India College of Technology, India. His research topic is signal processing, machine learning and information security. Dey is an Associate Editor of IEEE Access and is currently the Editor-in-Chief of the *International Journal of Ambient Computing and Intelligence*, Series Editor of Springer Tracts in Nature-Inspired Computing Book Series.

Surekha Borra is currently a Professor in the Department of Electronics and Communication Engineering and Chief Research Coordinator of K.S. Institute of Technology, Bangalore, India. She earned her doctorate in Copyright Protection of Images from Jawaharlal Nehru Technological University, Hyderabad, India. Her current research interests are image and video analytics, machine learning, biometrics, biomedical signal and remote sensing. She has filed one Indian patent, published 6 books, 12 book chapters and several research papers to her credit in refereed and indexed journals, and conferences at the international level. She is the recipient of several research grants and awards from professional bodies and Karnataka State Government of India. She has received the Young Woman Achiever Award for her contribution in copyright protection of images, Distinguished Educator and Scholar Award for her contributions to teaching and scholarly activities, Woman Achiever's Award from The Institution of Engineers (India) for her prominent research and innovative contribution(s).

Contributors

Mai Abdrabo
Information Systems Department
Faculty of Computers and Information
Suez Canal University
Ismailia, Egypt

Sherif Barakat
Information Systems Department
Faculty of Computers and Information
Mansoura University
Mansoura, Egypt

R. Bhavani
Department of Computer Science and Engineering
Government College of Technology
Coimbatore, Tamilnadu, India

Vinay Chowdary
Department of Electronics and Instrumentation Engineering
University of Petroleum and Energy Studies
Dehradun, Uttarakhand, India

Mario Cifrek
Faculty of Engineering and Computing
University of Zagreb
Zagreb, Croatia

Satya Ranjan Dash
School of Computer Applications
Kalinga Institute of Industrial Technology
Odisha, India

Vijaya R. Dirisala
Department of Biotechnology
Vignan Foundation for Science, Technology and Research University
Guntur, India

Nagwa Elaraby
Information System Department
Faculty of Computers and Information
Mansoura University
Mansoura, Egypt

Laura Elezabeth
Department of Computer Science and Engineering
Amity University Dubai
Dubai, UAE

Mohammed Elmogy
Information Technology Department
Faculty of Computers and Information
Mansoura University
Mansoura, Egypt

Ghada Eltaweel
Computer Science Department
Faculty of Computers and Information
Suez Canal University
Ismailia, Egypt

Heba El-Zeheiry
Information System Department
Faculty of Computers and Information
Mansoura University
Mansoura, Egypt

Kresimir Friganovic
Faculty of Engineering and Computing
University of Zagreb
Zagreb, Croatia

Rajani Reddy Gorrepati
Department of Computer Science and Engineering
Jeju National University
Jeju, Republic of Korea

Sitaramanjaneya Reddy Guntur
Department of Biomedical Engineering
Vignan Foundation for Science, Technology and Research University
Guntur, India

Mukul Kumar Gupta
Department of Electronics and Instrumentation Engineering
University of Petroleum and Energy Studies
Dehradun, Uttarakhand, India

Alan Jovic
Faculty of Engineering and Computing
University of Zagreb
Zagreb, Croatia

Kresimir Jozic
INA–industrija nafte, d.d.
Avenija Veceslava Holjevca
Zagreb, Croatia

Kunal Kabi
School of Computer Engineering
Kalinga Institute of Industrial Technology
Odisha, India

Vivek Kaundal
Department of Electronics, Instrumentation and Control Engineering
University of Petroleum and Energy Studies
Dehradun, Uttarakhand, India

Davor Kukolja
Faculty of Engineering and Computing
University of Zagreb
Zagreb, Croatia

K. Anitha Kumari
Department of IT
and
Department of CSE
PSG College of Technology
Coimbatore, Tamil Nadu, India

Ved P. Mishra
Department of Computer Science and Engineering
Amity University Dubai
Dubai, UAE

Bhabani Shankar Prasad Mishra
School of Computer Engineering
Kalinga Institute of Industrial Technology
Odisha, India

Amit Kumar Mondal
Department of Electronics and Instrumentation Engineering
University of Petroleum and Energy Studies
Dehradun, Uttarakhand, India

Jyotiprakash Panigrahi
School of Computer Engineering
Kalinga Institute of Industrial Technology
Odisha, India

Sachin Patil
Computer Science and Engineering Department
RIT Rajaramnagar
Maharashtra, India

Md. Fahim Shahrier Rasel
Department of Computer Science and Engineering
East West University
Bangladesh

Manas Kumar Rath
School of Computer Applications
Kalinga Institute of Industrial Technology
Odisha, India

Shamim H. Ripon
Department of Computer Science and Engineering
East West University
Bangladesh

G. Sudha Sadasivam
Department of Computer Science and Engineering
PSG College of Technology
Coimbatore, Tamilnadu, India

Md. Golam Sarowar
Department of Computer Science and Engineering
East West University
Bangladesh

Paawan Sharma
Department of Electronics, Instrumentation and Control Engineering
University of Petroleum and Energy Studies
Dehradun, Uttarakhand, India

Shefali Sonavane
Information Technology Department
WCE Sangli
Maharashtra, India

Varnita Verma
Department of Electronics and Instrumentation Engineering
University of Petroleum and Energy Studies
Dehradun, Uttarakhand, India

1 Big Data Mining Methods in Medical Applications

Laura Elezabeth and Ved P. Mishra

CONTENTS

CHAPTER AND SCOPE OF THE BOOK

The book *Medical Big Data and Internet of Medical Things: Advances, Challenges, and Applications* addresses advances in mining, learning, and investigation of huge volume of medical data coming about at a high rate from both ongoing technologies and disconnected frameworks. This book presents scientific categorizations, patterns and issues. For example, veracity in distributive, dynamic, various information gathering, information administration, information models, theories testing, preparing, approval, demonstrate building, improvement procedures and administration of therapeutic huge information gathered from different, heterogeneous IoT devises, systems, stages and technologies.

This chapter discusses data related to human health and medicine and how it can be stored, searched, shared, analysed and presented in ingenious ways. In this chapter, we will see how data mining technology is more convenient for integrating this medical data for a variety of applications. This chapter aims at analysing the existing data mining methodologies, resolving the existing drawbacks, investigating its future potential and proposing a multi-relational and accumulative employment for mining data in a multi-relational format.

1.1 INTRODUCTION

Big data is a term describing structured and unstructured data sets of large volume, which are evaluated computationally to reveal progressions, associations and patterns specifically relating to human interactions and behaviour. To look for significant or germane information from large data sets, data mining is used, which is the method of digging through the data, involving relatively intricate search operations that return definite and explicitly targeted results.

In the medical field, huge amounts of data are generated, from the patient's personal information to medical histories, and genetic data to clinical data. This medical big data is stored not simply for the sake of storing, but contains valuable information, which if and when analysed and methodized properly, can aid in understanding the 'concepts' of illness and health and thus bring about major breakthroughs in the medical field, especially in the areas of disease diagnosis and prevention.

With the aid of computers and technology, this medical data can be analysed faster in a less cumbersome manner. We can thus draw meaningful and reliable conclusions regarding the health of a person. Data mining technology has opened a new door to disease diagnosis. Similarly, in order to provide effective treatment for a disease's triennial prevention, data mining can be used. If the assumptions are true and the results are reliable, this could be the beginning of the absolute prevention or even eradication of diseases. Medical data can also be used for healthcare and drug administrative purposes. Big data mining can aid in analysing medical operation indicators of hospitals to help hospital administrators provide data support for medical decision-making.

Dealing with huge amounts of data has many issues related to data integrity, security and inconsistency. Data mining methods and their applications in the medical field is a new concept, although data mining methods have been applied in other fields for quite a while. Therefore, we face issues of practicality. Also, if the medical assumptions deduced from the data are wrong, all this work would be futile. Therefore, in this case, science and technology must go hand in hand.

This chapter aims at analysing the existing data mining methodologies, resolving the existing drawbacks, investigating future potentials and proposing a multi-relational and accumulative employment for mining data in a multi-relational format.

1.2 RISE OF BIG DATA

1.2.1 Growth Rate

Since the emergence of the digital age, which has been marked by the rise of computer technology and the Internet, there has been an outburst of data. There has also been a dramatic increase in our capability to capture and store these data in various formats. Data is not a new gimmick. The archived files and records of old are data themselves.

But with today's technology like computer databases, spreadsheets and so on, data can be retrieved and accessed easily with just the touch of a finger.

Every day we create large amounts of data, and this keeps increasing. We leave a digital footprint with every digital activity. By 2020, the amount of digital information available will have grown from around 4.4 zettabytes to 44 zettabytes.

This outpouring of data has surpassed our ability to process, store, understand and analyse these datasets. Therefore, our existing methodologies and data mining tools and software are unable to handle this huge amount of data. This is where big data comes into the picture. Big data is simply the collection of this enormous quantity of data and its manipulations.

1.2.2 Benefits of Big Data

As we have more information about something, the data becomes more reliable and we can draw meaningful conclusions by observing large data patterns. We can conceptualize patterns and relationships among these data to make smarter decisions.

In big data, data isn't confined to simply spreadsheets and databases but to everything from photographs to sound recordings.

Although big data gives us remarkable insights and opportunities, it also raises concerns and questions that must be addressed such as data privacy, data discrimination and data security [1].

1.2.3 Big Data Analytics Usage Across Industries

Big data is applicable in banking, technology, manufacturing, energy, consumer healthcare and so on.

Figure 1.1 shows the big data analytics usage across industries.

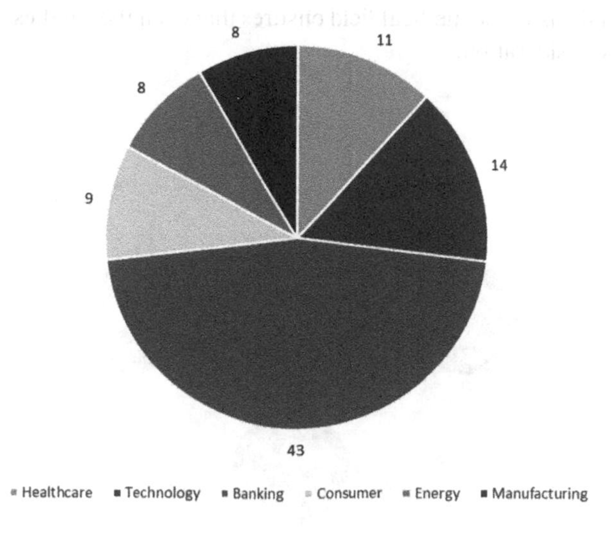

FIGURE 1.1 Big data analytics usage across industries. (From Big data analytic survey research, Sudhaa Gopinath.)

1.3 MEDICAL DATA AS BIG DATA

Big data is defined by its large volume, velocity, variety and veracity. The data recorded in the medical field is large in terms of volume as large amounts of patient details are stored; also, the data is in high velocity in terms of large amounts of data coming in at high speeds, such as from the constant monitoring of patients condition. This data also has variations and in terms of large amounts of medical data from different age groups, or high variability of veracity in terms of incomplete patient records and so on [2].

But medical big data has slightly different features compared to big data from other fields. Medical big data is hard to frequently access, it is somewhat structured in comparison and also has legal complications and issues associated with its use [3].

Basically, medical big data comprises data on human genetics, medical imaging, pathogen genomics, routine clinical documentation, pharmacokinetics, digital epidemiology, course assessment and so on.

Figure 1.2 shows the composition of medical big data.

1.3.1 Benefits of Medical Big Data

Medical big data can be used to improve healthcare quality, predict epidemics, increase analytical abilities, cure disease, build better health profiles, improve quality of life, improving outcomes, avoid preventable deaths, build better predictive models and reduce resource wastage.

Big data can be used in understanding the biology of a disease by integrating the available large volumes of data to build meaningful relational models. Medical big data drives the changes behind models of treatment. Thus, with this kind of technology we can understand more about patients, be informed about their lives as early as possible, collect warning signs of serious diseases at an earlier stage for faster and cheaper treatment.

Big data analysis in the medical field ensures that even the smallest of details will be taken into consideration.

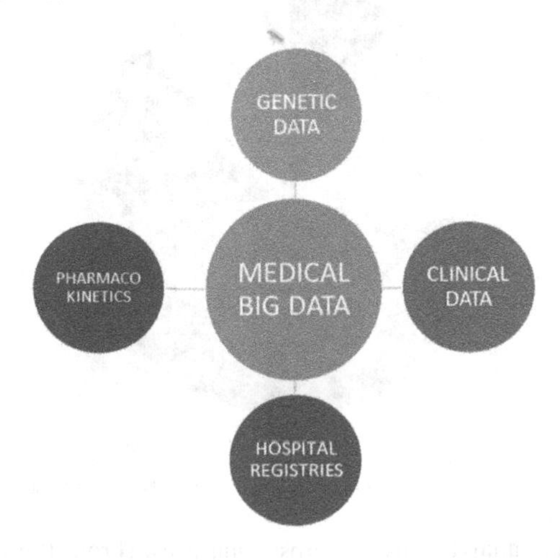

FIGURE 1.2 Composition of medical big data.

Many devices are collecting medical data continuously. In addition, various efforts explore improving the monitoring of patients through activity recognition, and thus, body area networks (BAN) have been developed in the recent past [4].

1.4 BIG DATA MINING PROCESS

Data mining is the technique of extraction of desired and targeted data from a very large dataset. The technique involves traversing through the huge volume of data using methods of association, classification, compilation and so on [5].

Association creates interrelationships between two or more data sets to identify a pattern. Association enables the deduction of general tendencies among data sets.

An important part of data mining techniques is classification and clustering. Classification assigns data into particular target classes to precisely predict what will occur within the class [6–8]. And, clustering is the process of grouping similar data records.

By identifying similar and frequent occurrences of data, we can have sequential patterns which can aid the data mining method.

1.4.1 Data Mining Techniques

Figure 1.3 Different techniques of data mining.

Data mining methods have been present for a long time, but it is more prevalent now with the emergence of big data. Big data has caused a need for more extensive mining methods.

The method is a mixture of the fields of statistics and artificial intelligence, with a touch of database management. Methods of data mining can be employed on large amounts of data in an automated matter [9].

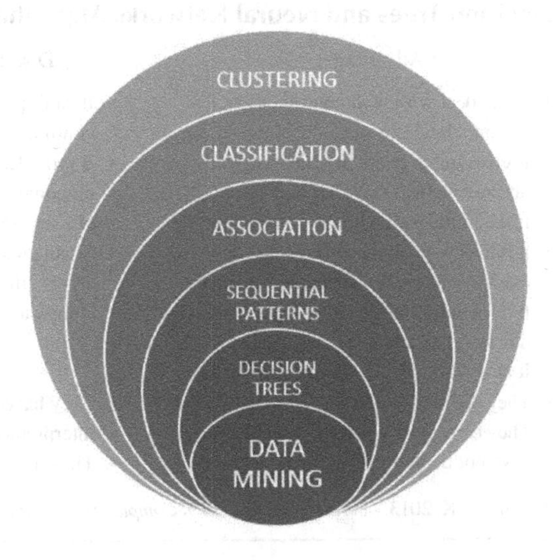

FIGURE 1.3 Different techniques of data mining.

- *Association*: Association, or also known as a relation, is one of the most used and simple data mining techniques. Correlations are made between data items and patterns in them. Building association or relation-based data mining tools can be achieved simply with different techniques [5].
- *Classification*: Classification is used as a feeder too, or the result of, other techniques. For example, you can use decision trees to determine a classification [5].
- *Clustering*: At a basic level, clustering is using at least one quality as a reason for recognizing a group of associated outcomes. Clustering is valuable to recognize distinctive data since it associates with different illustrations, so you can see where the likenesses and ranges concur [5].
- *Sequential patterns*: Sequential patterns are a helpful technique for identifying trends, or regular occurrences of similar events and are often used on longer-term data [5].
- *Decision trees*: The decision tree can be utilized either as part of the determination criteria or to help with the utilization and choice of particular information inside the general structure. Inside the decision tree, you begin with a straightforward inquiry that has two (or more) answers. Each answer prompts a further inquiry to help order or recognize the information with the goal that it can be arranged, or that an assumption can be made in light of each answer [5].

1.4.1.1 Comparative Analysis of Data Mining Techniques

Tables 1.1 through 1.3 compare the advantages and disadvantages of various algorithms and Table 1.4 shows the evolution of big data mining techniques to big data analytics.

TABLE 1.1
Comparison of Decision Trees and Neural Networks Algorithms

Algorithm	Advantages	Disadvantages
Decision Tree	• It can deal with both continuous and discrete data • It gives quick result in classifying obscure records • It functions well with redundant attribute • It furnishes great outcomes with small size tree. Results are not affected by anomalies • It does not require a preparation method like normalization • It also works well with numeric data	• It can't predict the value of a continuous class attribute • It provides error prone outcomes when too many classes are used • Small change in data can change the decision tree completely
Neural Networks	• They are well suited for continuous value • They can classify pattern on which they have not been trained	• They have poor interpretability • They have long training time

Source: Garg, S. and Sharma, A.K. 2013. *International Journal of Computer Applications* 74(5), pp. 1–5 [11].

TABLE 1.2
Comparison of CART and C4.5 Algorithms

Algorithm	Advantages	Disadvantages
CART	• It easily handles outliers • It is nonparametric • It does not require variable to be selected in advance	• It may have unstable decision trees • It splits only by one variable
C4.5	• It uses continuous data • It handles training data with missing and numeric values • It avoids over-fitting of data • It improves computational efficiency	• It requires that target attribute will have only discrete values

Source: Garg, S. and Sharma, A.K. 2013. *International Journal of Computer Applications* 74(5), pp. 1–5 [11].

TABLE 1.3
Comparison of K-NN and K-Means Algorithms

Algorithm	Advantages	Disadvantages
K-Nearest Neighbor	• It provides more accurate results • It performs better with missing data • It is easy to implement and debug • Some noise reduction techniques are used that improve the accuracy of a classifier	• It has poor run time performance • It is sensitive to irrelevant and redundant feature • It requires high calculation complexity • It considers no weight difference between samples
K-Means	• It is reasonably fast • It is a very simple and robust algorithm. It provides the best results when data sets are distinct	• It requires that target attribute will have only discrete values • It can't work with non-linear data sets • It can't handle noisy data and outliers

Source: Garg, S. and Sharma, A.K. 2013. *International Journal of Computer Applications* 74(5), pp. 1–5 [11].

1.4.2 Data Mining Process

Rather than merely a set of tools, data mining can be considered as a process following the subsequent steps: sample, explore, modify, model and assess (SEMMA).

The five SEMMA stages can be further explained as:

1. Sample: to retrieve data that can be represented statistically
2. Explore: to implement exploratory, statistical and imaging techniques

TABLE 1.4
Evolution of Big Data Mining Techniques to Big Data Analytics

Data Mining Task	Technique	Development in Big Data Analytics	Dimensions Handled
Classification	K-Nearest Neighbour	Yes	Volume and veracity
	Decision trees	Yes	Volume, velocity and veracity
	Support vector machines	No	Nil
	Naïve Bayes classifier	No	Nil
	RIPPER	No	Nil
	Neutral networks	Yes	Volume
Association	FP Growth	Yes	Velocity
	Apriori	Yes	Volume and velocity
Clustering	K- Medoids	No	Nil
	K-Means clustering	Yes	Volume
	Agglomerative	No	Nil

Source: Fawzy, D. et al. 2016. *Asian Journal of Applied Sciences* 4(3), pp. 1–11 [12]; Yadav et al. 2013. *Journal of IJCSN, IJCSN (International Journal of Computer Science and Network)*, 2(3) [13]; Aloisioa, G. et al. 2013. Scientific big data analytics challenges at large scale. *Proceedings of Big Data and Extreme-Scale Computing (BDEC)*. pp. 2–4 [14]; Wu, X. 2014. *IEEE Transactions on Knowledge and Data Engineering*, 26(1), pp. 97–107 [15].

3. Modify: to choose and manipulate the predictive variables of data
4. Model: to model the predictive outcome
5. Assess: to evaluate the accuracy of a model

1.4.3 Data Mining Tools

Data mining methods are evolving day by day. The techniques and tools used are constantly revolutionised. Some of the tools presently being used are:

- *Database analysis*: Data mining algorithms are intended to extract information from various heterogeneous databases and anticipate evolving patterns;
- *Text analysis*: The text processing algorithms are intended to discover meaningful information from a given text; and
- *Seeking out incomplete data*: Data mining intimately depends on the information available. Thus, it is fundamental to find out if there are inadequate organization of information in the dataset [16].

Some tools being used currently in data mining are:

- *Rapid miner*: It is a readymade, open source, no-coding-required software, which produces advanced analytics;

- *WEKA*: This is a JAVA-based customization tool, which is free to use. It incorporates representation and prescient investigation and displaying strategies, grouping, affiliation, relapse and order; and
- *Knime*: Essentially utilized for information pre-handling – that is, information extraction, change and stacking, Knime is a powerful tool with GUI that shows the network of data nodes.

1.5 DATA MINING IN MEDICAL DATA

The key techniques of medical data mining involve analysis of medical information, combination and comparative study of various example, quick and powerful mining calculations and unwavering quality of mining outcomes. Methods and applications of medical data mining based on computation intelligence such as artificial neural networks, fuzzy systems, evolutionary algorithms, rough sets, and association rules have been introduced [17].

1.5.1 Application of Big Data Mining in Medical Field

Previously, in the field of business and marketing, the application of data mining was implemented and might have been ahead of healthcare. But this is not the case now. Successful mining applications have been implemented in the medical field, some of which some are described below.

- *Identifying health risks in patients*: With the help of medical big data, robust mining methods and model building solutions, we can identify patients with high-risk health conditions. This information can be harnessed by doctors and medical staff to identify the condition to take steps to improve healthcare quality and to prevent health problems in the future [18].

  Figure 1.4 Steps of identifying health risk using big data mining.

  For example, cancer is a serious illness which can be prevented and cured with the help of big data analytics. Cancer quickly devastates individuals all over the world. Big data can help battle this disease even more viably. Healthcare suppliers will have an upgraded capacity to recognize and analyse infections in their early stages, selecting more adequate treatments in view of a patient's genetic makeup, and directing medication protocols to limit symptoms and enhance viability. Big data can likewise be a fantastic help in parallelization and help to map the 3 billion DNA (Deoxyribonucleic acid) base sets [19].

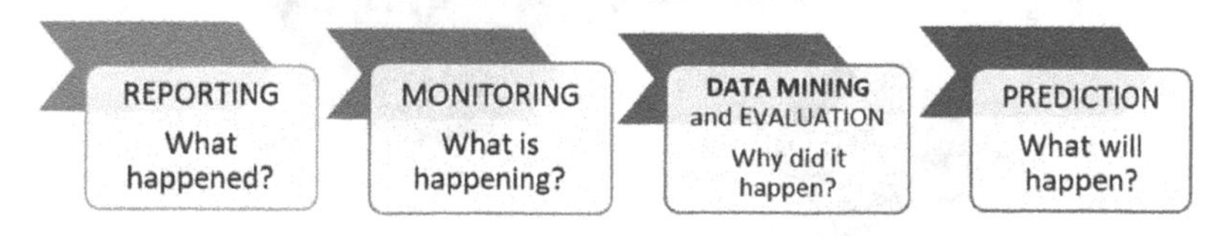

FIGURE 1.4 Steps of identifying health risk using big data mining. (From McKinsey Big Data Value Demonstration team.)

Advances in medical imaging innovation bolstered by software engineering have improved the understanding of medical images to a significant degree. The establishment of computer-supported analysis technology in health centres helps pathologists to assess their outcomes' viability, and in addition, gives additional clues to empower specialists to confirm their outcomes. An organized structure can possibly improve the nature of medical care and democratize its access.

There has been a recent trend of applying profound learning strategies to medical data examination for analysis, development and forecast in ophthalmology applications such as diabetic retinopathy, glaucoma and so on [20].

Medical image analysis plays a noteworthy part in distinguishing and determining various illnesses. As of late, analysts are intrigued with biomedical image analysis. Mostly, techniques of machine analysis including artificial neural networks (ANNs) have attracted the interest of several specialists. Computer aided diagnosis (CAD) is viewed as a fast and dynamic tool with the assistance of present day computer-based strategies and new medical imaging modalities [21].

- *Outbreaks and early recognition of epidemics require constant surveillance*: To identify high-risk patients, possible cases and detect the deviation in the occurrence of predefined events, we can use the aid of computer-assisted surveillance research. Using the medical big data already on hand, we can use powerful mining tools to deduce patterns and correlations to understand the health pattern of an area. The surveillance system utilizes data mining techniques to recognize a new and fascinating potential in contamination control. The technology cooperatively utilizes the regional data and patient health information to create monthly data displays that are checked by a specialist in contamination control. Figure 1.5 shows the method of surveillance. Developers of these systems have concluded enhancing infection control using data mining systems is

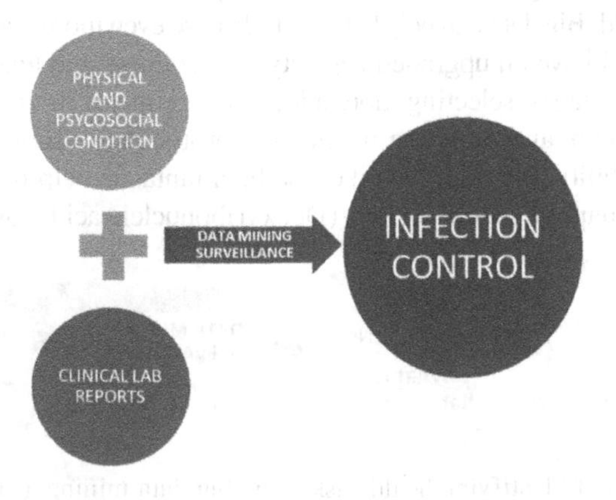

FIGURE 1.5 Method of surveillance.

more sensitive than traditional control surveillance, and significantly more specific. This has been developed and implemented at the University of Alabama [18].

- *Patient monitoring*: The use of big data makes the work easier for doctors and staff, and enables them to work all the more proficiently. Sensors are utilized and set up near the beds of patients to persistently screen pulse, blood pressure, and respiratory rate. Any change is detected immediately and the staff is alerted [1].

  Wireless innovation improvements have expanded quickly because of their ease-of-use and reduced costs compared with wired applications, especially considering the favourable environment offered by wireless sensor network (WSN)-based applications. Such applications can exist in many spaces including medical, restorative, mechanical and home computerization. A locally established remote ECG checking framework making use of the 'Zigbee' technology is considered. Such technology can be valuable for observing individuals in their own home and additionally for punctual checking by doctors for suitable medical services, enabling individuals to live in their home longer. Well-being observing systems can continually screen for numerous physiological flags and offer leads for further investigations and translations [22].

  One way in which big data can be used to aid in monitoring patient's vitals is using electronic health records (EHR). Big data analytics is widely used in EHR. Significant patient information can be obtained like clinical history, lab reports and other relevant statistics among others. The data assembled in these EHRs enable professionals to precisely study the patient's history and encourage better and convenient conveyance of healthcare services. Another notable point is that patients can now screen their lab test results without delay. Along these lines, frequent hospital visits of patients can be thus avoided [23,24].
- *Telemedicine*: Digitization, cell phones, remote devices and online video data have set the stage for the possibility of remote clinical administrations. Telemedicine is of great importance and uses today, due to these technological advancements. The information gathered from these devices can be effortlessly shared, which simplifies diagnosis considerably. Aside from remote patient checking, big data additionally helps in foreseeing serious medical conditions and preventing the decline in patients' conditions. Another significant manner in which big data has changed telemedicine is by gathering ongoing information originated remotely utilizing robots [25].
- *Informed strategic planning and predictive analytics*: Quick and smart decisions ought to be taken when treating patients encountering complex conditions. The role big data plays here is altogether more basic in this condition. Big data yields bits of information which empower professionals to reach informed decisions and upgrade general treatment processes. These bits of information furthermore help in the insightful examination, as it provides much more accurate understanding.

  Anticipating the occurrence of illness can be exceptionally valuable for clinical help. The PARAMO platform is utilized for building computational models for clinical diagnosis. To proficiently process parallel jobs, independent jobs are executed in parallel as MapReduce jobs, which breaks down information for large medical data that can be handled in reasonable time [26,27].

- *Fraud prevention*: Healthcare fraud is a national issue. In the course of the most recent decade, human services misrepresentation has soared with billions of dollars being paid on improper claims. The National Healthcare Anti-Fraud Association conservatively assesses that 3% of all health service spending, or $60 billion, is lost to healthcare fraud. Different estimates put this number closer to $200 billion [25].
- *Hierarchical medical system based on big data*: In this model, medical information can be recorded and shared among clinics at various levels. The critical fundamental patient data including physical examination documents, medical records, lab results, imaging results, prescription records, and other applicable data can be transmitted to providers of health services, which would enable patients to receive consistent health treatment when they seek medical services. Additionally, without being limited by space and time, primary hospitals or clinics can access resources of larger hospitals. Medical specialists of advanced health facilities can actively take part in the medical activities of lower level health facilities online and thus help the medical specialists at these facilities to improve their quality of health services. As a result of this, specialists across different facilities can build up relationships with each other, which can prove to be beneficial. Moreover, patients would have the advantage of getting consistent medical treatment at various establishments [28].
- *Bioinformatics applications*: The part of big data strategies in bioinformatics applications is to provide data archives and data computing and manipulating tools to accumulate and break down biological data.

  These days, a sequencing machine can deliver a large quantity of short DNA sequencing information in one run. The sequencing information should be mapped to reference genomes with the specific end goal to be used for further examination. CloudBurst is a parallel read-mapping model that facilitates the genome mapping process [29]. CloudBurst parallelizes the short-read mapping procedure to enhance the versatility of perusing substantial sequencing information.

  Various instruments have been produced to recognize errors in sequencing information: SAMQA distinguishes such mistakes and guarantees that extensive scale genomic information meets the base quality gauges [30]. Initially developed for the National Institutes of Health Cancer Genome Atlas to naturally distinguish and report mistakes, SAMQA incorporates an arrangement of specialized tests to discover information anomalies containing empty reads. For organic tests, analysts can set an edge to channel peruses which could be incorrect and report them to specialists for manual assessment [26].

  The utilization of big data stages generally requires a solid handle on distributed figures and systems administration learning. To enable biomedical analysts to grasp big data innovation, novel techniques are expected to coordinate existing big data advances with easy to use activities.

Hydra is a versatile proteomic web index using the Hadoop-distributed computing system. Hydra is a product bundle for preparing huge peptide and spectra databases, executing a distributed computing condition that supports the adaptable seeking of enormous measures of spectrometry information.

The proteomic search in Hydra is partitioned into two stages:

1. Producing a peptide database.
2. Scoring the spectra and recovering the information. The technology is fit for performing 27 billion peptide scorings in around 40 minutes on a 43-hub Hadoop bunch [31].

- *Imaging informatics applications*: Imaging informatics is widely utilized for enhancing the productivity of image processing work, for example, stockpiling, recovery and interoperation. PACS (Picture Archiving and Communication System) are prevalent for conveying pictures to nearby display stations, which is achieved fundamentally through DICOM (Digital Imaging and Communications in Medicine) protocols in the radiology department. Numerous online medical applications have been created to get to PACS, and more noteworthy, the utilization of big data technology has been enhancing their execution [32].

Notwithstanding big data innovations in view of the execution of cloud stages with PACS, a Hadoop-based restorative image retrieval technology separates the qualities of therapeutic pictures utilizing a Brushlet transform and a local binary pattern algorithm. At that point, the HDFS (Hadoop Distributed File System) stores the image information, trailed by the execution of MapReduce. The assessment outcomes showed a diminished error rate in images in contrast to the outcome without homomorphic separating [26,33,34]. These were used to evaluate the results of the comprehensive evaluation value for the synthesized image [35,36].

PACS basically provide an image data archiving and analysis work process at single sites. Radiology bunches working under a dissimilar delivery model confront huge difficulties in an information sharing environment.

Super-PACS, a technology that empowers a radiology bunch which serves multiple sites and has different PACS, RIS (Remote Installation Services), and other vital IT technology to view these sites from another, virtually and by making use of a virtual desktop for efficient completion of work.

Super PACS have two methodologies:

- The combined approach, in which every single patient datum remain local, and
- The merged approach, in which the information are put away halfway by a solitary operator.

The operator can:

- Give an interface to DICOM, HL7, HTTP (Hypertext Transfer Protocol), and XDS standard and non-standardized information;
- Synchronize metadata on neighbourhood PACS and RIS;

- Reserve pictures and information got from local PACS, RIS, any information device, or another specialist;
- Give worklists, organizers, and directing rationale for picture and non-image information;
- Circulate picture information through a web server; and
- Get to local and remote information from a SuperPACS web customer [26,36].

Soft computing based medical image analysis displays the principal procedures of soft figuring in medical image examination and handling. It incorporates image enhancement, segmentation, classification-based soft computing, and their application in diagnostic imaging. It includes a broad foundation for the advancement of sharp technologies in view of delicate registering utilized as a part of medical image investigation and preparing [37].

- *Mental health management*: Messages posted via web-based networking media could be utilized to screen for and conceivably recognize depression [38]. Their investigation depends on past research of the relationship between depressive issue and dull contemplations conduct. Big data mining tools assume a vital part in their work by mining shrouded behavioural clues and hints given in messages, or 'tweets', posted on Twitter. Inside these tweets, we might have the capacity to recognize symptoms, which is a formerly concealed side effect. Furthermore, Dabek and Caban [39,40] introduced a neural system able to anticipate the probability of mental conditions, for example, tension, behavioural disturbances, anxiety and depression issues. They tested the adequacy of their model against a group of 89,840 patients, and the outcome demonstrated a precison of 82.35% for all conditions [26].
- *Capsule endoscopy*: Conventional wired endoscopic devices are utilized for stomach and colon examinations. The capsule endoscopy (CE) is a noteworthy medical gadget for inspecting the small intestine, where the wireless CE (WCE) allows for a visual examination of the entire gastrointestinal (GI) tract. The WCE has transformed the detection procedure for bowel sicknesses. Consequently, it has become the main screening technique for the entire GI tract. The WCE is considered as a promising technology superior to conventional diagnosing gear because of links and the ineptitude of examining digestive system segments. In order to improve the WCE, computational methods can be executed for exact case restriction and follow. Precise information on the WCE position shows the variation from the norm condition has a key part for various reasons. This clinically guarantees practicable insightful calculations to diminish the indicative mistakes, to limit the variations from the norm, to evaluate intestinal motility and also yield improved video quality and WCE restriction. From now on, with the gigantic volume of WCE-produced data, big data and cloud advances will become basic requirements for productive information extraction [41].

1.6 CHALLENGES FACED IN APPLICATION OF BIG DATA IN THE MEDICAL FIELD

Although big data applications are a major breakthrough in the medical field, there are a few challenges that need to be overcome (Figure 1.6).

The potential for learning and information stored in medical databases require us to create particular devices for retrieving, examining, analysing, and using this knowledge; however, because of the expansion of information volume, there is much difficulty in extricating helpful data for decision support. Conventional manual data analysis is inefficient.

Critical issues arising from the quickly expanding datasets are :

- The creation of standard terminology, vocabularies and organizations to help multi-linguist and sharing of information
- Standards for the deliberation and representation of information
- Integration of heterogeneous information
- Standards for interfaces between various sources of information
- Reusability of information, learning and tools

A considerable lot of the environments still needs measures that block the use and analysis of information on an extensive variety of worldwide information, restricting this application to informational indexes gathered for analytic, screening, prognostic, observing, treatment bolster or on the other hand other patient administration purposes [17,42].

- *Issues related to data structure*: Big data applications ought to be easy to understand, straightforward, and menu-driven. The greater part of data in medical services is unstructured, for example, from natural language processing, it is frequently divided, scattered, and very unstandardized.

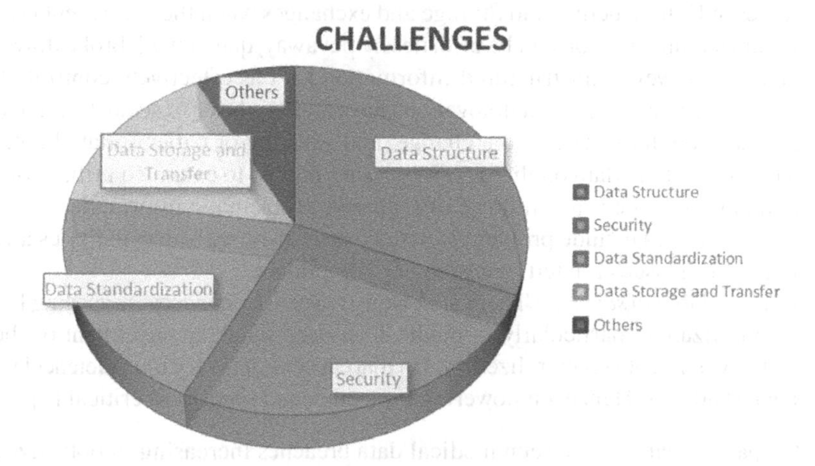

FIGURE 1.6 Summarises the challenges. (Adapted from Eysenbach, G. 2016. *JMIR Med Inform* 4(4), e38, pp. 1–11 [43].)

The EHRs don't share well crosswise over authoritative lines, yet with unstructured information, even inside a similar association, unstructured information is hard to total and investigate. Medical big data is more heterogeneous compared to other big data from other fields. Big data also needs to solve metadata transparency issues [43].

- *Issues related to security*: There are extensive security concerns regarding big data utilization, particularly in medical services given the institution of the Health Insurance Portability and Accountability Act (HIPAA) enactment in the United States. Information made accessible on an open source is accessible and, thus, much less secure. Furthermore, because of the sensitivity of medical data, there are noteworthy concerns identified with protection of privacy. Additionally, this data is centralised, and in that capacity, it is very much defenceless against attacks. Hence, improving protection and security is critical [43].
- *Issues related to standardisation of data*: In spite of the fact that the EHRs share information inside a similar association, intra-authoritative, EHR stages are divided, best case scenario. Data is put away in groups that are not compatible with all applications and innovations. Issues are caused by the exchange of data and, in addition, due to the absence of standardisation of data. It makes obtaining and purging date confusing.

  Restricted interoperability represents a huge challenge for big data due to the rare standardisation of data. This creates issues with the exchange of information within institutionalized organizations, as well as worldwide sharing. With the globalization of information, big data should manage an assortment of guidelines, hindrances of dialect, and distinctive wordings.
- *Issues related to storage and transfer*: When data is created, the expenses related with securing and storing it is high in contrast to the cost of generating it. Expenses are additionally brought about by exchanging data and breaking down it. Research has shown we have possessed the capacity to join the topics of Data structure and Storage and exchanges when they represent how organized information can be effortlessly put away, questioned, broke down, etcetera, however, unstructured information isn't as effectively controlled. Cloud-based medical data innovations have added a layer of security-related issues through the extraction, change, and stacking of patient-related data. The usage of big data ought to convey issues related to extended utilizations and moreover the transmittance of secure or problematic information.

  Other issues include problems related to accuracy, real-time analytics and managerial issues related to regulatory compliance.
- *Issues related to security*: There are extensive security concerns regarding big data utilization, particularly in medical services since the enactment of the HIPAA. This data is centralized, and in that capacity, it is very muchdefenceless against attacks. Hence, empowering protection and security is critical [7].

In the past 5 years, we've seen medical data breaches increasing in both size and recurrence, with the biggest breaches affecting more than 80 million individuals. Healthcare data thefts frequently expose sensitive data, ranging from personal data

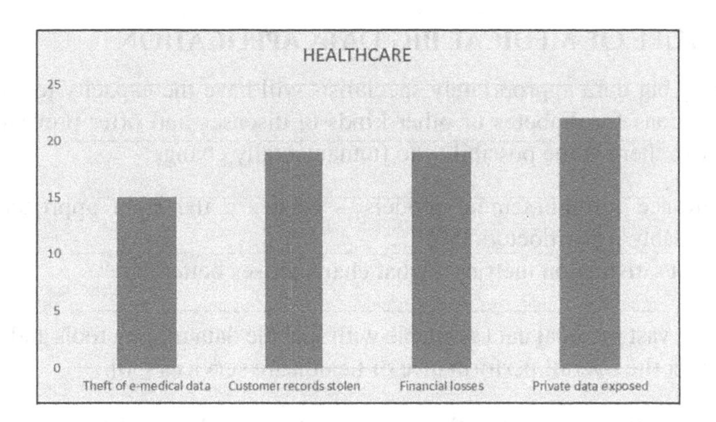

FIGURE 1.7 Most frequent types of cybercrime in healthcare sector. (From Mishra, V.P. and Balvinder, S. 2017. *Process Mining in Intrusion Detection—The Need of Current Digital World*. Springer Nature Singapore Pte Ltd., CCIS 712, pp. 238–246 [45].)

such as social security numbers, names and addresses to health information such as Medicaid ID numbers, medical coverage data, and health histories [44–48].

With the larger part of human services suppliers announcing that they were affected by a ransomware assault in the previous year, it isn't shocking that nine out of the ten biggest breaches in 2017 were caused by hacking or IT episodes. Put something aside for one announced occurrence, the biggest information breaks originated from ransomware assaults, unapproved server access, and PC infections [44] (Figure 1.7).

Although assailants will dependably look for approaches to utilize recently created developments and set up stages against healthcare industry to impact people and associations from turning into the cause of all their own problems.

Table 1.5 shows the top ten healthcare data breaches as given by the U.S. Department of Health and Human Services Office for Civil Rights [49].

TABLE 1.5
Top Ten Healthcare Data Breaches

1.	Anthem Blue Cross	78.8 Million Affected	January 2015
2.	Premera Blue Cross	11+ Million Affected	January 2015
3.	Excellus BlueCross BlueShield	10+ Million Affected	September 2015
4.	TRICARE	4.9 Million Affected	September 2011
5.	University of California, Los Angeles Health	4.5 Million Affected	July 2015
6.	Community Health Systems	4.5 Million Affected	April–June 2014
7.	Advocate Health Care	4.03 Million Affected	August 2013
8	Medical Informatics Engineering	3.9 Million Affected	July 2015
9.	Banner Health	3.62 Million Affected	August 2016
10.	NewKirk Products	3.47 Million Affected	August 2016

Source: Lord, N. 2017. Top 10 Biggest healthcare dat breaches of all time. Retrieved from https://digitalguardian.com/blog/top-10-biggest-healthcare-data-breaches-all-time (Accessed on 10th March 2018) [49].

1.7 FUTURE OF MEDICAL BIG DATA APPLICATION

By utilizing big data appropriately specialists will have the capacity to estimate at risk populations for diabetes or other kinds of diseases and offer preventive care. Additionally, there is the possibility to fundamentally change:

- Insurance reimbursement models – what are the right approaches to reasonably repay doctors [50]?
- Patient satisfaction metrics – what characterises better care?

Using the vast medical data available with suitable data mining tools and methods, we can better the overall performance of healthcare services [50].

- Risk identification and mitigation – how to bring down settlement rates
- Staffing and arranging
- Medication and antibody advancement

Organizations and even political movements have been utilizing scientific devices to exploit this immense measure of data in new and profitable routes, running from identifying extortion to more adequately spending publicizing dollars. Medical services have customarily fallen behind these different parts because of worries around understanding classification, however, that may change soon.

In the medical field, this could mean better alternatives to treat, forestall, and foresee ailments or better distinguish patients in danger. For instance, unique medications may work with changing levels of effectiveness on various patients. Giving access to these instruments that use more than the information and are implanted inside a doctor's work process can help encourage better choices. A considerable measure of these achievements is as of now happening, yet it is extremely just being starting. What's more, soon enough big data will simply be as common as normal data.

1.8 DATA MINING TOOL: HADOOP

Hadoop is utilized as a part of a wide range of utilization like Facebook and LinkedIn. The potential for big data and particularly Hadoop in human services and overseeing healthcare information is energizing, however—starting at yet—has not been completely figured out. This is primarily claiming Hadoop isn't surely known in the medical industry and halfway in light of the fact that human services don't exactly have the enormous amounts of information seen in different enterprises that would require handling power available in Hadoop.

Although medical analytics haven't yet been hampered by not utilizing Hadoop, it never hurts to look forward and think about the potential outcomes. Hadoop is a crucial device for effectively storing and handling extensive amounts of information. It has the capacity to offer better approaches for utilizing and investigating medical information and to enhance patient care at a lesser expense.

Hadoop is an open-source distributed data storage and analysis application. It was developed by Yahoo, in light of research papers distributed by Google [10]. Hadoop actualizes Google's MapReduce calculation by dividing expansive queries into many parts, sending those particular parts to a wide range of handling hubs, and afterward joining the outcomes from every hub.

Hadoop additionally is a collection of other significant data tools like Hive, Pig, HBase and Start.

Hadoop was composed from the earliest stage as a distributed processing and storage platform. This moves essential obligation regarding managing equipment failure into the software, streamlining Hadoop for use on vast bunches of ware equipment.

Regardless of whether existing database applications could suit these substantial databases, the cost of ordinary venture equipment and plate stockpiling is plainly restrictive. Hadoop was outlined from the onset to keep running on commodity hardware with frequent failures. This significantly diminishes the requirement for costly equipment with Hadoop. Since Hadoop is open source application there are no authorizing charges for the product either, which is another significant saving.

Hadoop is the underlying innovation utilized in numerous medical service analysis stages. This is to say that Apache Hadoop is the correct application to deal with colossal and complex healthcare data and to adequately manage the difficulties it creates for the medical services industry.

A few reasons for utilizing Hadoop with big data in healthcare are:

- *Hadoop makes information stockpiling more affordable and more accessible*: At present, 80% of all medical data is unstructured information. This incorporates doctors' notes, medical reports, lab results, X-rays, MRI pictures, vitals and monetary information among others. Hadoop gives specialists and analysts the chance to discover experiences from informational indexes that were prior difficult to deal with;
- *Capacity limit and data handling*: Most health services associations can store close to 3 days of information for each patient, restricting the open door for examination of the created information. Hadoop can store and handle humongous measures of information, making it perfectly possible to do this; and
- *Hadoop can fill in as an information coordinator and furthermore as an investigation apparatus*: Hadoop enables specialists to discover relationships in informational indexes with numerous factors, a difficult undertaking for people. This is the reason it is a suitable system to work with healthcare information.

Sending the Hadoop cloud stage can be a major test for scientists who don't have a software engineering foundation. CloudDOE is a product bundle that gives a straightforward interface to sending the Hadoop cloud claiming the Hadoop stage is regularly excessively complex for researchers without software engineering ability and additionally comparative specialized aptitudes. CloudDOE [51] is an easy-to-use device for examining high-throughput sequencing information with MapReduce, epitomizing the confused systems for arranging the Hadoop cloud for bioinformatics analysts. A few bundles are coordinated with the CloudDOE bundle (CloudBurst, CloudBrush, and CloudRS), and its task is additionally streamlined by wizards and realistic UIs [26].

1.9 CONCLUSION

In the field of medicine, an enormous quantity of data is created, from patient's personal data to health history, from hereditary information to clinical information. This medical big data contains valuable data. This information, when investigated and methodized legitimately, can help in the understanding of ailments and health and hence help realize significant leaps forward in medical sciences, particularly in the fields of diagnostics and disease prevention. There are many useful applications that have already been implemented, and many more potential applications in this field exist such as disease diagnosis, disease detection, infection control, telemedicine, fraud prevention and so on. But there also are challenges that need to be overcome such as issues related to data structure, security, standardization, storage and transfer. With the use of technology and potential data mining tools like Hadoop, big data can be investigated more quickly and in a less cumbersome way. With the large quantities of medical data available, we feel confident that we can uncover important and solid information regarding people's health. If this presumption is valid and the analysis outcomes are dependable, this could be the start of a new phase of total illnesses prevention or even eradication.

REFERENCES

1. Dey, N., Hassanien, A.E., Bhatt, C., Ashour, A., and Satapathy, S.C. 2018. *Internet of Things and Big Data Analytics toward Next Generation Intelligence*. Springer International Publishing, pp. 3–549.
2. Raghupathi, W. and Raghupathi, V. 2014. Big data analytics in healthcare: Promise and potential. *Health Information Science and Systems*, Springer, 2(3), pp. 2–10.
3. Lee, H.C. and Yoon, H.-J. 2017. Medical big data: Promise and challenges. *Kidney Research and Clinical Practice*, 36(1), pp. 3–11.
4. Kamal, M.S., Dey, N., and Ashour, A.S. 2017. Large scale medical data mining for accurate diagnosis: A blueprint. In Samee U. Khan, Albert Y. Zomaya, Assad Abbas (Eds) *Handbook of Large-Scale Distributed Computing in Smart Healthcare*. Springer, Cham, pp. 157–176.
5. Brown, M. 2012. Data mining techniques. Retrieved from www.ibm.com/developerworks/library/ba-data-mining-techniques/index.html (Accessed on 25th Feb 2018).
6. Acharjya, D. and Anitha, A. 2017. A comparative study of statistical and rough computing models in predictive data analysis. *Int. J. Ambient Comput. Intell.* 8(2), pp. 32–51.
7. Mahmoudi, S., Belarbi, M.A., and Belalem, G. 2017. PCA as dimensionality reduction for large-scale image retrieval systems. *Int. J. Ambient Comput. Intell.* 8(4), pp. 45–58.
8. Wang, D., Li, Z., Cao, L., Balas, V.E., Dey, N., Ashour, A.S., McCauley, P., Dimitra, S., and Shi, F. 2017. Image fusion incorporating parameter estimation optimized Gaussian mixture model and fuzzy weighted evaluation system: A case study in time-series plantar pressure data set. *IEEE Sensors Journal*, 17, pp. 1407–1420.

9. Kamal, S., Dey, N., Ashour, A.S., Ripon, S., Balas, V.E., and Kaysar, M.S. 2017. FbMapping: An antomated system for monitoring Facebook data. *Neural Network World* 1/2017, pp. 27–57.
10. Crapo, J. Hadoop in Healthcare. Retrieved from https://www.healthcatalyst.com/Hadoop-in-healthcare (Accessed on March 2018).
11. Garg, S. and Sharma, A.K. 2013. Comparative analysis of data mining techniques on educational dataset. *International Journal of Computer Applications* 74(5), pp. 1–5. 0975-8887.
12. Fawzy, D., Moussa, S., and Badr, N. 2016. The evolution of data mining techniques to big data analytics: An extensive study with application to renewable energy data analytics. *Asian Journal of Applied Sciences* 4(3), pp. 1–11. ISSN: 1996–3343.
13. Yadav, C., Wang, S., and Kumar, M. 2013. Algorithm and approaches to handle large data-A survey. *Journal of IJCSN, IJCSN (International Journal of Computer Science and Network)*, 2(3), pp. 1–5.
14. Aloisioa, G., Fiorea, S., Foster, I., and Williams, D. 2013. Scientific big data analytics challenges at large scale. *Proceedings of Big Data and Extreme-Scale Computing (BDEC)*. pp. 2–4.
15. Wu, X. 2014. Data mining with big data. *IEEE Transactions on Knowledge and Data Engineering*, 26(1), pp. 97–107.
16. Kamal, S., Sarowar, G., Dey, N., and Amira, S. Self-organizing mapping-based swarm intelligence for secondary and tertiary proteins classification. *International Journal of Machine Learning and Cybernetics*, pp. 1–30. 10.1007/s13042-017-0710-8
17. El-Hasnony, I.M., El Bakry, H.M., and Saleh, A.A. Data mining techniques for medical applications: A survey. *Mathematical Methods in Science and Mechanics*, pp. 205–212. ISBN: 978-960-474-396-4.
18. Obenshain, K.M. 2011. Application of data mining techniques to healthcare data. *Infection Control & Hospital Epidemiology*, 25(8), pp. 690–695. doi: 10.1086/502460
19. Big Data Applications in Healthcare. Retrieved from http://attunelive.com/big-data-applications-healthcare/ (Accessed on 25th Feb 2018).
20. Dey, N., Ashour, A.S., and Borra, S. (Eds.). 2017. *Classification in BioApps: Automation of Decision Making*, Vol. 26. Springer. pp. 3–447.
21. Chakraborty, S., Chatterjee, S., Ashour, A.S., Mali, K., and Dey, N. 2017. Intelligent computing in medical imaging: A study. *Advancements in Applied Metaheuristic Computing*, pp. 143–163.
22. Dey, N., Ashour, A.S., Shi, F., Fong, S.J., and Sherratt, R.S. 2017. Developing residential wireless sensor networks for ECG healthcare monitoring. *IEEE Transactions on Consumer Electronics*, 63(4), 442–449.
23. Greely, T.H. and Kulynych, J. 2017. Clinical genomics, big data, and electronic medical records: Reconciling patient rights with research when privacy and science collide. *Journal of Law and the Biosciences*, 4(1), pp. 94–132, https://doi.org/10.1093/jlb/lsw061
24. Dey, N., Hassanien, A.E., Bhatt, C., Ashour, A., and Satapathy, S.C. (Eds.). 2018. *Internet of Things and Big Data Analytics toward Next-Generation Intelligence*. Springer. pp. 3–549.
25. Wang, L. and Alexander, C.A. 2015. Big data in medical applications and health care. *Current Research in Medicine*, 6(1), pp. 1–8.
26. Luo, J., Wu, M., Gopukumar, D., and Zhao, Y. 2016. Big data application in biomedical research and health care: A literature review. *Biomed Inform Insights*. 19(8), pp. 1–10. doi: 10.4137/BII.S31559
27. Ng, K., Ghoting, A., Steinhubl, S.R. et al. 2014. PARAMO: A parallel predictive modeling platform for healthcare analytic research using electronic health records. *J Biomed Inform*, 48, pp. 160–70. doi: 10.1016/j.jbi.2013.12.012

28. Wan, X., Kuo, P., and Tao, S. 2017. Hierarchical medical system based on big data and mobile internet: A new strategic choice in health care. *JMIR Medical Informatics*, 5(3), e22, pp. 1–6, http://doi.org/10.2196/medinform.6799
29. Schatz, M.C. 2009. CloudBurst: Highly sensitive read mapping with MapReduce. *Bioinformatics*. 25(11), pp. 1363–1369.
30. Robinson, T., Killcoyne, S., Bressler, R. et al. 2011. SAMQA: Error classification and validation of high-throughput sequenced read data. *BMC Genomics*, 12: 419, pp. 1–7.
31. Lewis, S., Csordas, A., Killcoyne, S. et al. 2012. Hydra: A scalable proteomic search engine which utilizes the Hadoop distributed computing technology. *BMC Bioinformatics*. 13: 324, pp. 1–6.
32. Silva, L.A., Costa, C., and Oliveira, J.L. 2012. A PACS archive architecture supported on cloud services. *International Journal for Computer Assisted Radiology and Surgery*, 7(3), pp. 349–358.
33. Yao, Q.A., Zheng, H., Xu, Z.Y. et al. 2014. Massive medical images retrieval system based on Hadoop. *Journal of Multimedia*, 9(2), pp. 216–222.
34. Jai-Andaloussi, S., Elabdouli, A., Chaffai, A. et al. 2013. Medical content based image retrieval by using the Hadoop technology. *2013 20th International Conference on Telecommunications (ICT)*, Casablanca, Morocco. IEEE.
35. Wang, D., Li, Z., Cao, L., Balas, V.E., Dey, N., Ashour, A.S., and Shi, F. 2017. Image fusion incorporating parameter estimation optimized Gaussian mixture model and fuzzy weighted evaluation system: A case study in time-series plantar pressure data set. *IEEE Sensors Journal*, 17(5), 1407–1420.
36. Benjamin, M., Aradi, Y., and Shreiber, R. 2010. From shared data to sharing workflow: Merging PACS and teleradiology. *European Journal of Radiology*, 73(1), pp. 3–9. doi: 10.1016/j.ejrad.2009.10.014. Epub 2009 Nov 14
37. Dey, N., Ashour, A.S., Shi, F., and Balasm, V.E. 2018. *Soft Computing Based Medical Image Analysis*. Elsevier, pp. 1–292.
38. Nambisan, P., Luo, Z., Kapoor, A. et al. 2015. Social media, big data, and public health informatics: Ruminating behavior of depression revealed through twitter. *2015 48th Hawaii International Conference on IEEE System Sciences (HICSS)*, Honolulu, Hawaii. IEEE.
39. Dabek, F. and Caban, J.J. 2015. A neural network based model for predicting psychological conditions. In: Guo Y., Friston K., Aldo F., Hill S., Peng H. (eds) *Brain Informatics and Health*. BIH 2015. Lecture Notes in Computer Science, vol 9250. Springer International Publishing, Cham, pp. 252–261.
40. Guo, Y., Friston, K., Aldo, F., Hill, S., and Peng, H. 2015. *Brain Informatics and Health*. Springer, pp. 3–459.
41. Dey, N., Ashour, A.S., Shi, F., and Sherratt, R.S. 2017. Wireless capsule gastrointestinal endoscopy: Direction-of-arrival estimation based localization survey. *IEEE Reviews in Biomedical Engineering*, 10, pp. 2–11.
42. Liewis, S., Han, L., and Keane, J. 2013. Understanding low back pain using fuzzy association rule mining. *IEEE Conference on Systems, Man and Cybernetics*, pp. 3265–3270.
43. Eysenbach, G. 2016. Challenges and opportunities of big data in health care: A systematic review. *JMIR Med Inform* 4(4), e38, pp. 1–11.
44. Snell, E. 2017. Health IT security. Retrieved from https://muckrack.com/elizabeth-snell/articles (Accessed on 26th Feb 2018).
45. Mishra, V.P. and Balvinder, S. 2017. *Process Mining in Intrusion Detection—The Need of Current Digital World*. Springer Nature Singapore Pte Ltd., CCIS 712, pp. 238–246.
46. Mishra, V.P., Shukla, B., and Bansal, A. 2018. Analysis of alarms to prevent the organizations network in real-time using process mining approach. *Cluster Computing*, pp. 1–8. https://doi.org/10.1007/s10586-018-2064-8 (https://link.springer.com/article/10.1007/s10586-018-2064-8)

47. Mishra, V.P., Yogesh, W., and Subheshree, J. 2017. Detecting attacks using big data with process mining. *International Journal of System Modeling and Simulation*, 2(2), 5–7.
48. Mishra, V.P. and Shukla, B. 2017. Development of simulator for intrusion detection system to detect and alarm the DDoS attacks. *2017 International Conference on Infocom Technologies and Unmanned Systems (Trends and Future Directions) (ICTUS)*, Dubai, United Arab Emirates, pp. 803–806. doi: 10.1109/ICTUS.2017.8286116
49. Lord, N. 2017. Top 10 Biggest healthcare dat breaches of all time. Retrieved from https://digitalguardian.com/blog/top-10-biggest-healthcare-data-breaches-all-time (Accessed on 10th March 2018).
50. Mirdamadi, A. 2015. The future of big data in healthcare. Retrieved from https://www.logianalytics.com/bi-trends/future-big-data-healthcare (Accessed on 15th Feb 2018).
51. Chung, W.C., Chen, C.C., Ho, J.M. et al. 2014. CloudDOE: A user-friendly tool for deploying Hadoop clouds and analyzing high-throughput sequencing data with MapReduce. *PLOS ONE*, 9(6), pp. 1–7.

2 Approaches in Healthcare Using Big Data and Soft Computing

*Kunal Kabi, Jyotiprakash Panigrahi,
Bhabani Shankar Prasad Mishra,
Manas Kumar Rath, and Satya Ranjan Dash*

CONTENTS

2.1 INTRODUCTION

Healthcare is a service which provides protection and improvement for everyone's health. But due to the growing complexity of healthcare, high-quality and affordable healthcare services have become a challenging task to maintain for all individuals. Healthcare services researchers are continually trying to evolve innovative approaches for organizations and health professionals to implement new technological ideas in order to improve healthcare. The quality of care includes different tools and techniques such as improving communication, making knowledge more readily available, assisting in calculations, providing better decision support, performing important controls in real time, assisting in monitoring and providing key information. The structure of organs can be analyzed through IT properly to find the fault and bring a clear patient-centred description at the point of health maintenance. Healthcare professionals use a transparent, and swift way to achieve most of the knowledge through the use of IT. Simple, smart and fast use of IT in the field of healthcare was initially difficult and complex, however, new technologies like small handy devices are now being widely used. Portable devices like laptops, smart phones, tablets and wearable devices like smart watches are used by healthcare professionals and patients to provide care to patients. These devices give the patient's health data, or other electronically stored instructions, in numerous media like text, figure, voice or video, all of which helps healthcare professionals.

2.2 THEORETICAL STUDY

2.2.1 Big Data

Big data refers to large volumes of structured and unstructured data, which may be analyzed computationally. Big data contains five types of V's; velocity, value, veracity, variety and volume. In Figure 2.1 we list the five V's present in big data.

However, many authors have different definitions of big data. According to Gartner, data is growing at the rate of 59% every year. This growth can be depicted in terms of the following four V's: volume, veracity, velocity and variety. Shan et al. defines three types of common V's for big data; variety, velocity and volume. According to IBM, big data is being generated by nearly everything possible thing around the world at all times at an alarming velocity, volume, and variety. But in theories suggested by many scientist basically three V's are important in big data: variety, velocity and volume.

a. The volume represents the capacity of the data in petabyte or zettabyte or greater;
b. Velocity represents the processing high speed in real-time processing;

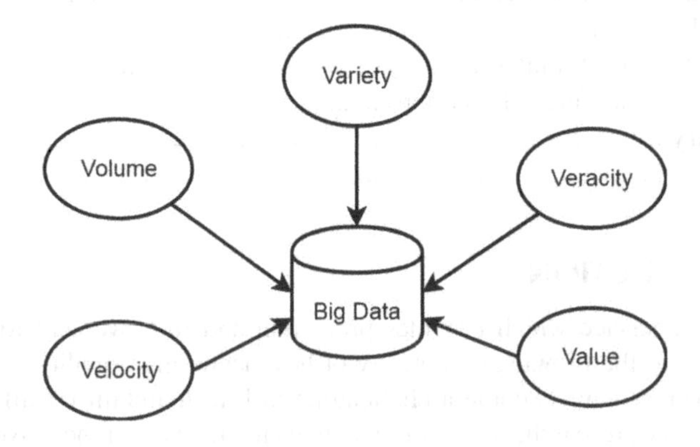

FIGURE 2.1 Different types of V's in big data.

c. Variety represents the different formats of data. There are three formats of data: structured, semi-structured and unstructured;
d. Value represents the value the collected data adds to knowledge creation. There is some valuable information somewhere within the data; and
e. Veracity includes two types: Data consistency (or certainty) indicating data trustworthiness. Data inconsistency (or uncertainty) indicating data can be doubtful, incomplete, ambiguous and deceitful.

In healthcare, big data analytics is used for storing, analysing and many more things. Electronic health data is very bulky in size and complex in nature, and it is problematic to manage with traditional data management methods, software, and hardware. Wearable devices produce data in less time than conventional devices. Like data collected from wearable or implantable biometric sensors like blood pressure, heart rate or body temperature, which is often processed and analysed in real time. Data in healthcare gives a wide idea about brief analysis. Big data analytics offer great possibilities to operate on massive amounts of data and are capable of solving hidden problems that are discovered. This distinguishing approach can be implemented to reduce the costs of processing time for massive amounts of data.

2.3 SOFT COMPUTING

In computer science and IT, soft computing is a collection of different techniques which provide solutions that are unpredictable, uncertain and between 0 and 1. Old mathematical methods take more time to process the data, which may be fatal to life of a patient. Soft computing deals with cases that are uncertain and approximate in nature. The different soft computing techniques are

1. Fuzzy Logic
2. Genetic Algorithm (GA)
3. Support Vector Machine (SVM)
4. Simulated Annealing (SA)
5. Perceptron
6. Bayesian Network
7. Particle Swarm Optimization (PSO)
8. Ant Colony Optimization (ACO)
9. Cuckoo Search (CS)
10. Artificial Bee Colony (ABC)
11. Bat algorithm
12. Artificial Neural Network (ANN)

2.3.1 Machine Learning

Formal or informal language can be implemented through different methods of AI. A system can treat automatically these explanations in precise languages by the usage of some predefined logical inference rules. This process is known as the knowledge-based approach in artificial intelligence and these pre-defined data are

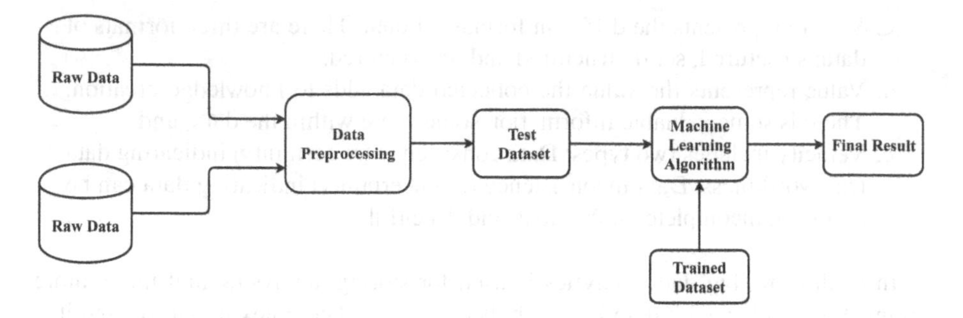

FIGURE 2.2 Machine learning module.

fed manually. It is an unwieldy process. People struggle to come up with formal rules with enough complexity to accurately describe all the activities. AI systems have the ability to acquire their own cognition, by clipping patterns from raw data. This whole mechanism is known as machine learning. Machine Learning trains the system about the problem, which helps to take a decision.

In plain words, machine learning is a part of AI, where the computer system has stored some predefined trained datasets and the system compares it with results given by test dataset. Machine learning has different types of algorithms like linear regression, logistic regression, support vector machine, neural network, deep neural network, naive Bayes, nearest neighbour, deep learning (a new aspect of machine learning) and so on.

In healthcare, machine learning is very useful for disease detection, diagnosis, proper treatment and medication. In all the processes the trained datasets are the patient's current information. The test case will be a new patient's disease symptoms and causes will be compared with the training data and the name, type and medication process will be predicted according to the result. In Figure 2.2 a common machine learning process model is presented.

2.3.2 Internet of Things (IoT)

IoT runs the interface between humans and objects present in the physical world, in conjunction with data and virtual environments, merging them with each other. In general, we can say Internet on connected devices. Over the past years, there has a been tremendous increase in interest about the IOT. Organizations have opened the path to familiarize the public with various IoT based devices and services which have been making the headlines. The application area for IoT technology is as wide as is distinct, as IoT are applied in all areas. The purpose of the IoT is to design a smart environment and simple life by saving time, energy and money. Through this technology, the expenses in different industries can be reduced. The most applied areas of application include, for example, healthcare, smart industries and so on. IoT uses three major communication protocols:

a. Device to Device (D2D): Enables communication between small devices like mobile phones

b. Device to Server (D2S): Enables data communication between devices to the server. In case of cloud processing, this protocol is used
c. Server to Server (S2S): It enables communication between two different servers.

The advancements in IoT-based medical devices and sensors, therapeutic care and healthcare are some of the research areas with the most potential. These are health-supervising systems, designated as healthcare industrial IoT (HealthIoT). This system has significant potential to reach the goal of analysing patients' healthcare data in critical condition, and continuously collect the data through wearable sensors [14].

2.3.3 Cloud Computing

Cloud computing provides a shareable architecture between system resources with minimal management over the Internet. The system includes both systems of client and server. It also minimizes the IT infrastructure cost. With less maintenance and advanced managing skills, it adjusts the resources to the unpredictable IT environment. A subpart of cloud computing is fog computing. In fog computing, data is handled at the network level in smart devices like mobile phones. In Figure 2.3 a common cloud architecture is presented.

2.3.4 Augmented Reality

Combination of the physical and non-physical world which demonstrate improvement to a blended reality and transverses between the fully physical world and the fully non-physical world. As such, blended reality is the 'middle ground' between the non-physical world or environment/reality and real environments, or it is the link between the real and the virtual world. Blended reality is made up of augmented virtuality and augmented reality (AR). Augmented virtuality is the non-physical surroundings in which everything is virtual and is augmented by real-world objects, while in AR the

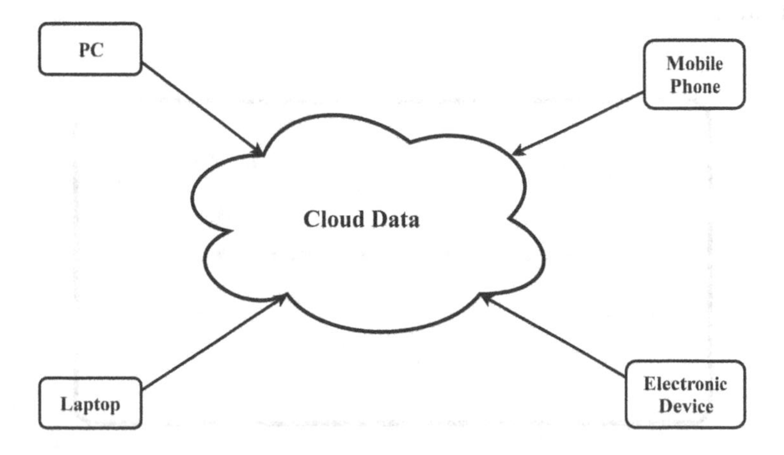

FIGURE 2.3 Cloud architecture.

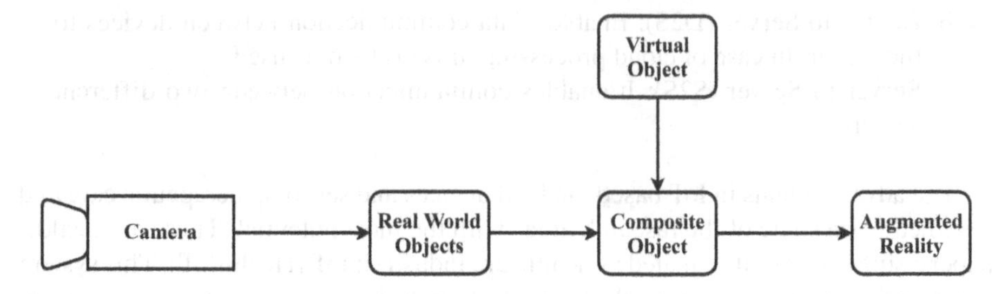

FIGURE 2.4 Augmented reality.

physical surroundings in which everything is real is augmented by virtual objects. In Figure 2.4 we present common processes of augmented reality.

2.3.5 Organs-on-Chips

These are microchips called 'organs-on-chips' [106–109], which are engineered to replicate the organ's architecture and act like living human organs, such as the lungs, intestines, kidneys, skin, bone marrow and blood cells. This technology offers a better alternative to traditional animal drug testing, which is very painful for animals. Because the medicine tested may or may not be effective in humans this new technology will help pharmaceutical researchers get a better idea about the effects and side effects of the medicine. Each individual organ-on-chip is constructed with a transparent amenable synthesized polymer, about the size of a computer memory stick, which is composed of hollow microfluidic channels coordinate by living human cells interfaced with a human cell-lined artificial vasculature. A bit of mechanical force can be applied to recreate the physical activity of living organs, like the breathing motion in lungs and the digestion activity going through the intestine. Because the microdevices are visible, they provide a clear vision of the inner workings of human organs. Figure 2.5 is an example of a simple 'organ-on-chip' architecture.

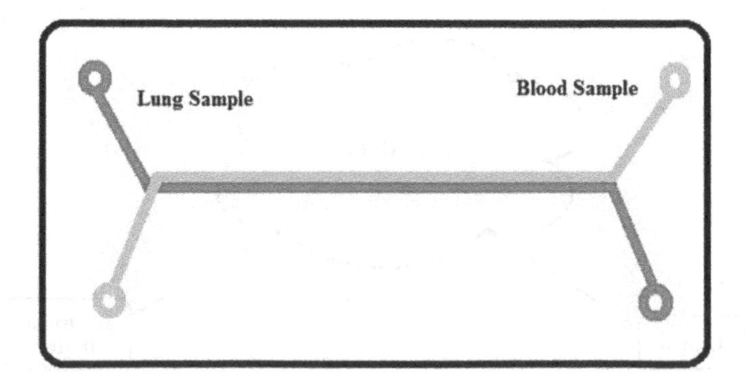

FIGURE 2.5 Design of organ-on-chip.

2.3.6 Personalized Drug

Every patient has either the same or a different type of disease. Thus accordingly, the medication parameters may or may not be same. So personalized drug dosing [110–115] provides a better medication protocol to each individual patient. For instance, if one patient is taking 2 to 3 pills per day, sometimes the preferred dose does not match with the prescribed drug. The personalized drug protocol for medication is the patient has to eat only one pill at a time. It helps also by decreasing the number of pills or injections by taking only one medicine combining them all. Thus it becomes less irritant to patient, relatives and the medical staff.

2.4 LITERATURE SURVEY

In 2017, Ahmad et al. [1] proposed a highly flexible electrocardiogram (ECG) as a wearable device. ECG devices are commonly used for heart and cardiovascular diseases. A conventional ECG is very bulky in size and immobile in nature. Constant use can cause discomfort to the patient. It is also very expensive and requires punctual maintenance. So monitoring with a wearable ECG provides better heart monitoring and treatment. The author used a reduced graphene oxide as a cotton fabric patterning and circuit design. Graphene oxide is a good conductor of heat and electricity. It is also stretchable so that it is fabricated in cotton. Graphene has carbon particles tightly packed in a 2D honeycomb shape.

In 2017, Hallfors et al. [2] proposed a reduced graphene oxide (rGOx)-based nylon ECG sensor based on IoT as a self-powered wearable device measuring heart pulsations. This new approach is a nominal energy electronic systems, or energy efficient electronic systems (E3S), built into IoT devices, systems and usage. In order to inspect the effectiveness of the rGOx fabric electrode, the author compared it with the clinical Ag/AgCl electrode. Data was collected using the rGOx fabric electrodes at the wrists and neck, and simultaneously at the same locations with the Ag/AgCl wet electrodes for result comparisons.

In November 2017, Verma et al. [3] proposed an IoT-based cloud-centric disease diagnostic framework. The conceptual framework of an IoT-based medical health monitoring system consists of three stages [20–28].

Stage 1: Users' health data are converged from medical devices and sensors. The picked up data is relayed to a cloud subsystem implementing a gateway or local processing unit (LPU);

Stage 2: The medical measurements are used for medical diagnosis systems to make intelligent decisions related to personal health; and

Stage 3: An alert is generated and sent to the parents or caretakers about the person's health. Furthermore, if any emergency situation occurs then an alert is also sent to the nearby hospital to deal with the medical emergency [3].

In 2018, Firouzi et al. [4] proposed an IoT-based smart healthcare system to provide an advanced and better healthcare facility to the patients. In 2018, Adame et al. [5] proposed an upgraded hybrid IoT-based checking system using both radio frequency identification (RFID) and wireless sensor network (WSN) technologies to measure the healthcare properties (using some involved and uninvolved RFID tags), region, and fitness of patients (using an active wristband monitoring body heat, heartrate [30] and movement). For real-time measurement the positioning system is connected to the wristband, which measures the body's vital parameters, the whole architecture is joined as a pattern, which gives feedback to the back-end server. The overall process has been implemented under a hospital environment.

In 2018, Farahani et al. [6] has done a survey on IoT [30] architecture for eHealth and mHealth. The author extensively interpreted the hazards present in healthcare across the globe. In its response, the article suggests to transform healthcare services from a medical-centered model to an individual-centered model with the support of IoT devices. The authors proposed an integrated, miscellaneous-layer IoT ecosystem for eHealth that is carried by three layers including edge devices, fog nodes, and cloud computing. The article also enumerated the different obstacles for such IoT ecosystem and proposed some conceivable solutions. The study involved advanced eyeglasses and smart textiles, which are presented to give a general idea about the capability of the proposed IoT ecosystem for eHealth.

In 2017, Sood et al. [3] designed a fog assisted cloud-based healthcare system to diagnose and prevent the outbreak of the chikungunya virus. The state of chikungunya virus outbreaks are determined by temporal network analysis at the cloud layer using proximity data.

In 2017, Babar et al. [8] proposed energy harvesting-based IoT and big data analytics [31–34] for a smart health monitoring system. Energy harvesting is the process using sustainable energy like solar, thermal, wind and so on to generate electricity in a lossless manner. Some special circuit boards are designed as wireless modules for energy harvesting. A Hadoop server is used to save, pre-process and test the data. The IoT sensors are attached to gain energy from human body movement or pressure generated from different positions. For instance, while running, energy-harvesting devices are in the soles of shoes, or on the legs and arms. In this way, the sensors get their power and measure the body's vital signs like temperature, blood pressure and heartbeat.

In 2017, Lavanya et al. [9] proposed an IoT [35] device to intelligently measure heartbeat and blood pressure. This device consists of three main stages,

a. *Smart medical service*: Here, the patient has to place his finger on the sensor to measure the heartbeats per minute (average 72 per minute). A Raspberry Pi is used to analyse the heartbeat, which is sent so that the doctor decides whether the patients need medication or not;
b. *Medication management*: If the patient takes the medicine at the wrong time, or the patient takes too much medicine or too little medicine, it will cause severe health issues to the patients. Real-time tracking and inspecting of vital signs is needed for early-detection. It helps doctors and family members to check whether the patients are following their prescribed treatment on time using real time clock (RTC) and RFID tags, which are connected to a Raspberry Pi. It sends an SMS to the patients, doctors and family members if there is any abnormal behaviour occurring. It will improve the user experience and service efficiency; and

c. *Cloud integration*: The heartbeat sensor reads and feeds the timing history on the cloud storage for future reference. By doing this, the doctor can examine the full sensor readings and timing history of patients and check whether the patient is taking the medicine on time or not, and whether they are following the prescribed routine or not on the cloud storage.

In this way, the patient can monitor his health by using this device. An SMS will be sent to both the doctors and patients so they can check the data whenever it's necessary.

In October 2016, Kamal et al. [36] proposed an fbMapping for dynamic monitoring of data from social media on Facebook. In this proposed work, multiple machine learning algorithms were designed for a system for monitoring the social networking website Facebook data, where at first data was assembled from web pages and Facebook user pages. These data contained different types of noise and irrelevancy, and the data was then pre-processed to remove it. The authors proposed an algorithm named Fisher's discrimination criterion (FDC) [37–40] to calculate on large datasets. The authors used this algorithm because it is efficient for dealing with large datasets like Facebook, Twitter and so on. It has the ability to do an iterative operation in a limited manner. So it helps in the important aspect of removing the noise, classifying the data, mapping control, a grouping of large data sets. Table 2.1 lists the author's results using this approach.

The author then compares this result with the naive Bayes filtering and without filtering technique. In Figure 2.6 the author gives a comparison between FDC, naive Bayes, and without using any filtering technique. From this, the author concludes the FDC requires the least time compared to the other methods.

In 2017, Wang et al. [41] presented a hybrid mechanism mixed with a parameter estimation optimized Gaussian mixture model (GMM), and a fuzzy weighted evaluation system as an image fusion [42–47] for plantar pressure dataset calculation. In this proposed mechanism there are frameworks involved in three phases; the preprocessing, data handling and result interpretation. The parameters present in the GMM were adjusted according to fuzzy operations [29,49]. The author used the data set of a foot from software Footscan 7.0. The author used it to get root mean square error (RMSE). The author compares his proposed algorithm with other algorithms like max-min, mean-mean, and up-down. In Figure 2.7 the author gives a comparison of these four algorithms.

TABLE 2.1
fbMapping Result

Attributes	Counts (Number of Persons/Hours)
Teenagers	223
Adults	227
Males	265
Females	185
Expend timings	4 hours 290 minutes
Wandering interests	209
Cuisine interests	231

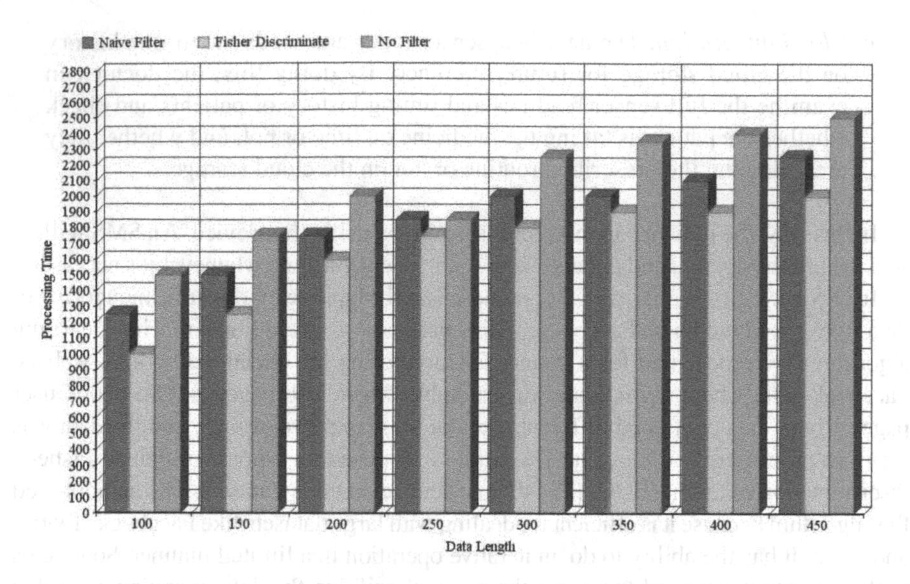

FIGURE 2.6 Algorithmic comparison between FDC, Naive Bayes, and no filter.

From the above result, the author concludes three results:

i. A GMM was more impressive when examined in contrast to the other blended methods for the plantar imaging fusion through indices system
ii. The synthesized image interpretation indices were added and separated into positive and negative indices
iii. The fuzzy interpretation system was more meticulous than other methods for the individual indices, which helped weighted operations

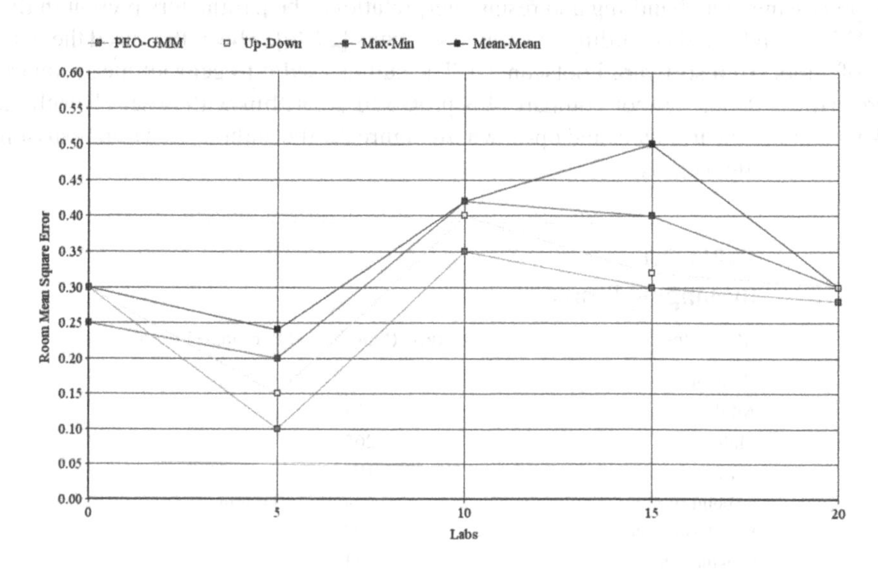

FIGURE 2.7 Comparison between PEO-GMM, Max-Min, Mean-Mean, and Up-Down.

In 2017, Matallah et al. [75] proposed a hybrid approach combining a big data Hadoop system and cloud computing [76–87]. The Hadoop environment provides a MapReduce model where it processes the data in a parallel and distributed manner. The cloud computing is used to treat the metadata in order to increase the objectives of performance and scalability. The hybrid approach provides consistency in the data management system as to not imperil metadata, as the author introduces a blended solution between centralization and distribution of the metadata service.

This hybrid approach provides a better result for huge data repositories. Nandi et al. [88] proposed a principal component analysis (PCA) [55–58,61,65] technique for medical image processing. Medical image processing [69–71] means analysing patient images collected from bio-metrics, gene expression, image sensors and so on. It assists doctors and other medical staff in patient diagnosis. Basically, computed tomography (CT scan), magnetic resonance imaging (MRI scan) [62, 67] and X-rays provide imagery data. But as advanced technologies are present, images like positron emission tomography (PET) and single-photon emission computed tomography (SPECT) furnish better diagnosis capabilities. PCA [50–54] can be used for the following medical image processing properties:

1. Medical image noise reduction
2. Medical image fusion [60,64,66,68,72–74]
3. Medical image segmentation [63]
4. Medical image feature extraction [48]
5. Medical image compression
6. Medical image classification
7. Medical image registration [59].

In Figure 2.8, the author gives a wide application of PCA for image processing properties.

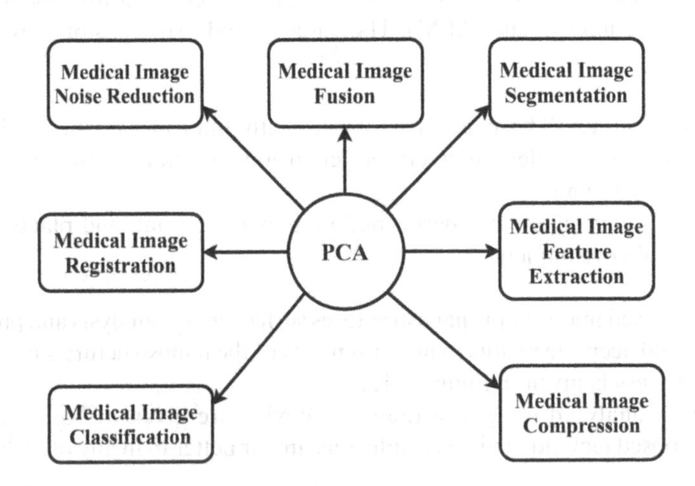

FIGURE 2.8 PCA application on medical image processing.

The PCA provides a better result than other medical image processing approaches because it is based on the statistical description of a random variable and a classical statistical method.

Moraru et al. [89] proposed a diffusion tensor-based MRI for human brain scanning. The diffusion tensor imaging (DTI) technique tracks the diffusion of water molecules through brain tissues, which are able to provide structural information of the brain. The author tested this technique on three patients (two men and one woman) aged between 45 and 50 years old. In Figure 2.9, the author shows a test medical image scan.

The result gives an excellent performance in comparison to conventional MRI scans, as it provides a clearer vision of the brain tissue.

Moraru et al. [90] proposed an analysis of the texture of parasitological liver fibrosis images [91,97,99]. Liver fibrosis is a liver infection disease, which is considered a fatal disease. If not diagnosed and medicated in time it can cause death. The authors' proposed algorithm is able to intelligently diagnose liver fibrosis [92]. In Figure 2.10, the image scans of liver fibrosis are shown.

In Table 2.2 the authors' give the complete comparison between IA, AC and LI

In Figure 2.11 the author graphs the results for liver fibrosis,

From the above table and figure, the authors reveal a local texture feature [96] called laminarity is a better texture descriptor and allows for a 100% confidence level in the discrimination stage. Global and local coherence of gradient vector approaches are efficient tools in microscopic image analysis tasks such as classification of liver image datasets [93–95] for integral anisotropy coefficient and laminarity, which are proposed based on the relative orientation of the pixel pairs.

In 2016, Fadlallah et al. [100] proposed a scanning electron microscopy (SEM) image processing technique for titanium alloys [101–105] in orthopedic and dental care. Titanium and its alloys are used for implantation or filling for bone damage repair. Titanium dioxide (TiO_2) NiTi (nitinol), Ti–6Al–4V and Ti–6Al–7Nb are titanium alloys used for the filling. The structure of titanium implant surfaces is determined by electron microscopes such as scanning electron microscopy (SEM) and atomic force microscope (AFM). The authors used two types of studies for the analysis:

1. *Invitro studies*: These are performed with microorganisms, cells, or biological molecules in a closed environment, not in the biological environment; and
2. *Invivo studies*: These are performed on human, animal, and plants in the biological environment.

The author used microscopic nanostructures surface image analysis and processing, which can yield accurate results in measurements of the nanostructure's morphology. The SEM process is given in Figure 2.12.

The author analysed images of titanium dioxide are given in Figure 2.13. The authors' proposed technique yields results that are far better than any other 3D-image analysis.

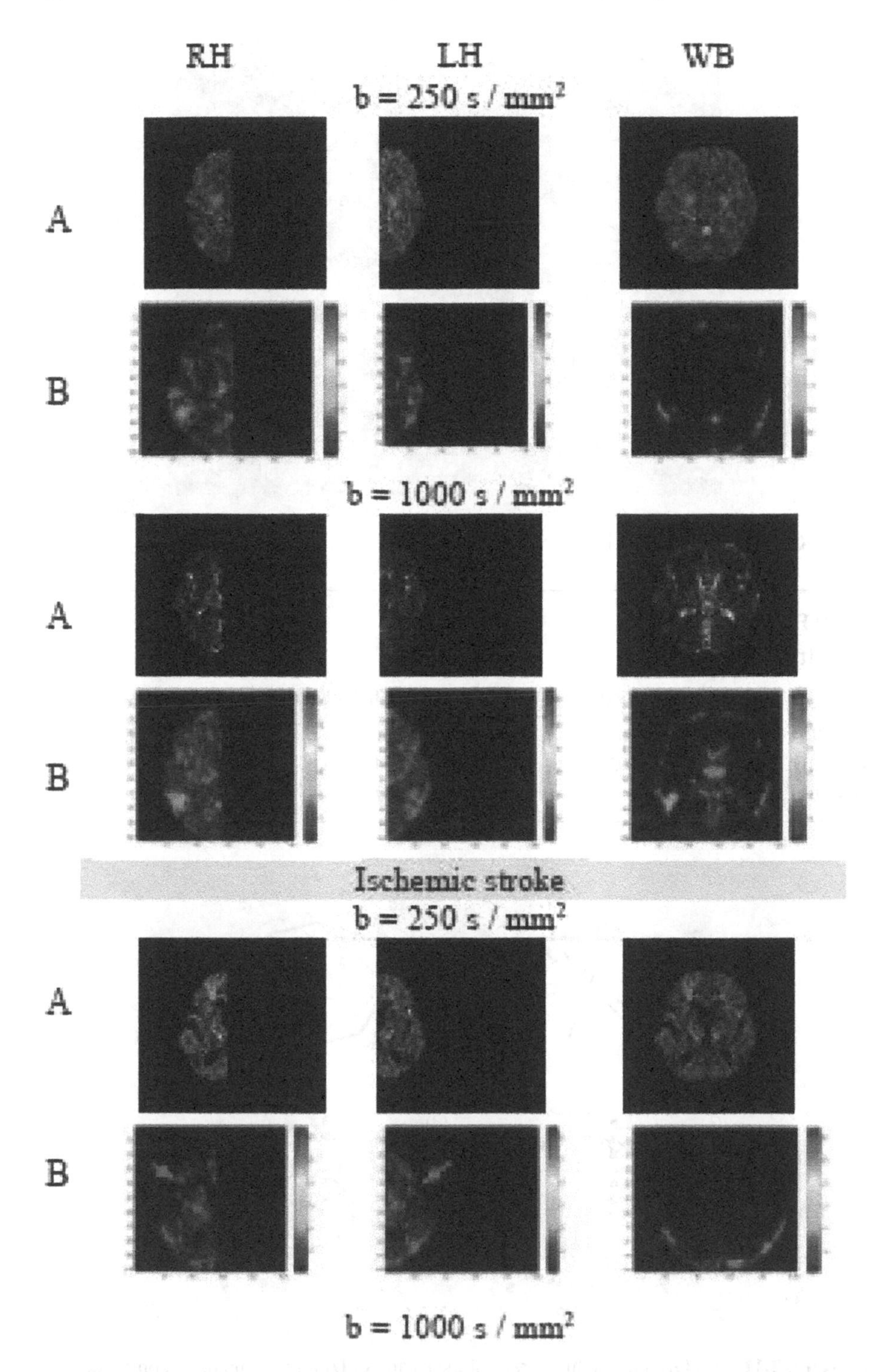

FIGURE 2.9 Brain scanned image.

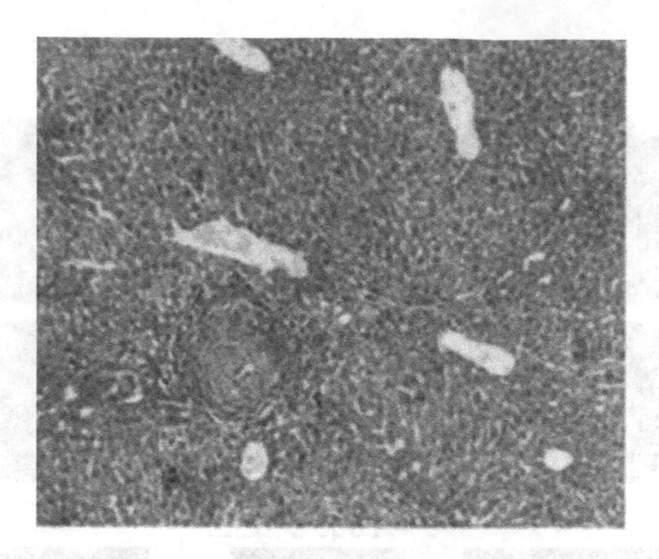

FIGURE 2.10 Liver fibrosis.

TABLE 2.2
Result Analyses

	Laminarity (LI)	Anisotropy Coefficient (AC)	Integral Anisotropy (IA)
NC-GRA	5,12E-05	0,0151	2,31E-04
FIB-GRA	1,04E-05	0,1364	2,54E-04
NC-FIB	1,46E-06	0,1137	2,03E-04

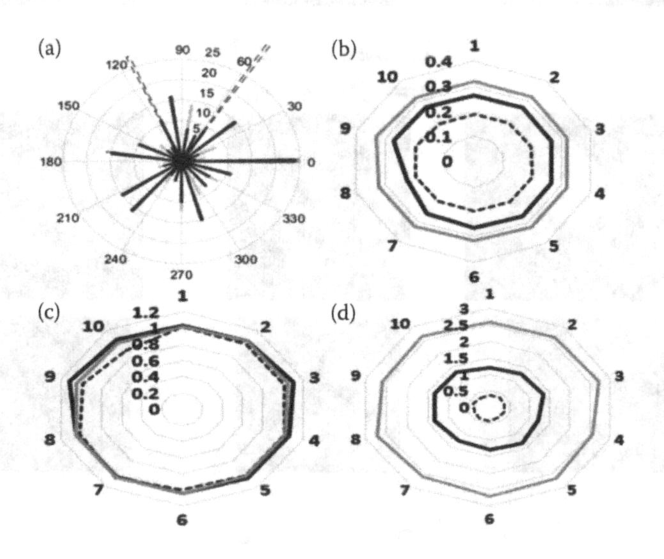

FIGURE 2.11 (a) Orientation histogram plots; (b and c) IA and AC features computed from the orientation histogram for healthy, fibrosis and granulomas liver; and (d) LI feature.

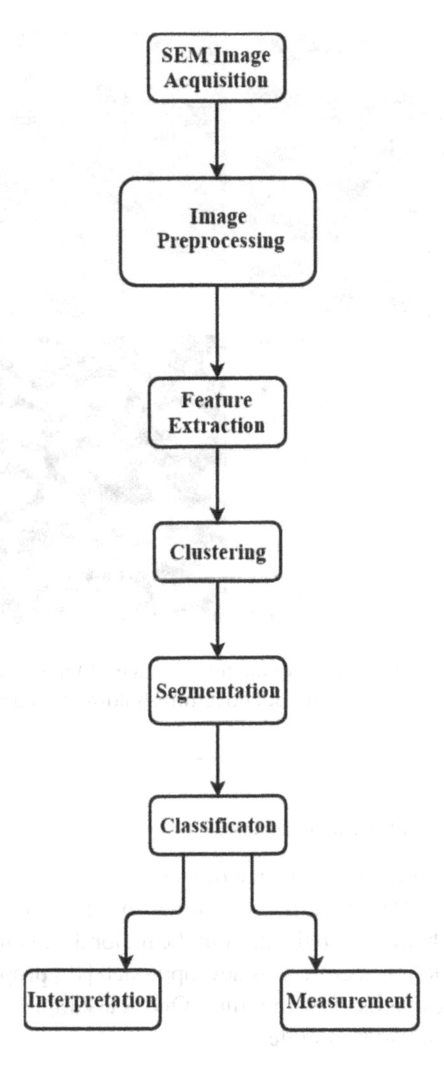

FIGURE 2.12 SEM image processing.

Machine Learning techniques like SVM and K-SVM also can be applied to automate a process in biomedical image application [120–121], which helps to detect the disease more efficiently. A wireless sensor framework for ECG observations can be used, in which the patient's vital signs like heart pulse are measured and scrutinized [119].

An IoT-based sensor framework for human body functioning such as a WBAN (Wireless Body Area Network) exists for healthcare purposes [122]. Some efficient machine-learning approaches exist for disease interpretation of fatal diseases like Alzheimer, cancer, and bacterial vaginosis [123]. A study of different soft computing approaches for medical imaging examination discuss medical imaging processes like PET, CT scan, MRI, X-rays and so on. [3].

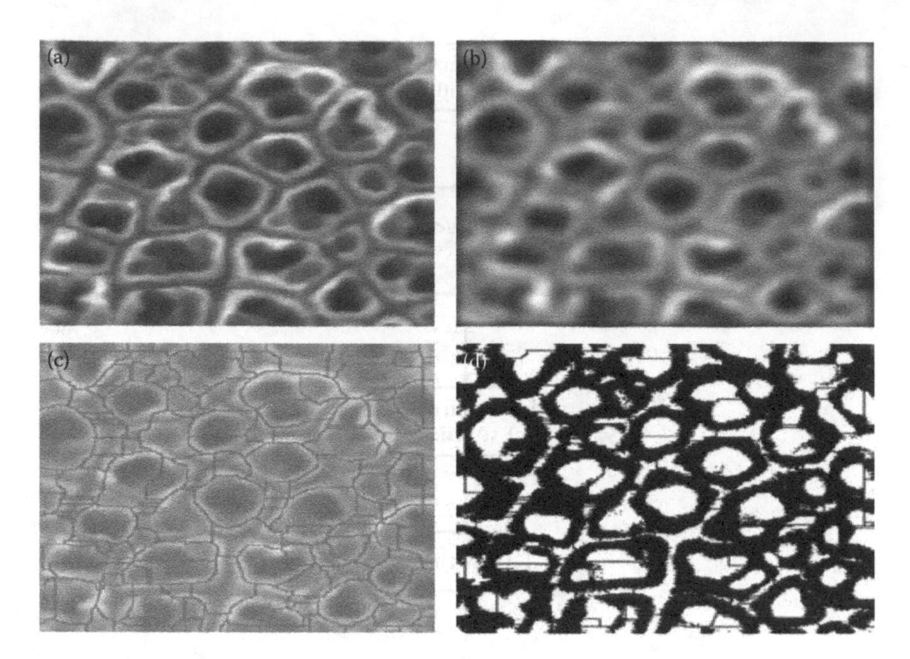

FIGURE 2.13 (a) Original titanium dioxide tubes image, (b) pre-processed (filtered) image, (c) watershed segmented image, and (d) adaptive thresholding for each segment.

2.5 CASE STUDY

2.5.1 Air Analysing Electronic Nose

In 2017, Murugan et al. proposed a device that can analyse air and chemicals present, called an electronic nose [18,19]. Some researchers have developed electronic devices which detect specific chemicals in the air, but the major disadvantage of these devices has been cost and prediction accuracy. A new approach [10] proposes to detect alcohol using a neural network that has two modules. One is a sample-collecting module and the other is the data analysing module.

i. *Sample collecting module*: Inside the sample-collecting module, gas sensors are present. These sensors are made of a metal oxide semi-conductor base. Both signal amplification and ADC (analogue-to-digital converter) are integrated in this module. Here a power amplifier amplifies both voltage-current ratios and fits the analogue signal to the converter. After converting the analogue signal to digital, the sample-collecting module transmits the sample data to the analysing module.
ii. *Analysing module*: The analysing module gathers the sample data and filters the unwanted or noise parts. In the next phase, the wave is converted to different wavelets and prepared for a data set. A feed-forward neural network is designed for analysis of these signals. In some cases, datasets are divided into two subparts. One is for a test set while another is for the trained set. There are four numbers of hidden layers to recognize or detect

the patterns, where the patterns are a virtual representation of the smell present in the air. This device can be implemented inside the hospital or inside an ICU chamber, in order to recognize the possibilities of infection and or bad smells that can irritate the patient. This system also requires a continuous synchronization of the hospital's own fog through which all the algorithmic decisions [118] can be sent back to the IoT board.

The algorithm for the electronic nose is

Algorithm 1: Artificial Nose

```
Input: Air sample
Output: Air harmfulness

Step 1
  Acknowledge sensor inside nostril
   If sensor=1
     Then Process_Air()
   Else throw an expression
Step 2
  Process_Air()
   Smell a sample of air from hospital through nostril sensor
   Synchronize the data from sensor
   Analyze the data
   Match()
Step 3
  Match()
   Check stored data about harmfulness
   If new data= harmful data
     Then hospital air is hazardous
   Else the air is good.
```

In Figure 2.14 we present a working model of the electronic nose

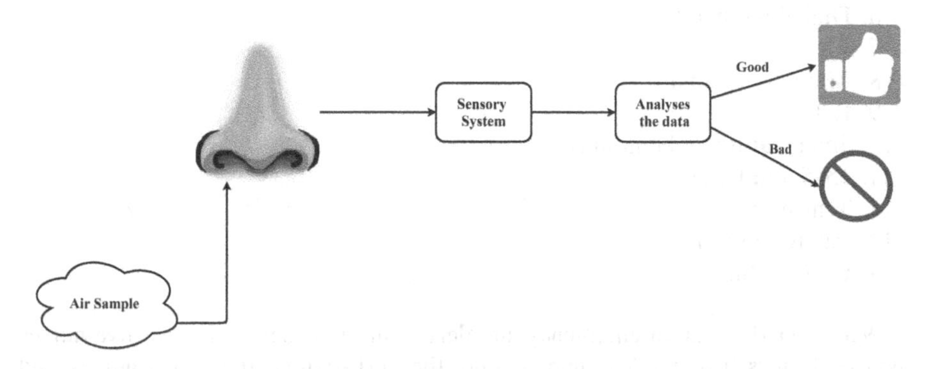

FIGURE 2.14 Working principle of electronic nose.

2.5.1.1 Networks inside a Hospital

Inside a hospital, a number of IoT boards and sensors can be implemented, but these IoT boards have little computing capacity. Some IoT boards like Arduino and Raspberry Pi have less computing power than our mobile phones, so these boards require high computing engines. These high computing machines can be replaced by a private fog with a number of servers. All IoT boards send the data they gather for processing in these servers and predictions, or the analysed data, are sent back to the IoT boards. Cost and risk can be eliminated through this process.

This network is connected to a 802.11 standard Wi-Fi connection. The number repeaters are also present to drive the Wi-Fi signal strength. In our case study, we suggest a new network protocol web socket, which is a full-duplex protocol. With this protocol, we can communicate with the server using two different channels, where one channel is dedicated for the request and another is dedicated for the response. In the comparison to of HTTP protocol, web socket is superior because it is full duplex and HTTP is half duplex communication.

2.5.2 Next Generation Medical Chair

A chair cum bed will be available for critical patients where the patients can sit or lie whenever necessary. A 3-kg oxygen cylinder will be attached to the chair, which will be regulated by a sensor. An emergency oxygen supply can thus be provided when patients need it. A motor will be attached to the wheel of the chair so it can move according to the requirements of the patients. A biodegradable pouch will be attached to the chair so it can collect stool and urine samples from the patient if he/she is paralyzed.

The chair is attached to a sensor module having 13 sensors in a 1-sensor kit. The following sensors will be attached to the module.

1. Active buzzer
2. Button
3. Microphone big
4. Reed switch
5. Heartbeat
6. Digital temperature
7. Joystick
8. Microphone small
9. Relay
10. Temperature and humidity
11. 15*5 mm LED
12. Temperature
13. Passive buzzer
14. GPS module

Whenever there is an emergency, an alert sound will be received to five nurses and two doctors. The wireless module sends the alert sound to the nearby nurses' and doctors' wristwatches with the patient's ID so that they know which patient needs

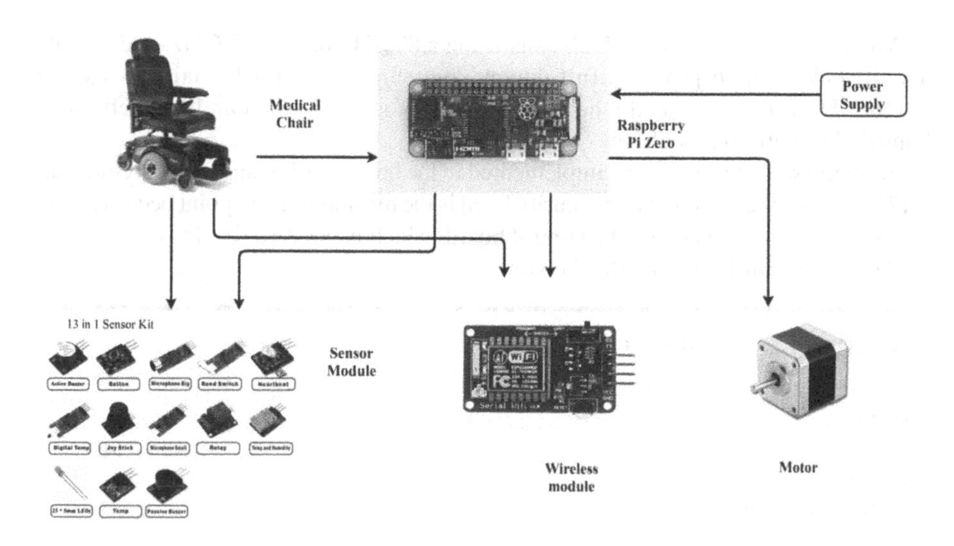

FIGURE 2.15 Next generation medical chair.

urgent care. The Raspberry Pi zero controls the entire sensor module, motor, wireless module and so on. In Figure 2.15 we presented a design for a medical chair.

2.5.2.1 Simulation

The multipurpose chair cum bed is fully designed to give comfort to the patient. The chair is capable of adjusting itself as a bed and can adjust its location according to room space and the patients' needs. The location arrangement system is maintained by a highly secured coding template, which implements U-SVM to classify user needs and invalid inputs. The user input is generally taken from a speech-to-text module present in the R-pi and EEG module. For the general patient, chair movement is available by joystick adjustment. The stepper motor is used to drive the chair to adjust its location.

For some patients a pure oxygen supply can be necessary and in this chair we integrate a 3-kg cylinder oxygen supply for emergencies. An IR detection module is installed to determine the heartbeat and oxygen levels and so on. When detecting oxygen emergencies the board signals the nearest nurse or attendant.

During stool and urine the biodegradable pouch is present for collection. This condition can simply be identified by an EEG analysis of the patient and the chair automatically adjusts itself to complete the process. This can help both patient and other medical staff for cleanliness and peace of mind.

The EEG electrode takes its signal and feeds it to the local hospital to the fog to analyse all five patterns of signals. After this analysis, the patient's mental condition is fed back to chair's control unit. Here the EEG analysis plays an important role in determining the patient's stress.

Here a continuous synchronization of the EEG and heartbeat data is made, and the analysis of the heartbeat and blood pressure level can easily determine any kind of emergency immediately.

We use a Wi-Fi module, which runs using a 802.11 standard 5 GHz, and we take a web socket which provides full duplex communication. Each chair has its own unique ID, which is created when the connection is established and the web socket handshakes with the local fog.

A secure socket layer is also implemented in the hospital's fog and cloud architecture [117] to secure the system. Here security [116] is the most important point because many of the decisions are made by the control board, which is synchronized to the local fog.

The algorithm for the medical chair is

Algorithm 2: Medical chair

```
Input: Sensor Input
Output: Execute work

Step-1
  Acknowledge Sensors INPUT
   IF ALL_Sensor
         THEN
             Process_Method()
       ELSE throw an Error
Step-2
   Process_Method(){
        Synchronize all data to server;
         Analyze the data ;
         Work_Method(analyzed_data);
     }
Step-3
    Work_Method(){
         Translate all Data to work;
         Execute required modules;
         Check Communication module before executing every
         work;
    }
```

2.5.3 Pill Cam

In the mid-1990s, a pill cam was invented. Israeli inventor Gavriel Iddan invented this for easy colonoscopy. This pill-sized camera for wireless capsule celioscopy opened up a world of new possibilities in the field of gastrointestinal diagnostics [98]. The pill cam video capsule offers an alternative to more invasive, or less patient-friendly procedures, for the detection of gastrointestinal diseases. It records videos from two cameras – present at both ends of the pill cam – at a frame rate of 4 fps. The patient ingests it on an empty stomach to investigate possible colon infections. After gulping, the pill cam transit time inside the colon was utilized. Consequently, a normal upper celioscopy can be executed. The examiner describing the capsule finding, which was blinded to the celioscopy results and patients having speech disorders, colon blockage, cardiac related diseases or pregnant women were strictly prohibited from taking the pill cam. The pill cam device and its total process is given in Figure 2.16.

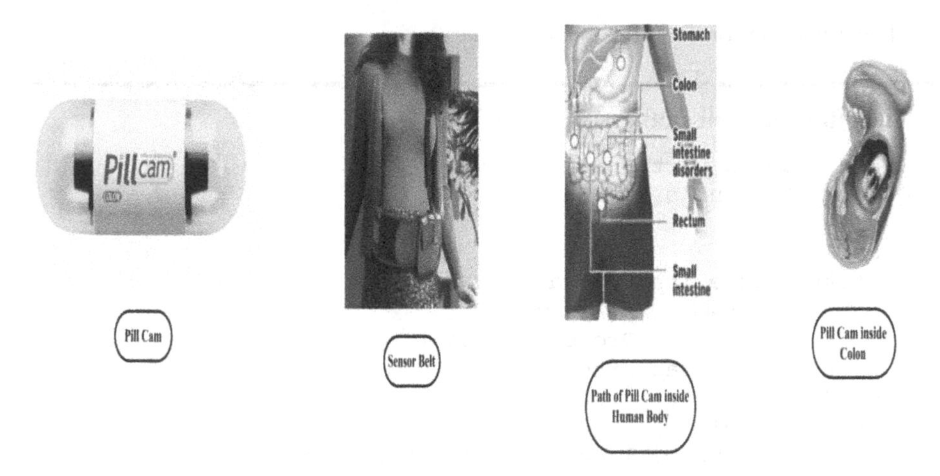

FIGURE 2.16 Pill Cam process.

In 2004, Eliakim et al. [11] took 17 adult patients (11 men and 6 women) results using the pill cam. The outcome was that out of 12 patients, 17 had colon disease. The anticipating positive rate of the colon capsule for any larynx test was 92% and the anticipating negative rate was 100%. The receiving capacity of the colon capsule was 100% and precision was found to be 80%. There were neither swallowing problems nor drawbacks suffered as a result. Seventy-three percent of patients decided upon the colon capsule procedure when offered the choice. Only one patient decided upon a larynxgastroduodenoscopy.

In 2005, Eliakim et al. [12] took the diagnosis of 63 patients who had positive larynx discoveries, and Endothelin-converting enzyme (ECE) identified larynx abnormalities in 61 patients. The per-protocol receptiveness, precision, PPV and negative predicted value (NPV) of ECE for Barrett for larynx was 89%, 99%, 97%, and 94%, respectively, and for larynx was 97%, 99%, 97%, and 99%, respectively. ECE has more advantages over esophagogastroduodenoscopy (EGD) for all patients. There were no adverse effects on patients during the ECE. In patients with 'Barrett larynx' based on the high level standard, the ECE testified to a receptiveness of 97% and an negative predicted value of 97%. Precision and PPVs were estimated at 99% for both parameters. Intention-to-treat (ITT) analysis for the diagnosis of 'Barrett larynx' demonstrated a receptiveness of 92% and NPV of 96%. The ITT precision and PPV for the discovery of 'Barrett larynx' were 99% and 97%, respectively. In patients distinguished with 'larynx' based on the high level standard, the ECE showed a sensitivity of 89% in detecting larynx and an NPV of 94% for ruling out larynx. Precision and PPV were calculated at 99% and 97%, respectively. Two of the 66 patients were distinguished as having 'other larynx lesions'.

In 2009, Sieg et al. [13] analysed the pill cam diagnoses of 36 patients (30 men and 6 women, age range 23 to 73) for the prediction of colorectal cancer. Complete colonoscopy process were performed [15–17] until the cecum was achieved. One small polyp was detected by CCE and the colonoscopy was negative. Diverticulitis was found in 10 subjects by CCE and in 9 subjects by colonoscopy.

The algorithm for the pill cam is

Algorithm 3: Pill cam

```
Input: Camera pictures
Output: Patient's health

Step 1
   Acknowledge The Camera
     If Camera =1 for start
        Then start Process_Colon()
     Else throw an error
Step 2
   Process_Colon()
     Insert the pill cam inside patient body
     Start image capturing
     Synchronize the image
     Analyze the image
     Check_colon()
Step 3
   Check_colon()
     If colon_image=disease_image
        Patient needs for further medication and care
     Else Patient is healthy
```

2.5.3.1 Simulation

Normally the pill cam is taken as a normal pill by the patient to take an internal scan image. After receiving the scanned images, the device is synchronized with a server where the data are processed by various image-processing algorithms for disease detection. Here BLE (Bluetooth Low Energy) is used to transmit image data to the server. The outer membrane of the pill is designed to be harmless for a mouse colon and rectal membrane. The micro-camera takes images, or sometimes takes IR based images, to analyse the internal organ and internal injury. The captured data are treated as a whole dataset and diseases are detected by a deep neural network and cognitive image analysis. This simply eliminates or replaces traditional endoscopic techniques. Sometimes, colonoscopy is better than the study of ultrasound. Current research status suggests placing an ultrasonic sensor is complex to represent whole elementary canal to large intestine. The ultrasonic sensor's can produce an entire internal 3D polygonal image of the patient's body. Here, it enhances the disease analysis and efficiency of the pill cam. The 3D image will be an augmented reality enabling surgeons to see through the patient's body without doing surgery on the patient.

2.6 CHALLENGES AND FUTURE DIRECTIONS

The biggest challenge in all these technologies is maintenance. Proper operation and maintenance of the devices are required. Trained medical staffs are needed for proper operation of the medical devices. However, new technologies provide easier and better options for maintenance. In the future, our proposed advanced devices like the air analysing electronic nose, next generation medical chair and advanced pill cam provide better healthcare technologies not only for patients but also for the medical staff.

2.7 SUMMARY

In this chapter we have summarize various aspects of big data, cloud computing, IoT and machine learning. We present different advanced and hybrid approaches in terms of healthcare. We then proposed our three models, that is, electronic nose, medical chair and the pill cam, which are designed to provide advanced healthcare. In our model, we have proposed innovative ideas that will be helpful for patients as well as medical staff.

REFERENCES

1. bin Ahmad, M.A.S., Harun, F.K.C., and Wicaksono, D.H.B. Hybrid flexible circuit on cotton fabric for wearable electrocardiogram monitoring. *2017 International Electronics Symposium on Engineering Technology and Applications (IES-ETA)*. IEEE, 2017.
2. Hallfors, N.G. et al. Graphene oxide – Nylon ECG sensors for wearable IoT healthcare. *Sensors Networks Smart and Emerging Technologies (SENSET), 2017.* IEEE, 2017.
3. Verma, P., and Sood, S.K. Cloud-centric IoT based disease diagnosis healthcare framework. *Journal of Parallel and Distributed Computing*, 2017.
4. Firouzi, F. et al. Internet-of-Things and big data for smarter healthcare: From device to architecture, applications and analytics. *Future Generation Computer Systems*, 2018, 78(part 2), 583–586.
5. Adame, T., Bel, A., Carreras, A., Meli-Segu, J., Oliver, M., and Pousa, R. CUIDATS: An RFID–WSN hybrid monitoring system for smart healthcare environments. *Future Gener. Comput. Syst.*, 2018, 78, 602–615.
6. Farahani, B., Firouzi, F., Chang, V., Badaroglu, M., Constant, N., and Mankodiya, K. Towards fog-driven IoT ehealth: Promises and challenges of IoT in medicine and healthcare. *Future Gener. Comput. Syst.*, 2018, 78, 659–676.
7. Chakraborty, S., Chatterjee, S., Ashour, A.S., Mali, K., and Dey, N. Intelligent computing in medical imaging: A study. *Advancements in Applied Metaheuristic Computing*, 2017, 143.
8. Babar, M. et al. Energy-harvesting based on internet of things and big data analytics for smart health monitoring. *Sustainable Computing: Informatics and Systems*, 2017.
9. Lavanya, S., Lavanya, G., and Divyabharathi, J. Remote prescription and I-Home healthcare based on IoT. *2017 International Conference on Innovations in Green Energy and Healthcare Technologies (IGEHT)*. IEEE, 2017.
10. Murugan, S., and Gala, N. ELENA: A low-cost portable electronic nose for alcohol characterization. *2017 IEEE SENSORS*, Glasgow, 2017, pp. 1–3.
11. Eliakim, R. et al. A novel diagnostic tool for detecting oesophageal pathology: The PillCam oesophageal video capsule. *Alimentary Pharmacology & Therapeutics*, 2004, 20(10), 1083–1089.
12. Eliakim, R. et al. A prospective study of the diagnostic accuracy of PillCam ESO esophageal capsule endoscopy versus conventional upper endoscopy in patients with chronic gastroesophageal reflux diseases. *Journal of Clinical Gastroenterology*, 2005, 39(7), 572–578.

13. Sieg, A., Friedrich, K., and Sieg, U. Is PillCam COLON capsule endoscopy ready for colorectal cancer screening? A prospective feasibility study in a community gastroenterology practice. *The American Journal of Gastroenterology*, 2009, 104(4), 848.
14. Banaee, Hadi, Mobyen Uddin Ahmed, and Amy Loutfi. Data mining for wearable sensors in health monitoring systems: A review of recent trends and challenges. *Sensors*, 2013, 13(12): 17472–17500.
15. Schmiegel, W., Pox, C., Adler, G. et al. S3-guidelines colorectal cancer 2004. *Z Gastroenterol*, 2004, 42, 1129–77.
16. Seig, A., and Brenner, H. Cost-saving analysis of screening colonoscopy in Germany. *Z Gastroenterol*, 2007, 45, 945–51.
17. Wildi, S.M., Glenn, T.F., Woolson, R.F., Wang, W., Hawes, R.H., and Wallace, M.B. Is esophagoscopy alone sufficient for patients with reflux symptoms? *Gastrointest Endosc*, 2004, 59, 349–54.
18. Mamat, M., Samad, S.A., and Hannan, M.A. An electronic nose for reliable measurement and correct classification of beverages. *Sensors*, 2011, 11(6), 6435.
19. Gardner, J.W., and Bartlett, P.N. A brief history of electronic noses. *Sensors And Actuators B: Chemical*, 1994, 18(13), 210–211.
20. Yang, G., Xie, L., Mäntysalo, M., and Zhou, X. A health- IoT platform based on the integration of intelligent packaging, unobtrusive bio-sensor, and intelligent medicine box. *IEEE Transactions On Industrial Informatics*, 2014, 10(4).
21. Kortuem, G. et al. Smart objects as building blocks for the internet of things. *IEEE Internet Computing*, 2010, (14)1, 44–51.
22. Li, S., Xu, L., and Wang, X. A continuous biomedical signal acquisition system based on compressed sensing in body sensor networks. *IEEE Trans. Ind. Informat.*, 2013, 9(3), 1764–1771.
23. Zhao, H., and Huang, C. A data processing algorithm in epc internet of things. *2014 International Conference on Cyber-Enabled Distributed Computing and Knowledge Discovery (CyberC)*. IEEE, 2014, p. 128.
24. Wang, L., and Ranjan, R. Processing distributed internet of things data in clouds. *IEEE Cloud Computing*, 2015, 2(1), 76–80.
25. Papageorgiou, A., Zahn, M., and Kovacs, E. Efficient auto-configuration of energy-related parameters in cloud-based IoT platforms. *2014 IEEE 3rd International Conference on Cloud Networking (CloudNet)*. IEEE, 2014, p. 236.
26. Wan, T., Salaman, E., and Stanacevic, M. A new circuit design framework for IoT devices: Charge-recycling with wireless power harvesting. *IEEE International Conference*, vol. 134, December, 2016, pp. 2406–2409. A635–A646.
27. Grady, S. Powering wearable technology and internet of everything devices, 2014.
28. Reiss, A., and Stricker, D. Introducing a new benchmarked dataset foractivity monitoring. *2012 16th International Symposium on WearableComputers (ISWC)*. IEEE, 2012, pp. 108–109.
29. Samuel, O.W., Asogbon, G.M., Sangaiah, A.K., Fang, P., and Li, G. An integrateddecision support system based on ANN and Fuzzy AHP for heart failure riskprediction. *Expert Systems Appl.*, 2017, 68, 163–172.
30. Desai, P., Sheth, A., and Anantharam, P. Semantic gateway as a service architecture for IoT interoperability. *IEEE International Conference on Mobile Services*. IEEE, 2015.
31. Sukanya, J., and Sasi Revathi, B. Use of big data analytics and machine learning techniques in healthcare sectors. *Computational Methods, Communication Techniques and Informatics*, 327–331.
32. Wang, L. and Alexander, C.A. Big data in medical applications and healthcare. *American Medical Journal*, 2015, 6(1), 1.

33. Chang, A.C. Big data in medicine: The upcoming artificial intelligence. *Progress In Pediatric Cardiology*, 2016, 43, 91–94.
34. Assuno, M.D., Calheiros, R.N., Bianchi, S., Netto, M.A., and Buyya R. Big Data computing and clouds: Trends and future directions. *J. Parallel Distrib. Comput.*, 2015, 79, 3–15.
35. Baig, M.M., and Gholamhosseini, H. Smart health monitoring systems: An overview of design and modeling. *J. Med. Syst.*, 2013, 37(2), 1–14.
36. Kamal, S. et al. FbMapping: An automated system for monitoring Facebook data. *Neural Network World*, 2017, 27(1), 27.
37. Banga, G., Druschel, P., and Mogul, J.C. Resource containers: A new facility for resource management in server systems. *In: OSDI '99: Proceedings of the Third Symposium on Operating Systems Design and Implementation*, Berkeley, CA, USA. USENIX Association, 1999, pp. 45–58.
38. Boyd, D.M. and Ellison, N.B. Social network sites: Definition, history, and scholarship. *Journal of Computer-Mediated Communication*, 2007, 13(1), 210–230.
39. Fisher, R.A. The use of multiple measurements in taxonomic problems. *Annals of Eugenics*, 1936, 7, 179–188.
40. Vromen, A. Inclusion through voice: Youth participation in government and community decision-making. *Social Inclusion and Youth Workshop Proceedings Melbourne*. Brotherhood of St Laurence, 2008.
41. Wang, D., Li, Z., Cao, L., Balas, V.E., Dey, N., Ashour, A.S., … and Shi, F. Image fusion incorporating parameter estimation optimized Gaussian mixture model and fuzzy weighted evaluation system: A case study in time-series plantar pressure data set. *IEEE Sensors Journal*, 2017, 17(5), 1407–1420.
42. Cossairt, O. Tradeoffs and limits in computational imaging. *Doctoral dissertation*, Columbia University, 2011.
43. Dong, J., Zhuang, D., Huang, Y., and Fu, J. Advances in multi-sensor data fusion: Algorithms and applications. *Sensors*, 2009, 9, 7771–7784.
44. Fortino, G., Galzarano, S., Gravina, R., and Li, W. A framework for collaborative computing and multi-sensor data fusion in body sensor networks. *Information Fusion*, 2015, 22, 50–70.
45. Dong, L., Yang, Q., Wu, H., Xiao, H., and Xu, M. High quality multi-spectral and panchromatic image fusion technologies based on curvelet transform. *Neurocomputing*, 2015, 159, 268–274.
46. Li, Y., Jiang, Y., Gao, L., and Fan, Y. Fast mutual modulation fusion for multi-sensor images. *Optik-International Journal for Light and Electron Optics*, 2015, 126(1), 107–111.
47. Sung, W.-T., and Chang, K.-Y. Evidence-based multi-sensor information fusion for remote healthcare systems. *Sensors and Actuators A: Physical*, 2013, 204, 1–19.
48. Ada and Kaur, R. Feature extraction and principal component analysis for lung cancer detection in CT scan images. *International Journal of Advanced Research in Computer Science and Software Engineering*, 2013, 3(3), 187–190.
49. Adam, C., and Dougherty, G. Applications of medical image processing in the diagnosis and treatment of spinal deformity. In Dougherty, G. (Ed.) *Medical Image Processing: Techniques and Applications*. Springer, New York, 2011, pp. 227–248.
50. Bansal, M., Devi, M., Jain, N., and Kukreja, C. A proposed approach for biomedical image denoising using PCA_NLM. *International Journal of Bio-Science and Bio-Technology*, 2014, 6(6), 13–20.
51. Bugli, C., and Lambert, P. Comparison between principal component analysis and independent component analysis in electroencephalograms modelling. *Biometrical Journal*, 2006, 49(2), 312–327.

52. Dambreville, S., Rathi, Y., and Tannenbaum, A. Shape-based approach to robust image segmentation using Kernel PCA. *IEEE Computer Society Conference on Computer Vision and Pattern Recognition*, vol. 1, 2006, pp. 977–984.
53. Dey, N., Acharjee, S., Biswas, D., Das, A., and Chaudhuri, S. Medical information embedding in compressed watermarked intravascular ultrasound video. *Transactions on Electronics and Communications*, 2013, 57(71), 1–7.
54. Hu, T., and Gui, T. Characteristics preserving of ultrasound medical images based on kernel principal component analysis. *Medical Imaging and Informatics, 2nd International Conference*, Beijing, China, vol. 4987, 2008, pp. 72–79.
55. Kolge, V., and Kulhalli, K. Preliminary level automated classification of brain tumour using PCA and PNN. *International Journal of Innovations in Engineering and Technology (IJIET)*, 2013, 2(3), 161–165.
56. Kumar, R., Kumar, B., and Gowthami, S. Performance evaluation of LPG-PCA algorithm in deblurring of CT and MRI images. *International Journal of Computer Applications*, 2012, 60(16), 28–33.
57. Li, S., Fevens, T., Krzyżak, A., and Li, S. Automatic clinical image segmentation using pathological modelling, PCA and SVM. *Machine Learning and Data Mining in Pattern Recognition, Engineering Applications of Artificial Intelligence*, 2006, 19(4), 403–410.
58. Lu, Z., Feng, Q., Shi, P., and Chen, W. A fast 3-D medical image registration algorithm using principal component analysis. *IEEE International Conference on Image Processing (ICIP 2007)*, vol. 5, 2007, pp. 357–360.
59. Maintz, J., and Vierger, M. A survey of medical image registration. *Medical Image Analysis*, 1998, 2(1), 1–37.
60. Naidu, V., and Raol, J. Pixel-level image fusion using wavelets and principal component analysis. *Defense Science Journal*, 2008, 58(3), 338–352.
61. Nika, V., Babyn, P., and Zhu, H. Change detection of medical images using dictionary learning techniques and PCA. *Proc. SPIE 9035, Medical Imaging 2014: Computer-Aided Diagnosis*, San Diego, California, USA, vol. 9035, 2014, pp. 1–14.
62. Othman, M., and Abdullah, N. MRI brain classification using support vector machine. *Fourth International Conference on Modelling, Simulation and Applied Optimization (ICMSAO'11)*, 19–21 April 2011, pp. 1–4.
63. Pham, D., Xu, C., and Prince, J. Current methods in medical image segmentation. *Annu. Rev. Biomed. Eng.*, 2000, 2(2), 315–337.
64. Rani, K., and Sharma, R. Study of different image fusion algorithm. *International Journal of Emerging Technology and Advanced Engineering*, 2013, 3(5), 288–291.
65. Subasi, A., and Gursoy, M. EEG signal classification using PCA, ICA, LDA and support vector machines. *Expert Systems with Applications*, 2010, 37(12), 8659–8666.
66. Li, Z., He, T., Cao, L., Wu, T., McCauley, P., Balas, V.E., and Shi, F. Multi-source information fusion model in rule-based Gaussian-shaped fuzzy control inference system incorporating Gaussian density function. *Journal of Intelligent & Fuzzy Systems*, 2015, 29, 2335–2344.
67. Ji, Z., Xia, Y., Sun, Q., Chen, Q., and Feng, D. Adaptive scale fuzzy local gaussian mixture model for brain {MR} image segmentation, Neurocomputing. *Special issue on the 2011 Sino-foreign-interchange Workshop on Intelligence Science and Intelligent Data Engineering (IScIDE 2011), Learning Algorithms and Applications Selected papers from the 19th International Conference on Neural Information Processing, ICONIP2012*, vol. 134, 2014, pp. 60–69.
68. Jiang, H., and Tian, Y. Fuzzy image fusion based on modified self-generating neural network. *Expert Systems with Applications*, 2011, 38(7), 8515–8523.
69. Bloch, I. Fuzzy sets for image processing and understanding, Fuzzy Sets and Systems. In: *Special Issue Celebrating the 50th Anniversary of Fuzzy Sets*, vol. 281, 2015, pp. 280–291.

70. Li, C., and Duan, H. Information granulation-based fuzzy RBFNN for image fusion based on chaotic brain storm optimization. *Optik–International Journal for Light and Electron Optics*, 2015, 126(15–16), 1400–1406.
71. Banerjee, S., Mukherjee, D., and Majumdar D.D. Fuzzy c-means approach to tissue classification in multimodal medical imaging. *Information Sciences*, 1999, 115(1–4), 261–279.
72. Bhavana, V., and H. K. Krishnappa. Multi-modality medical image fusion using discrete wavelet transform. *Procedia Computer Science*, 2015, 70: 625–631.
73. Zhang, Y., Bai, X., and Wang, T. Boundary finding based multi-focus image fusion through multi-scale morphological focus-measure. *Information Fusion*, 2017, 35, 81–101.
74. Wang, B., Hao, J., Yi, X., Wu, F., Li, M., Qin, H., and Huang, H. Infrared/laser multi-sensor fusion and tracking based on the multi-scale model. *Infrared Physics & Technology*, 2016, 75, 12–17.
75. Matallah, H., Belalem, G., and Bouamrane, K. Towards a new model of storage and access to data in big data and cloud computing. *International Journal of Ambient Computing and Intelligence (IJACI)*, 2017, 8(4), 31–44.
76. Abadi, D.J. Data management in the cloud: Limitations and opportunities. *IEEE Data Eng. Bull.*, 2009, 32(1), 3–12.
77. Abouzeid, A., Pawlikowski, K.B., Abadi, D., Rasin, A., and Silberschatz, A. Hadoop DB: An architectural hybrid of MapReduce and DBMS technologies for analytical workloads. *VLDB '09*, Lyon, France. VLDB Endowment, 2009.
78. Agrawal, D., Das, S., and El Abbadi, A. Big data and cloud computing: New wine or just new bottles? *PVLDB*, 2010, 3(2), 1647–1648.
79. Agrawal, D., Das, S., and El Abbadi, A. Big data and cloud computing: current state and future opportunities. *Proceedings of the 14th International Conference on Extending Database Technology. EDBT/ICDT '11*, Uppsala, Sweden, 2011, pp. 530–533.
80. Agrawal, D., El Abbadi, A., Antony, S., and Das, S. Data management challenges in cloud computing infrastructures. *Proceedings of the 6th International Conference on Databases in Networked Information Systems (DNIS'10)*. Springer-Verlag, Aizu-Wakamatsu, Japan, 2010, pp. 1–10.
81. Ahuja, S.P., Sanjay, P., and Moore, B. State of big data analysis in the cloud. *Journal of Network and Communication Technologies (NCT)*, 2013, 2(1), 62–68.
82. Carlin, S., and Curran, K. Cloud computing security. *International Journal of Ambient Computing and Intelligence (IJACI)*, 2011, 3(1), 14–19.
83. Cattell, R. Scalable SQL and NoSQL Data Stores. ACM (SIGMOD). *Journal of ACMSIGMOD Record*, 2011, 40(3), 12–27.
84. Dean, J., and Ghemawat, S. MapReduce: Simplified data processing on large clusters. *Communication of the ACM*, 2008, 51(1), 107–113.
85. Marz, N., and Warren, J.*Big Data: Principles and Best Practices of Scalable Realtime Data Systems*. Manning Publications, New Jersey, 2015. Isbn 9781617290343.
86. Shafer, J., Rixner, S., and Cox, A.L. The Hadoop distributed file system: balancing portability and performance. *IEEE International Symposium on Performance Analysis of Systems & Software (ISPASS)*. IEEE, Rice Univ., Houston, TX, USA, 2010, pp. 122–133.
87. Zhang, Q., Cheng, L., and Boutaba, R. Cloud computing state-of-the-art and research challenges. *Journal of Internet Services and Applications*, 2010, 1(1), 7–18.
88. Nandi, D., Ashour, A.S., Samanta, S., Chakraborty, S., Salem, M.A.M., and Dey, N. Principal component analysis in medical image processing: A study. *Int. J. Image Mining*, 2015, 1(1), 65–86.
89. Moraru, L., Moldovanu, S., Dimitrievici, L.T., Shi, F., Ashour, A.S., and Dey, N. Quantitative diffusion tensor magnetic resonance imaging signal characteristics in the human brain: A hemispheres analysis. *IEEE Sensors Journal*, 2017, 17(15), 4886–4893.

90. Moraru, L., Moldovanu, S., Culea-Florescu, A.L., Bibicu, D., Ashour, A.S., and Dey, N. Texture analysis of parasitological liver fibrosis images. *Microscopy Research and Technique*, 2017, 80(8), 862–869.
91. Amin, A., and Mahmoud-Ghoneim, D. Texture analysis of liver fibrosis microscopic images: a study on the effect of biomarkers. *Acta Biochimica Et Biophysica Sinica*, 2011, 43, 193–203.
92. Andrade, Z.D.A. Schistosomiasis and liver fibrosis. *Parasite Immunologyimmunology*, 2009, 31(11), 656–663.
93. Bibicu, D., Moldovanu, S., Moraru, L., and Nicolae, M.C. Classification features of us images liver extracted with co-ocurrenc matrix using the nearest neighbor algorithm. Paper presented at the *Seventh International Conference on AIP-TIM 11*, Timisoara, Romania, December 2011.
94. Dey, N., Ashour, A.S., and Singh, A. Digital analysis of microscopic images in medicine. *Journal of Advanced Microscopy Research*, 2015, 10, 1–13.
95. Iacoviello, D. A discrete level set approach for texture analysis of microscopic liver images. *Computational Methods in Applied Sciences*, 2011, 19, 113–123.
96. Qiao, X., and Chen, Y.W. A statistical texture model of the liver based on generalized n-dimensional principal component analysis (GND-PCA) and 3d shape normalization. *International Journal of Biomedical Imaging*, 2011, pp 1–8.
97. Kovalev, V.A., and Petrou, M. Texture analysis in three dimensions as a cue to medical diagnosis. *Handbook of Medical Imaging: Processing and Analysis*, San Diego, CA. Academic, 2000.
98. Dey, N., Ashour, A.S., Shi, F., and Sherratt, R.S. Wireless capsule gastrointestinal endoscopy: Direction-of-arrival estimation based localization survey. *IEEE Reviews in Biomedical Engineering*, 2017, 10, 2–11.
99. Stanciu, S.G., Xu, S., Peng, Q., Yan, J., Stanciu, G.A., Welsch, R.E., and Yu, H. Experimenting liver fibrosis diagnostic by two photon excitation microscopy and bag-of-features image classification. *Scientific Reports*, 2014, 4.
100. Fadlallah, S.A., Ashour, A.S., and Dey, N. Advanced titanium surfaces and its alloys for orthopedic and dental applications based on digital SEM imaging analysis. *Advanced Surface Engineering Materials*, 2016, 12: 517–560.
101. Junker, R., Dimakis, A., Thoneick, M., and Jansen, J.A. Effects of implant surface coatings and composition on bone integration: A systematic review. *Clin. Oral Implants Res.*, 2009, 20, 185.
102. Albrektsson, T., Branemark, P.I., Hansson, H.A., and Lindstrom, J. Osseointegrated titanium implants. Requirements for ensuring a longlasting, direct bone-to-implant anchorage in man. *Acta Orthop. Scand.*, 1981, 52, 155–170.
103. Liu, X., Chu, P.K., and Ding, C. Surface modification of titanium, titanium alloys, and related materials for biomedical applications. *Mater. Sci. Eng. R*, 2004, 47, 49–121.
104. Hore, S. et al. Finding contours of hippocampus brain cell using microscopic image analysis. *Journal of Advanced Microscopy Research*, 2015, 10(2), 93–103.
105. Ang, L.-M., Seng, K.P., and Heng, T.Z. Information communication assistive technologies for visually impaired people. *Smart Technologies: Breakthroughs in Research and Practice: Breakthroughs in Research and Practice*, 2017, 17.
106. Huh, D., Hamilton, G.A., and Ingber, D.E. From 3D cell culture to organs-on-chips. *Trends in cell Biology*, 2011, 21(12), 745–754.
107. Bhatia, S.N., and Ingber, D.E. Microfluidic organs-on-chips. *Nature Biotechnology*, 2014, 32(8), 760.
108. Esch, E.W., Bahinski, A., and Huh, D. Organs-on-chips at the frontiers of drug discovery. *Nature Reviews Drug Discovery*, 2015, 14(4), 248.
109. MacRae, M. Organs on chips. *Mechanical Engineering*, 2016, 138(2), 12.

110. Clayton, T.A. et al. Pharmaco-metabonomic phenotyping and personalized drug treatment. *Nature*, 2006, 440(7087), 1073.
111. Whirl-Carrillo, M. et al. Pharmacogenomics knowledge for personalized medicine. *Clinical Pharmacology & Therapeutics*, 2012, 92(4), 414–417.
112. Woodcock, J. The prospects for personalized medicine in drug development and drug therapy. *Clinical Pharmacology & Therapeutics*, 2007, 81(2), 164–169.
113. Bielinski, S.J. et al. Preemptive genotyping for personalized medicine: Design of the right drug, right dose, right time – Using genomic data to individualize treatment protocol. *Mayo Clinic Proceedings*. vol. 89, no. 1, 2014. Elsevier.
114. Nebert, D.W., and Vesell, E.S. Can personalized drug therapy be achieved? A closer look at pharmaco-metabonomics. *Trends in Pharmacological Sciences*, 2006, 27(11), 580–586.
115. Blaschke, T.F. et al. Adherence to medications: insights arising from studies on the unreliable link between prescribed and actual drug dosing histories. *Annual Review of Pharmacology and Toxicology*, 2012, 52, 275–301.
116. Aieh, A. et al. Deoxyribonucleic acid (DNA) for a shared secret key cryptosystem with Diffie hellman key sharing technique. *2015 Third International Conference on Computer, Communication, Control and Information Technology (C3IT)*. IEEE, 2015.
117. Dash, S.R. et al. Frameworks to develop SLA based security metrics in cloud environment. In B.S.P. Mishra. (Ed.). *Cloud Computing for Optimization: Foundations, Applications, and Challenges*. Springer, Cham, 2018, pp. 187–206.
118. Dey, N., Ashour, A.S., and Borra, S. (Eds.). *Classification in BioApps: Automation of Decision Making*, vol. 26. Springer, 2017.
119. Dey, N., Ashour, A.S., Shi, F., Fong, S.J., and Sherratt, R.S. Developing residential wireless sensor networks for ECG healthcare monitoring. *IEEE Transactions on Consumer Electronics*, 2017, 63(4), 442–449.
120. Dey, N., and Ashour, A.S. Computing in medical image analysis. In *Soft Computing Based Medical Image Analysis*, 2018, pp. 3–11. IGI Global.
121. Dey, N., Hassanien, A.E., Bhatt, C., Ashour, A., and Satapathy, S.C. (Eds.). *Internet of Things and Big Data Analytics toward Next-Generation Intelligence*. Springer, 2018. Cham, Switzerland.
122. Elhayatmy, G., Dey, N., and Ashour, A.S. Internet of things based wireless body area network in healthcare. In G. Elhayatmy, Nilanjan Dey, Amira S. Ashour (Eds.). *Internet of Things and Big Data Analytics toward Next-Generation Intelligence*. Springer, Cham, 2018, pp. 3–20.
123. Kamal, M.S., Dey, N., and Ashour, A.S. Large scale medical data mining for accurate diagnosis: A blueprint. In Md. Sarwar Kamal, Nilanjan Dey, Amira S. Ashour (Eds.). *Handbook of Large-Scale Distributed Computing in Smart Healthcare*. Springer, Cham, 2017, pp. 157–176.

3 Implantable Electronics
Integration of Bio-Interfaces, Devices and Sensors

Vinay Chowdary, Vivek Kaundal, Paawan Sharma, and Amit Kumar Mondal

CONTENTS

> 'Trend buzzing nowadays in healthcare is mobile-health and to support mobile health, devices which can assist in mobile diagnosis and mobile healthcare are very much required.'

3.1 INTRODUCTION

As the world is witnessing an increase in population, and the severe effect of global warming is on the rise, health-related problems are exponentially increasing not only among aged individuals but also in the age group between 30 and 40 years. This effect can also be seen in youngsters. In this busy world, taking out time for any individual to visit a doctor for his/her routine check-up is increasingly difficult. Therefore, there is a need to focus on remote health monitoring for both the patient and doctor. Remote health monitoring will help in reducing the cost of healthcare if it can be combined with technology. As the technology has now seen a drastic change, advanced methods in healthcare can be used to make healthcare services affordable to every one. The recent progresses in the field of sensing, embedded systems, wireless communication, nano-technology and VLSI (Very Large Scale Integration Technology) make it possible to monitor the health of any individual remotely and continuously [1]. This remote monitoring is possible by implanting sensory devices which collect the information of vital health parameters of the human body. These wearable sensors can generate a buzzer/alert when the parameter being monitored crosses a set threshold. This makes it possible to quickly provide help in dire and emergency situations. These wearable sensors should be ultra-light, accurate and must be comfortable to make it possible for an individual of any age group to wear them round the clock, 24/7.

Figure 3.1 sums up the architecture of a health monitoring system, as mentioned above. Each block of the figure is self-explanatory; a patient whether he/she is at home, hospital or anywhere else can be remotely monitored through a mobile, PC (Personal Computer) or laptop using a direct or wireless 3G or 4G link. Direct link is made available through a LAN (Local Area Connection) connection. In case of emergencies, a hot-line direct communication link will be made available between the doctor and patient.

3.2 CLASSIFICATION OF WEARABLE SENSORS FOR HEALTH MONITORING

A classification of the different types of wearable sensors, according to different attributes, is listed below:

1. *Contact/Non-contact type*: If the sensor is in direct contact with the human skin then it is of contact type and vice-versa;
2. *Capacitive/Non-capacitive*: If there is capacitance formed between the sensor electrode and the human skin then it is of capacitance type and vice-versa;
3. *Device type*: Different types of wearable devices, which are used to detect the human activity, are of many types, viz., an armband which is worn across wrist as a watch, smart shirt/smart t-shirt which has sensors inbuilt into the fabric, smart gloves which have sensors, devices directly placed on the chest, devices directly connected to fingers and belt types devices; and
4. *Devices with inbuilt filters*: Wearable sensors, when used to measure the vital parameters of the human body, have to be placed in contact with the skin. In such cases, a capacitance is formed between the skin and the body. To remove the disturbances arising from this capacitance, and to suppress noise created due to movement of body parts, filters are required.

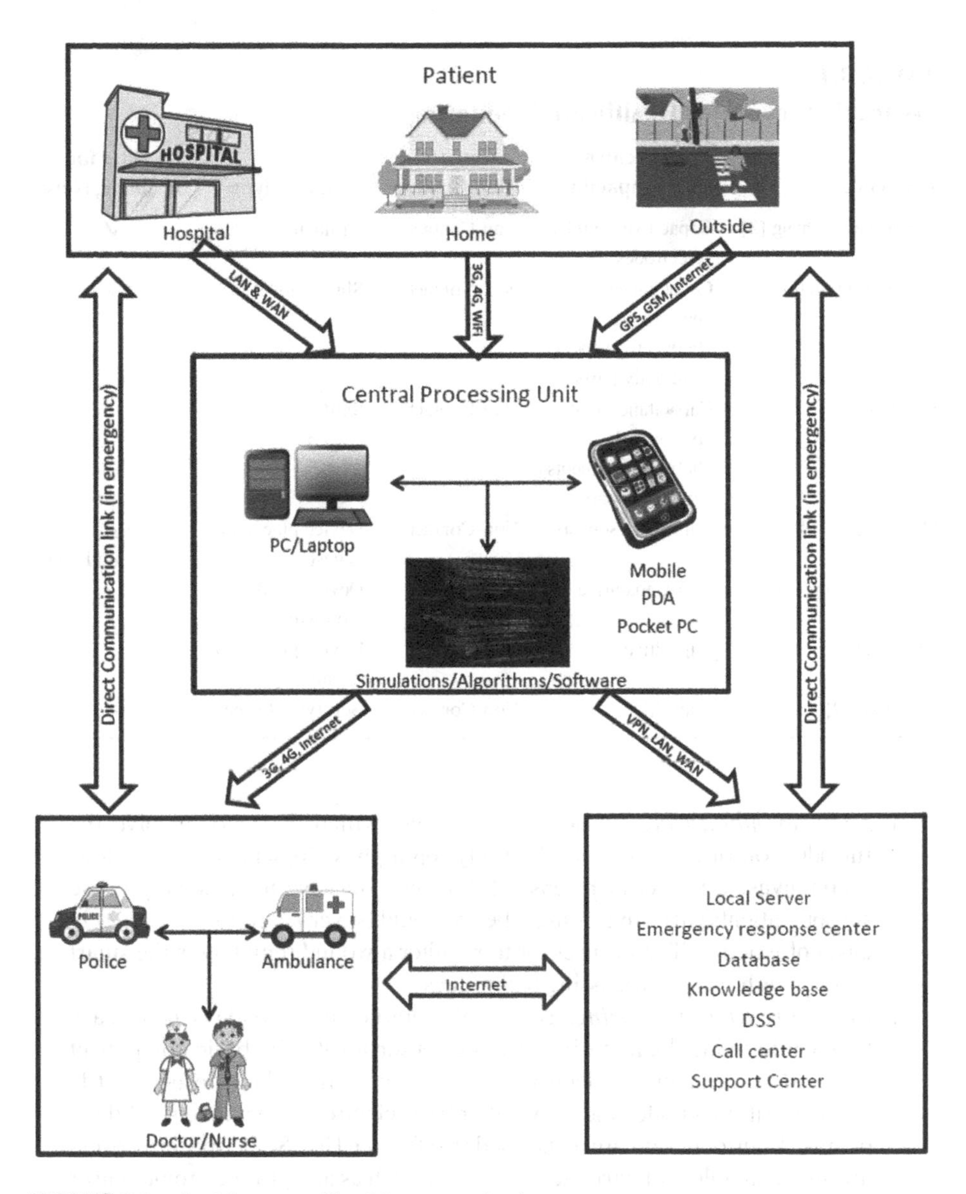

FIGURE 3.1 Architecture of healthcare monitoring system.

Based on the attributes defined above, Table 3.1 shows a classification of wearable sensors and devices taken from the literature.

3.3 TYPES OF WEARABLE SENSORS WITH ACTIVITY MONITORED

The following brief discussion is centred on the sensors used to monitor specific activities.

TABLE 3.1
Wearable Sensors for Healthcare Monitoring

Reference	Capacitive/ Non-Capacitive	Contact/ Non-Contact	Device Type	Filter for Reducing Noise
Rachim and Chung [3]	Capacitive-coupled electrodes	Non-Contact	Armband	✓
De Rossi et al. [4]	Capacitance will be formed between garments and body parts	Non-Contact	Shirt, Gloves	×
Pacelli et al. [5]	Capacitance will be formed between garments and body parts	Non-Contact	Shirt	×
Oehler et al. [6]	Capacitive sensors	Non-Contact	Device placed on chest	✓ (Software Filter)
Lee and Chung [7]	Non-Capacitive	Contact	Device connected to finger	×
Park et al. [8]	Capacitive	Non-Contact	Device placed on t-shirt	×
Lee et al. [9]	Capacitive	Non-Contact	Belt-type device	✓

1. *Diabetes monitoring*: A non-invasive (one which does not involve the introduction of a sensor into the body) opto-physiological sensor, such as a non-invasive micro-strip sensor [2], or non-invasive microwave glucose sensor. Patients suffering from diabetes should be very careful, especially in cases of wounds. The best sensor to monitor a wound remotely is the smart camera with image processing techniques;
2. *For heart rate monitoring*: Heart rate monitoring nowadays is so easy that any individual can easily design a sensor for it. The basic component required is a capacitive electrode, conducting wires, photo-resistor, LED (Light Emitting Diode) and a few other basic electronic components. All the details of sensor design are mentioned in reference [10]. Some special sensors are also available for heart rate monitoring, such as an opto-electronic sensor mentioned in reference [11];
3. *Blood pressure monitoring*: Any wearable sensor able to monitor the flow rate of blood. On such type is discussed in reference [12]. Also there are many other sensors available that can easily and accurately monitor the rate of blood pressure;
4. *Body temperature monitoring*: Although the thermometer is the best device to monitor body temperature, it does not have decision-making capabilities. One can design a system using LM35 (a temperature sensor) integrated with a controller so that it can give an alert as soon as the body temperature crosses a certain threshold;

5. *Fall detection*: In order to make sure that an emergency treatment can be made immediately available whenever an elderly person falls, one needs to detect this fall. For this an accelerometer sensor along with a gyro sensor can be used. One such fusion device is MPU6050, which is a 3-axis analogue gyroscope and accelerometer;
6. *For rehabilitation*: Once the patient undergoes medication and returns home, he/she needs to be continuously monitored in his/her rehabilitation phase. Sensors in the form of smart shirts, smart gloves, smart fabrics combined with wireless sensor networks (WSN) can be used. One such method for ubiquitous monitoring is presented in reference [13];
7. *Cough and asthma monitoring*: Smart cameras with built-in microphones can be combined with signal processing techniques in order to differentiate between a normal cough and an asthma-related cough. Flow rate of the air coming from the cough can be used to differentiate between both; and
8. *Physiological or psychological indications*: Electrical conductance of skin is the most accurate indicator of emotional stimulation. Galvanic skin response (GSR) sensor is used to measure the skin conductance, which can help in physiological or psychological estimations.

3.4 FAST EVOLVING WEARABLE SENSORS IN HEALTHCARE

Figure 3.2 pictures the fast evolving wearable sensors in the healthcare industry, which can be used for remote monitoring and activity tracking.

Figure 3.2a is the wearable wrist sensor in the form of a smart watch used for monitoring heart rate and heart rate variability (HRV), not only in day-to-day activities of any individual of any age, but also to track the heart rate and HRV during aerobics, daily exercises and in the gym.

Figure 3.2b is a band especially for smartphones. It is an activity-tracking device for daily routines, which counts the number of steps and calories burnt. At present, it is compatible with all android and iPhone devices.

Figure 3.2c represents a device useful for people of all ages with diabetes. It helps as an aid in therapies requiring insulin micro-infusions. Mirco-infusion therapy is a method of treatment mostly used by people who take insulin for diabetic care. This device monitors the insulin taken; it is an invasive type of device which needs to be inserted under the skin.

Figure 3.2d shows a device called as Smart Socks, which can be used by anyone of any age, even infants. In infants, they can track the heart and oxygen rate while infants are asleep. It is a device which can be worn as an ankle bracelet by adults to track their jogging speed, distance covered and calories burnt. It is a type of textile sensor, which comes with a conductive fabric.

3.5 PLACEMENT OF WEARABLE SENSOR

The placement of wearable sensors mostly depends on the parameter the sensor is measuring or monitoring. For example, an ECG sensor is best fitted on the chest, thus a sensor implanted in textile should be placed on the upper part of

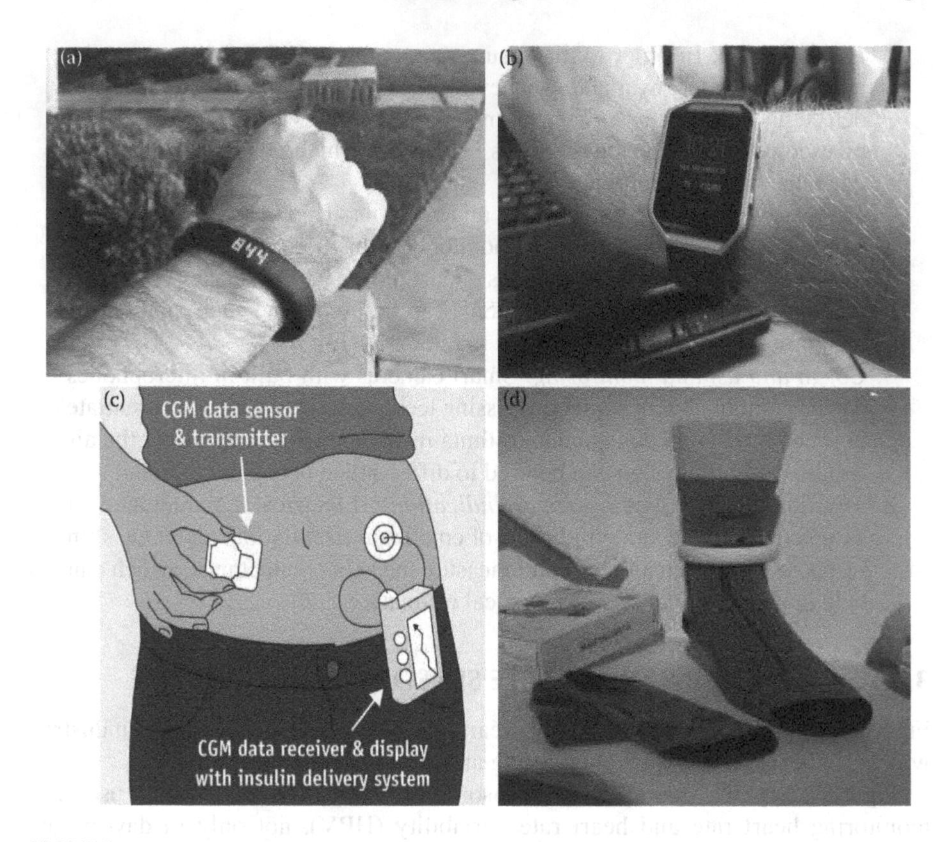

FIGURE 3.2 Fast evolving implantable sensors in healthcare.

the body if it is a shirt or t-shirt. Similarly, sensors embedded into socks, for tracking and counting the number of steps covered, should be placed below the ankle. Figure 3.3 shows the possible placement positions of wearable sensors on the human body.

3.6 BIOSENSORS IN HEALTHCARE

A biosensor is composed of a bio-element and a transducer that generates signals proportional to the analyte being measured. These biosensors consist of three main parts:

1. A detector unit
2. A transducer
3. A signal processing unit

A detector unit is used to detect the presence and type of a molecule present at its input, and to generate a corresponding stimulus signal at its output.

A transducer is a device which converts the signal from one form to the other. Here, the transducer is used to convert the stimulus into an electronic signal.

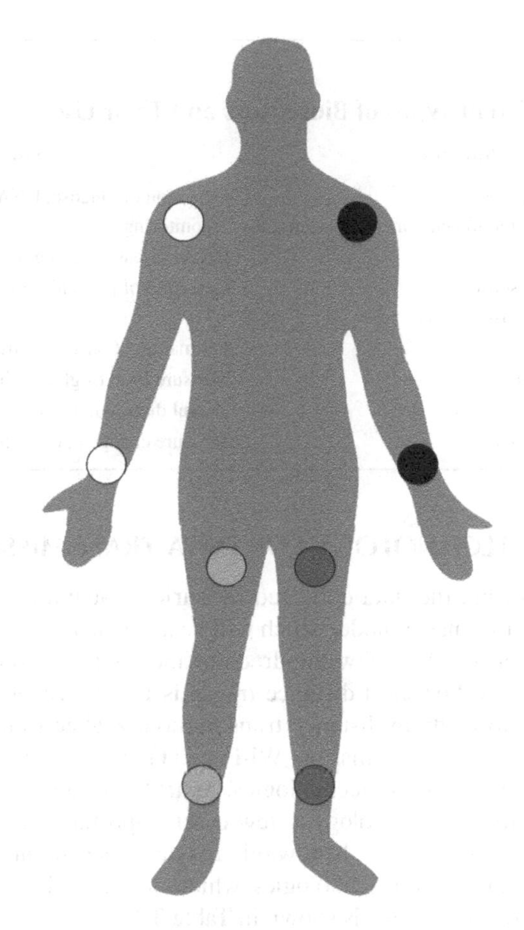

FIGURE 3.3 Possible sensor placement position.

A signal-processing unit is used to process the output of the transducer so that it can be presented in an appropriate form. Figure 3.4 shows the parts of a biosensor. Article [14] gives the details of BioApps used in wearable technology employing biosensors.

Table 3.2 gives an overview of the different types of biosensors and their use.

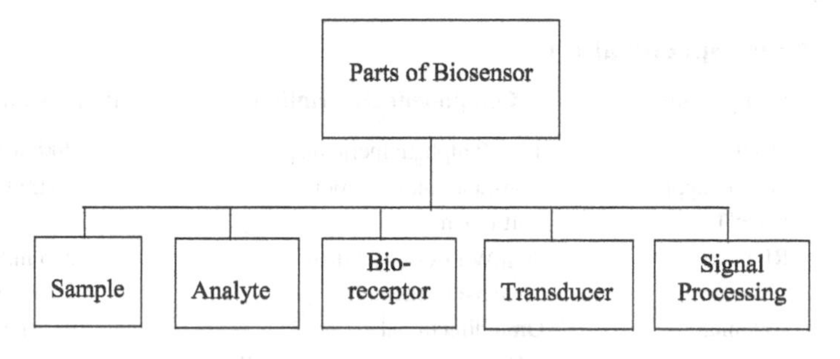

FIGURE 3.4 Parts of biosensor.

TABLE 3.2
Overview of Different Types of Biosensors and Their Use

S.No	Biosensor	Use
1	Electrochemical biosensor (Types can be potentiometric and amperometric)	Detection of glucose, DNA/RNA analysis, monitoring
2	Optical biosensor	Detect disease-causing bacteria, viruses in body
3	Piezoelectric biosensor (Also called as mass sensor)	Detection of pesticides in food
4	Calorimetric biosensor	Estimation of bacteria, virus and pesticides
5	Blood glucose biosensor	Measure level of glucose in blood
6	Conduct metric biosensor	Useful during dialysis
7	Fiber-optic biosensor	Measure change in oxygen concentration level

3.7 WIRELESS TECHNOLOGY FOR DATA TRANSMISSION

For remote monitoring, the data collected by various wearable sensors has to be transmitted locally to a master node, which will be a few meters away and connected remotely to a server, maybe a few hundreds of meters away from the individual wearing these sensors. For short distance transmission the Bluetooth signal is the best, and for short to medium distance transmission, Zigbee transmission is most suitable. For long distance transmission, Wi-Fi and GSM (Global System for Mobile Communication) are the suitable technologies. Apart for distance, when selecting the best suitable transmission technology, a few other important attributes to consider are data rate, power consumption, bandwidth and data encryption. A comparison of different wireless transmission technologies, which can be used for data transmission in remote healthcare monitoring, is shown in Table 3.3.

3.8 INTERFACING OF SENSOR IN BAN (BODY AREA NETWORK)

The BAN (Body Area Network) helps to continuously monitor the human body. The sensors for blood pressure, pulse oximetry, heartbeat and pedometer are integrated in

TABLE 3.3
Wearable Node Specification

S.No	Components	Component Description	Power Consumption
1	Display	16 × 2 alphanumeric display	200 mA
2	Power supply	Constant voltage supply	10,000 mAh
3	8 SPST	Soft touch	–
4	RF based Xbee	50 mW (Tx) −102 dBm (receiver sensitivity)	295 mA (Tx) 50 mA (Rx)
5	Antenna	Omnidirectional	2 dBi gain
6	SD module	8 GB memory extend to 32 GB	–

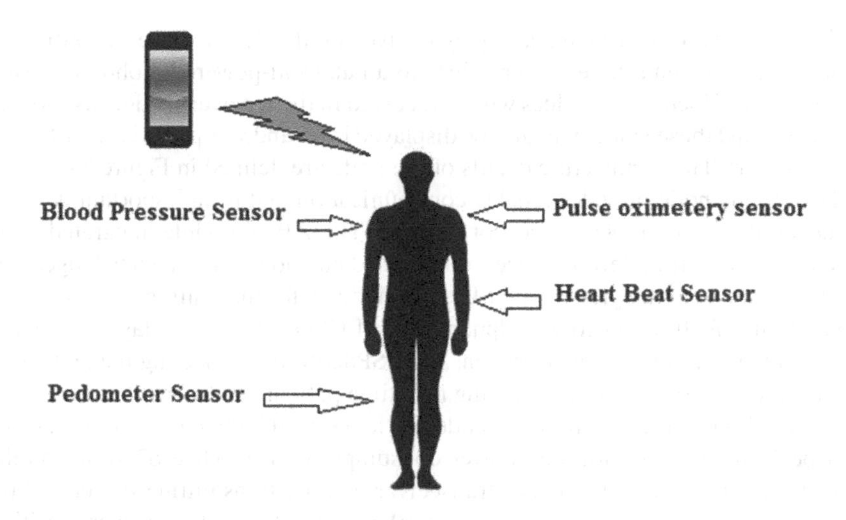

FIGURE 3.5 Sensors integrated in the human body.

the body to continuously measure health, as shown in Figure 3.5, and the measured data is sent to a mobile device or to a cloud server.

The monitoring system is integrated using a physiological parameters acquisition module and an ESP8266 module. The ESP8266 module is used to upload the data to the cloud server. The ESP8266 is a low-cost solution with months to years of battery life and featuring very low design complexity.

The daily activity of the people using the device is recorded, and the data can be viewed directly by the wearers, or be supervised and analysed by the doctor (physiotherapist). The main factors to be measured are heart rate count, step count, oxygen level in the hemoglobin and so on. The proposed system is efficient enough to do the entire task at once and has a very low cost and low complexity. In this section, the block diagram of the system is explained below. The transmitter section in Figure 3.6 is powered by a 5 V supply given to Arduino UNO. The pedometer sensor gives a digital output, the pulse oximetry sensor gives a serial output and the blood pressure/heart rate count sensor gives an analogue output. All these sensors

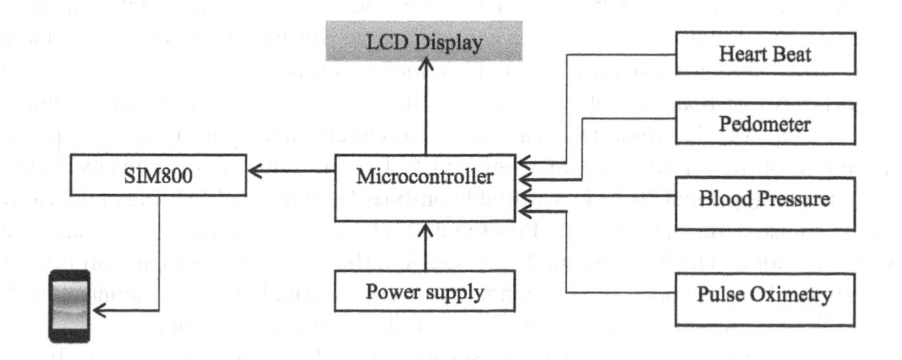

FIGURE 3.6 Transmitter section of fitness monitoring system.

pin into the Arduino UNO and a display is given on the device. After recording, the device is able to send these sensor values to an android-powered mobile for use by the supervisor/doctor. The values will be received in the receiver section, as shown in Figure 3.6, and these values will also be displayed in an android-powered smartphones (cloud server). The architecture details of the node are defined in Figure 3.6.

The mobile node provides reliable communication and is an important device in the designing of wearable devices for patients. The XBee module integrated with a node is used as a transceiver device to send the data and receive acknowledgements. The nodes are also integrated with SD card modules for maintaining data in a local server. It also features an 16 × 2 alphanumeric LCD for better display, a 2600 mAh power supply for continuous operations, an ESP8266 for connecting the node to the cloud server and the sensors for getting real-time patient conditions.

Xbee-S2 series is a new transceiver device and is best known for its minimal cost, high performance and minimal power consumption. The Xbee-S2 series module transmission range is 2.4 km. The transceiver module transmitting power is 2 mW at 3 dBm, with a receiver peak current of 40 mA at 3.3 V and a receiver sensitivity of −96 dBm. Additionally, the Xbee-S2 Pro module is the advanced version of Xbee-S2 with a transmitting power of 50 mW at 17 dBm and 10 mW at 10 dBm. The receiver sensitivity for the module is −102 dBm. The receiving peak current is 45 mA.

The Xbee module is able to create a point-to-point and mesh network. The network comprises a coordinator node, router node and end device node, and is programmed using API commands set mode using X-CTU (testing and configuration software); however, when it is used with a microprocessor/microcontroller, it will switch to AT command mode.

The Xbee-S2 module is very secure to design a wireless network, since it has advanced security features. The Xbee-S2 is able to form 6500 unique addresses using DSSS (Direct Sequence Spread Spectrum).

The detailed hardware/control architecture of a wearable node is shown in Figure 3.7. The main central processing unit is an AT Mega 328P. These boards are connected with the Xbee-S2 using the ZigBee protocol, and are driven with a 2600 mAh battery source (extendable). The battery source is connected via an A to B type cable. The RescOp (rescue operation) nodes can also be powered through AA batteries. Using IC 7805 and LM 1117 voltage regulators, the output voltage is regulated to +5 and +3.3 V, respectively. A separate SPI-based interface is available on-board for boot loading purposes (note: sometimes processors are not boot-loaded properly). An on-board FTDI port is also available to load the firmware on the node. The RescOp node has digital switches and LEDs to check the pull-up and pull-down of the pins of the ATMEGA 328P using an f-to-f connector. A 16 columns by 2 rows LCD display (optional 20 × 4) is available on-board and the backlighting of the LCD can be adjusted via a pre-set. The RescOp nodes have on-board pre-sets to check the ADC operation. The Xbee S2 module is fitted to the on-board shield and all pins of the Xbee-S2 remain open so as to connect the pins externally to a microcontroller. A micro SD card module is also clipped to fetch the required body data.

On the transmitter side, analogue sensors (blood pressure, heartbeat), digital sensor (pedometer) and serial output sensor (pulse oximetry) are connected with the

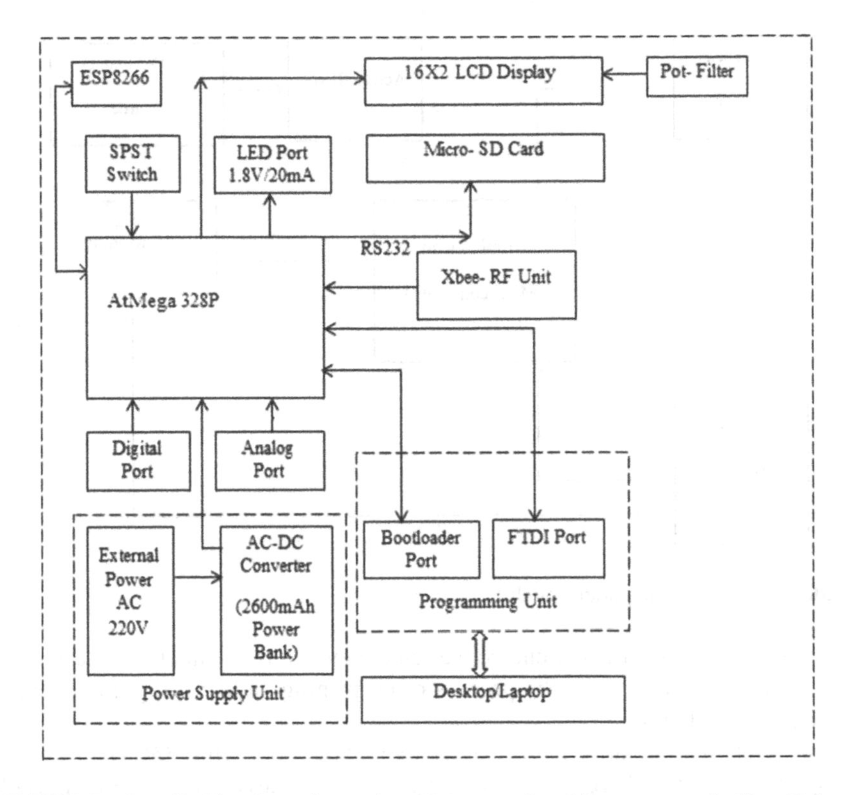

FIGURE 3.7 Detailed hardware/control architecture of mobile sensor node (RescOp).

ADC conversion port, digital PORT, RX and TX of the Arduino UNO, respectively, as shown in Figure 3.8. Data collected from the above sensors is processed and transmitted to the cloud server and then to a mobile app, which has a unique IP address. On the receiver side, the values from these sensors will be received and can be further analysed by a doctor or physician. In the prototype model, we have used the Arduino UNO, but the algorithm used can be implemented in any microcontroller and processor. The sending commands/values and receiving commands/values of the sensors are displayed in the LCD.

Figure 3.8 shows the prototype design of a BAN. This prototype is on the laboratory level and the data is successfully uploaded to the mobile data server (in mobile application). The prototype is a fully automated system, and the doctor or

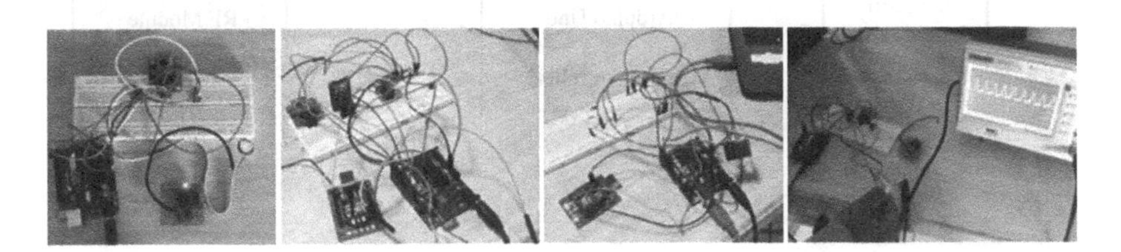

FIGURE 3.8 Prototype of BAN (body area network).

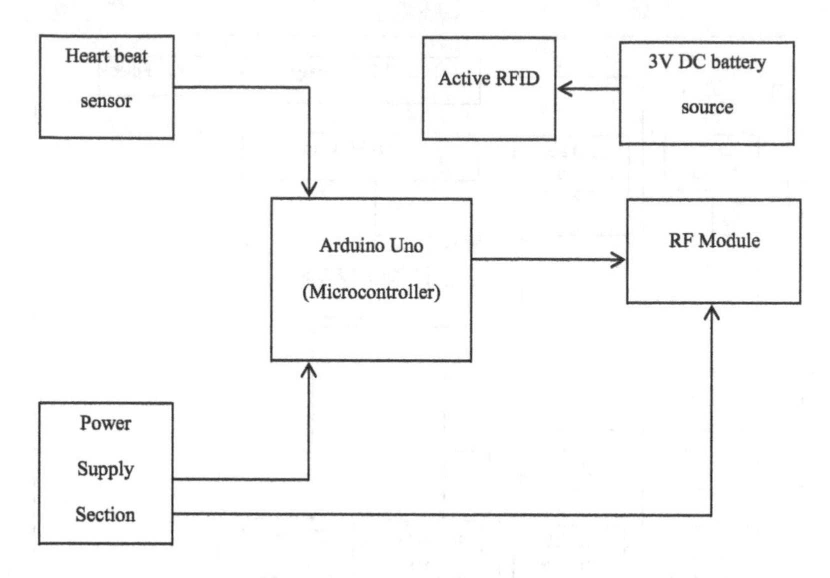

FIGURE 3.9 Wearable patient module.

physician can examine the patient from a remote location. The doctor can even track the patients. To elaborate on the process of examining and tracking, the operation details are given below.

A wearable module attached to the hands is shown in Figure 3.9. The module is equipped with active RFID for tracking purpose, and carries a unique node ID for every patient. This unique ID is helpful in tracking the patient through the whole wireless network. The data collected by the module will be collected in the central node along with the unique ID of the patient.

Once gathered, the patient module data will be transferred to the data-receiving unit, as shown in Figure 3.10. Figure 3.10 shows the central collection unit which

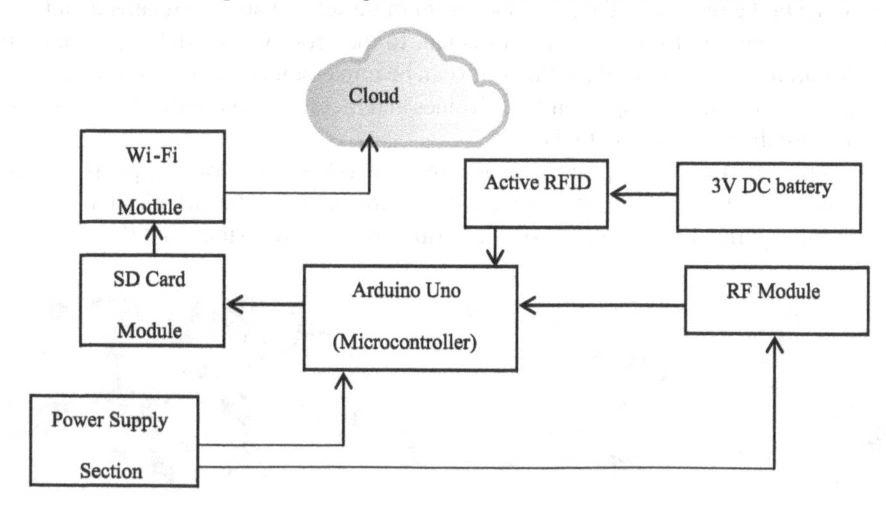

FIGURE 3.10 Central data collection unit.

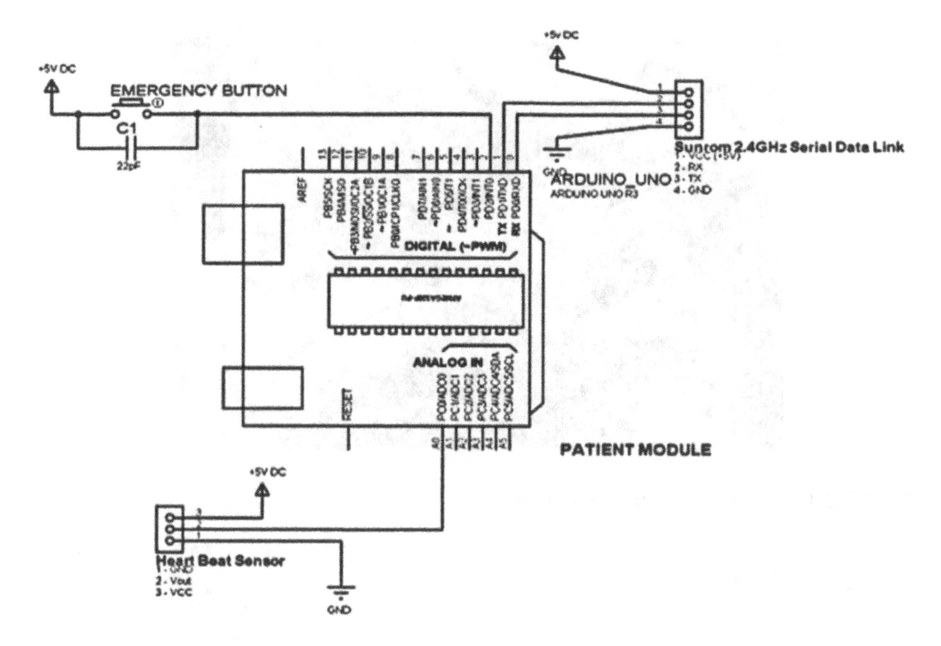

FIGURE 3.11 Patient module (wearable).

carries the data from the patients and stores it in the same SD card module equipped with the microcontroller. The data collected in the SD card will be uploaded to the cloud using the Wi-Fi module ESP8266 on the cloud server. The ESP8266 module has a free-of-cost cloud server with a public domain IP address.

The circuits of the patient monitoring module and central unit, where all the data is collected, are shown in Figures 3.11 and 3.12.

The patient monitoring module prototype is shown in Figure 3.13, and data in the ESP8266 cloud server is shown in Figure 3.14.

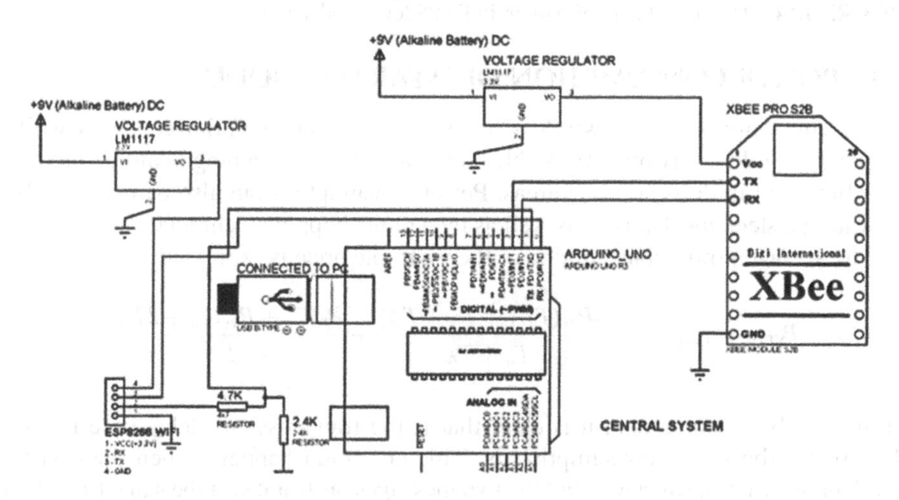

FIGURE 3.12 Central unit with ESP 8266 module.

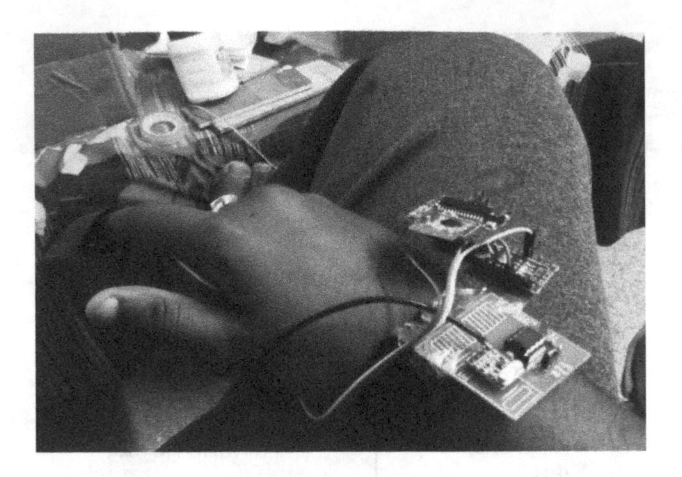

FIGURE 3.13 Patient monitoring system.

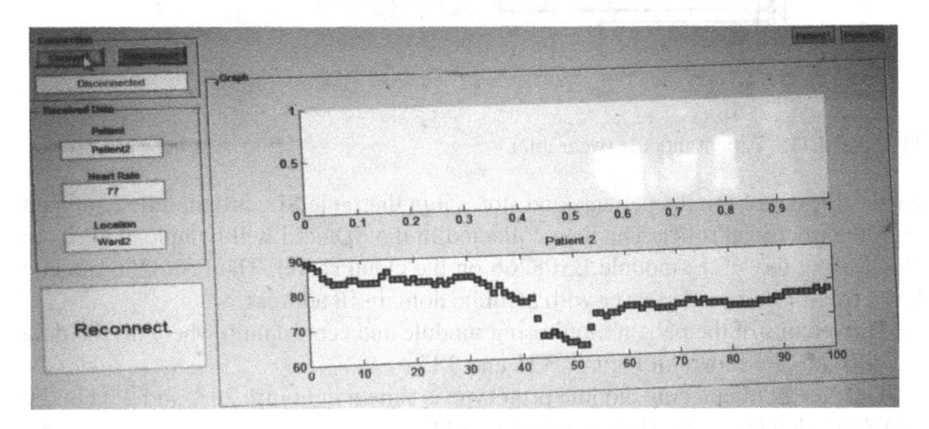

FIGURE 3.14 Heart beat data of patient in ESP8266 cloud server.

3.9 POWER CONSUMPTION OF WEARABLE NODES

Power consumption of the wireless node is one of the essential concerns while designing wireless networks. Wearable nodes are designed keeping in mind that they are often used in disaster prone areas. Power consumption can also be managed by activating a sleep mechanism, as well as by a multi-hopping technique.

The average power consumption of the wearable node is expressed as:

$$P_{\text{XBee}} = P_{\text{sleep mode}} + \frac{P_{Rx}(T_{s-f} + T_{fs} + T_{cp})}{T_{st} + T_{fs}} + \frac{P_{Tx}T_{cp} + P_{Rx}(T_d + 2T_t)}{l}$$

From the above expression, it is clear that if the transmissions delay time is less, the greater the power consumption is. This condition happens when there is an overlapping, that is, when we send the two messages (as frames) at the same time then the power consumption of the nodes increases. Most of the time, there is a request for resending the same frame. This leads to power wastage in the wireless network. This has to be rectified by proper balancing between the transmission delay and the power consumption, while maintaining the quality of the wireless channel.

The other type of solution for less power consumption is through writing the firmware in a wireless node. This can be addressed as:

1. Code length should be small
2. No use of else-if loops, as these commands consume more power than dedicated if loops
3. Debouncing problem to be rectified that will execute the loops in the firmware for one time only, thus having less power consumption. The firmware written in wireless nodes takes care of above-mentioned points

3.10 BATTERY LIFE OF NODES

Wearable nodes receive message packets, as discussed earlier, and are equipped with lithium-ion 2600 mAh power sources. These wearable nodes along with other wearable nodes fitted onto human bodies send a total of 295,651 message packets to the receiving nodes in a mesh network, for which a wearable node should be continuously operating for 13.65 hours with a 2600 mAh Lithium ion power source, as shown in Figure 3.15.

3.11 IMPLANTABLE DEVICES

Implantable medical devices are increasingly becoming popular due to rapid technological advancements. Such devices include, but are not limited to, hearing

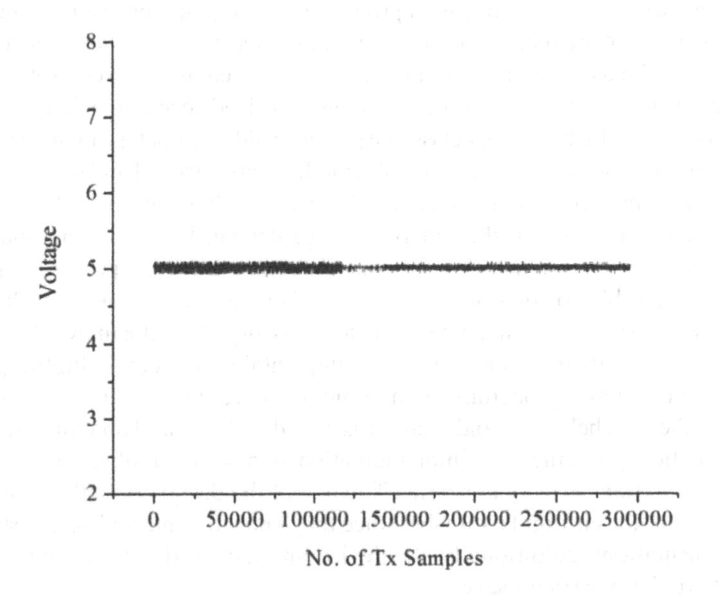

FIGURE 3.15 Battery performance.

aids, retinal implants, glucose monitoring, endoscopy, diaphragm pacemakers and so on. Recent developments in big data analysis have further fuelled the progress and demand for such implantable devices. Devices at present are small sized, power efficient and highly upgradable. The major challenges faced in developing newer devices are data security threats and large design-to-launch time.

Retinal implants involve specific electrical stimulation inside the eye using specially designed electrodes. Prominent research groups in this field are the Second Sight Medical Products, Retina Implant AG, EPI-RET Project, Pixium Vision, Bionic Vision Australia, Boston Retinal Implant Project and so on. Diaphragm pacing involves implant devices stimulating the phrenic nerve (which is important for breathing) to control the diaphragm pattern, thereby refining the breathing function in patients with respiratory problems.

Glucose monitoring is popularly used as a tool to monitor diabetes. A lot of development has taken place in glucose biosensors considering the remarkable demand for glucose monitoring. Cochlear implants involve inserting electrodes in ears. Pacemakers are used for stimulating normal heart function by transmitting a train of cardiovascular pulses synchronized with the R wave. Electrical stimulation therapies are also present for treating Parkinson's disease, wherein an electrical stimulus is provided to the brain area responsible for motor function. Similar devices are available for treating epilepsy. Many systems have been designed to store and deliver fixed doses of relief medicines into specific areas of the body, thereby removing the need for daily injection of drugs. Wireless endoscopy systems are available, such as capsule endoscopy, which can be used to visualize the internals of the body. Patients ingest a wireless endoscopy capsule similar to a pill. This capsule contains a miniature camera to capture internal images as it moves through the inner tract.

In most of the devices mentioned above, there is a short-range communication involvement, hence making the system prone to eavesdropping or unauthorized access to the system. There are major concerns regarding privacy of the external mechanisms of the system. The associated vulnerabilities are classified as either control or privacy. In the control class, there is a presence of an unauthorized access to gain control of the implant's operation. In the privacy class, the patient health parameters are made available to unauthorized persons. Such types of vulnerabilities are harmful and must be avoided.

The privacy and security of data emerging from such devices is still a challenge. The major challenge lies in the practical usage design. Data hacking and control are the most worrisome aspects in this situation. Devices present inside the body are vulnerable to unethical usage. The laws associated with such privacy concerns are not rigid and often are completely missing. With the unavoidable spread of IoT devices, their unauthorized use in implantable devices is highly possible. Various cryptographic procedures can come to the rescue under such conditions. Encrypting the data helps to avoid security issues, data leaks and unauthorized device accesses. Policy planning and implementation is also a possible and mandatory solution for tackling security concerns. The role of the designers in the formation of rules and regulations is highly valuable since they know the internal design structures and implementations. Additionally, the device interface available to the user is also crucial to avoid any wrong usage.

A brief listing of various implantable devices is shown in Table 3.4.

TABLE 3.4
A Brief Listing of Various Implantable Devices

S.No	Device	Manufacturing Companies	Use
1	Argus II Retinal Prosthesis	Second sight medical products	Used to provide visual perception to blind
2	Wide-View BVA (Bionic Vision Austrilia) system	Bionic Vision Australia	Used to provide visual perception to blind
3	Boston retinal implant project	Bionic eye technologies Inc. and Visus technologies, Inc.	Used to provide visual perception to blind
4	Sensimed triggerfish	Sensimed	Monitors pressure on eyes
5	Breathing Pacemaker System	Avery biomedical devices	Provides respiratory support
6	NeuRx Diaphragm Pacing System	Synapse Biomedical Inc.	Provides aid to people with injuries in spinal cord
7	Implantable Glucose Sensor	Glysens	Monitors glucose level
8	Lifestyle Cochlear Implant Systems	Advanced Bionics	Provides hearing aid
9	Maestro Cochlear Implant System	Med-El	Provides hearing aid
10	Transcatheter Pacing System	Medtronic Micra	Helps people with slow heart rate
11	Bladder Control Therapy Interstim System	Medtronic	Helps people with bladder related problems
12	St. Jude Medical Infinity DBS (Deep Brain Stimulation) System	Abbott	Aid to people with brain related problems and with Parkinson disease
13	RNS (Responsive Neurostimulation) Stimulator Neurostimulator	Neuropace Inc.	Aid to people with brain related problems

3.12 COMMON CHARACTERISTICS OF MEDICAL SENSORS

Common characteristics of medical sensors can be classified as either being static or dynamic. Static characteristics are transient free, this means these are characteristics measured after stabilizing all the effects of transient (which are of very short time interval) either to its steady state or final state. Characteristics which include transient properties on the other hand, are called dynamic characteristics. Brief definitions of a few, but not all, static and dynamic characteristics are given below:

Static characteristics:

Accuracy: It is defined as the ability of sensors to produce the true value of measure and error. It is the difference between the true value and the measured value by a sensor;

Precision: It is how carefulness how the measured value can be read, for example, the number of decimal places the measure and can be treated and read. It provides the quality and the fact of being accurate;

Resolution: It is the smallest change in the input quantity that can be measured;

Dynamic characteristics:

Speed of response: It is basically the rapidity (speed without any delay) with which the sensors can respond to changes in the measured quantity;
Measurement lag: It is the delay of the sensor in responding to changes in a measured quantity;
Dynamic error: Also called measurement error, it is defined as the difference between the actual value of the quantity being measured and the measured value with respect to changes in time; and
Fidelity: It is defined as the degree to which a measurement system indicates changes in measured quantity without any dynamic error.

3.13 SENSOR EVALUATION METRICS

In order to evaluate the performance of wearable sensors used in healthcare, the metrics to be considered are repeatability, accuracy and availability, which can be defined as shown below:

Repeatability: If the sensor is providing same reading for 'n' measurements with acceptable deviations, then the sensor is repeatable
Accuracy: If there is not much deviation of measured value from the actual value then the sensor is accurate
Availability: The sensor is available if it is providing measurements anywhere and anytime.

3.14 DESIGN AND IMPLEMENTATION OF AN EXAMPLE SYSTEM

The following is an example of a system which consists of wearable sensors, devices and a communication system. This system is designed and implemented to measure the fat percentage, body mass index (BMI), in the human body. The designed and fabricated body fat monitoring system basically determines the amount of cholesterol present inside the body in order to obtain a measure of BMI, fat and so on. Thus, a system is designed to measure the different parameters of the body through certain pre-designed fabrication methods of fat monitoring circuitry. In order to measure these fats, the following technology system is entirely located in a belt housing a controller and Bluetooth (low energy) which can act like like a pedometer, calorie counter, activity tracker, steps counter. It can also determine BMI (body mass index), body fat percentage, time spent sitting/standing and measure waist size. This works by being connected to a smartphone via Bluetooth, which could operate for a minimum duration of one week with a limited battery power supply. The following system does not interfere or add any new wearables on a person apart from their clothing, as the design is fully contained within a waist belt which looks no different than a normal one, but yet has all the capabilities to achieve what a basic wearable can do.

Being a serious health issue, increased body cholesterol can lead to various diseases. To enable continuous monitoring over such factors, numerous commercially available products are available and such as wrist bands, rings, abdominal belts, smart clothing and sometimes even smartphones. All these products have a certain aesthetic look or branded look, which attempts to add beauty and style, while performing basic operations such as a pedometer, calorie counter and activity tracker connected to smartphones via Bluetooth. There are many devices available on the market for tracking the physical activity of people who are experiencing physiological problems because they can't find the time to consult a doctor, or because of certain financial and availability issues. In the current situation, due to present-day food habits and many other factors, we are consistently watching the problem of unsaturated fat or cholesterol in the human body. Unsaturated fat deposited in our body creates numerous problems such as cardiac arrest, hypertension and so on, and it has been observed that 60%–70% of body fat is accumulated in the lower abdominal region. Also, it tends to alter our physical appearances.

Seating in improper postures, or being still for long periods of time, could increase the possibilities of spinal cord disc slips and so on, thus current research has been targeted at helping people avoid these mistakes; thus, providing them with a feedback via sound on their smartphones to improve their health. The key factors for a healthy and disease free life are management of diet and exercise. In the United States, the Center for Disease Control and the American College of Sports Medicine recommend 30 min/d of moderate activity on most week days. Fat distribution inside the human body is considered a primary risk factor for diseases related to obesity. An excessive amount of abdominal fat is always considered a prime factor of cardio metabolic disease. Increased intra-abdominal obesity is also somehow associated with diabetes.

3.14.1 Principle of Operation

It is correct that prevention is always better than the cure, so keeping this in mind this design is inspired to prevent people from spending too much time sitting, such as during an ordinary desk job. The objective of this study is to determine the body fat percentage in men. This technology should mainly comprise:

1. *Hardware*: Accelerometer which helps to determine the activities of the person whether the person is standing or sitting as well as walking
2. *Software*: Coding for the android app and interfacing with the sensor using a microcontroller
3. *Communication*: The communication between software and hardware is carried out via Bluetooth low energy, which is operated as a slave.

Body fat is the responsible factor for increased weight in the human body. Body fat is caused because of fatty (adipose) tissues, and the adipose is the proportion of your total body weight which includes weights of bone, organs, fluid, muscles, and so on.

Figure 3.16 shows the relationship between body fat percentage and body mass index for various observations. Two cases were considered: one where outliers were taken into the measurement, and the other where the outliers were removed.

There is one more method by which we can detect body fat: a method measuring the sugar contained in the blood. This experiment uses the method of weight measurement to detect body fat. Since the device is attached into the belt, it hence measures the weight

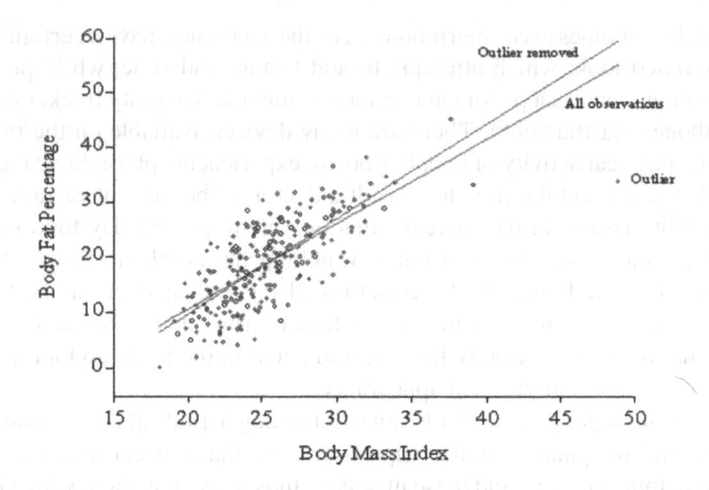

FIGURE 3.16 Amount of body fat with the variation of BMI.

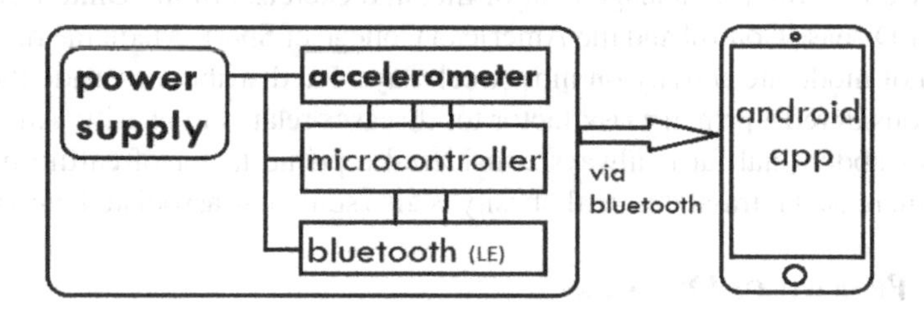

FIGURE 3.17 Block diagram of the proposed example system.

of the lower abdominal. Fat is the major cause of increased weight, thus by measuring the weight, the presence of fat in the lower abdominal region can be calculated. Once the fat is measured in the lower abdominal region, the device measures proportionally the fat of entire body by multiplying it with a defined constant, which yields an approximate measure of complete body fat. The fat is calculated through an algorithm which uses a relation of proportionality between weight and fat.

The accelerometer attached with the belt determines the exact position of the person wearing it. The accelerometer describes the physical activity of the person wearing the belt, such as standing, moving or doing some physical work. The calculated parameter is displayed on a smartphone, and the interfacing between the device and the smartphone is carried out by a microcontroller. The microcontroller fulfils the purpose of a transmitter, transmitting the information obtained by the device (in the belt) to the smartphone using Bluetooth. The block diagram of the example system is shown in Figure 3.17.

3.14.2 Methodology

The methodology used in this technique is basically an isolated circuitry, designed to measure the weight through a device installed on a belt. The belt is designed basically for men, where an accelerometer is connected which helps in determining the physical activities, or changes of body position. An algorithm is designed, which measures the body fat corresponding to the weight measured by the device. As per the algorithm, there are some certain variables which should be known to acquire the desired result. These variables are age, weight and height of the person. Bluetooth is interfaced with a microcontroller which connects the wearable device attached to the belt to a smartphone. The data produced by the accelerometer is transmitted by microcontroller via Bluetooth, and is then sent to the display unit.

3.14.3 Algorithm

The algorithm used for designing this belt is described using a flow chart as shown in Figure 3.18.

Based on the algorithms, these are some of the equations used for calculating body fat and other physical parameters for men.

```
(body weight *1.082) + 94
Waist measurement*4.15
Lean body mass= 1 - 2
Body fat weight= body weight - lean body mass
Body fat percentage= (body fat weight *100) / total body
weight.
```

For waist size determination, data from different specimens should be observed such as weight (in kg), height (in cm) and age to determine the fat percentage and BMI. This helps in determining the waist size of the person. After getting all the data, the microcontroller sends the information to a smartphone. An app can be designed based on the algorithm. After getting the result, the app does the final work and shows the result on a display.

3.14.4 Discussion

The design of the system should be simple to use and at the same time be a robust design to help the common person interact with the device in a user-friendly environment. The body parameters detected by the device should be accurate and precise, and should thus help save time, as the system reduces the need for consulting a doctor on a regular basis. The most important thing about the device should be its energy efficiency with respect to a given power source. The future scope of this project is that this belt can be used as a power bank to charge phones, which makes the belt handy for continuous monitoring. Adding an EEPROM and Bluetooth 4.0 will save the data and also reduce battery consumption since Bluetooth 4.0 is a low energy protocol. It can also help monitor our daily food and water intake and monitor calories burned per day. The body mass index (BMI) is a very important parameter used as a standard to differentiate between lean and obese persons. A BMI value of 37 or more is considered obese while a value of 23 or less is considered as lean/

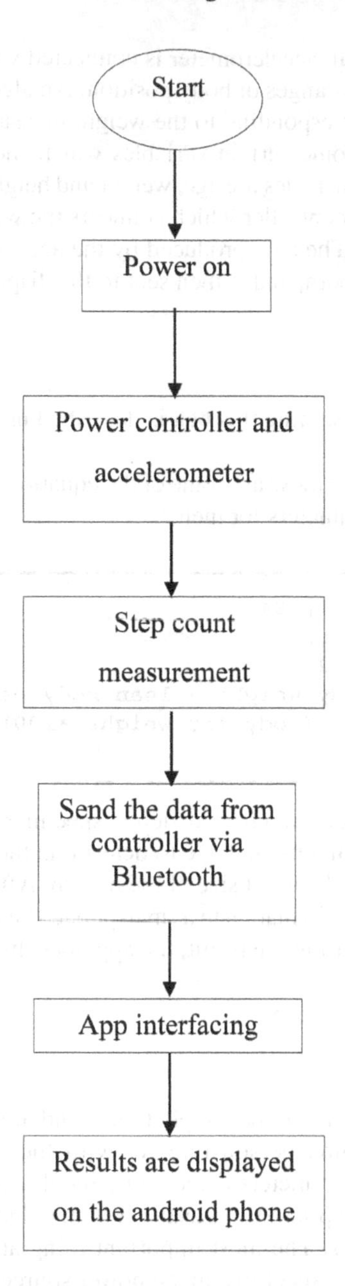

FIGURE 3.18 Flow chart of algorithm.

fit. Table 3.5 given below [15] lists the standard values and thresholds of various evaluation metrics of different body parameters. These threshold values act as cut-offs for an obese person or a lean person. These values are measured under standard assumptions and may vary slightly from person-to-person, but can nevertheless be used as standard values.

TABLE 3.5
Values of Different Evaluation Metrics for Lean and Obese Subjects

Evaluation Metrics	Cut-off for a Lean Human Being	Cut-off for an Obese Human Being
Body Mass Index measured in kgm^{-2}	23.0	37.0
Body weight measured in kilogram or pounds	71	116
Percentage of fat	15	32
Total fat (includes saturated and unsaturated fat measured in kg)	10	37
Fat that lies under the skin especially under stomach measured in kilograms	9	32
Belly fat measured in kilograms	4.3	12.3
Fat located inside the peritoneal area measured in kilogram	1.9	5.1
Intraperitoneal (IP) measured in kilogram	1.1	3.5
Retroperitoneal measured in kilogram	0.8	1.6

3.15 CHALLENGES AND FUTURE SCOPE

The basic challenge in the designs of wearable devices for the measurement of vital health parameters is the isolation of noise. When a wearable device is attached to a patient's body, noise can be generated in the measured parameters through even the slightest of movements, or external disturbances. Noise introduced in such cases will affect the output generated and may yield misleading data. In order to eliminate the noise generated, filters are designed suppressing this noise; thus, for instance, an analogue filter having a 0.04 Hz High Pass Filter and 40 Hz Low Pass Filter along with a Notch Filter can be employed.

Apart from the above-mentioned applications, wearable sensors can also be used in

1. Health monitoring through ECG signals for residential purpose by employing wireless sensor networks [16]
2. Applying soft computing techniques for data collected through these sensors for medical analysis [17]
3. IoT and big data technologies can be combined with wearable sensor for addressing the requirements of remote health monitoring. One serious issue related to this field is security and privacy, as addressed in references [18] and [19]
4. Wearable sensors can be combined with magnetic resonance imaging characteristics of the human brain to measure signals coming from the brain [20]
5. Disease investigations can be carried out by combining wearable sensors with machine learning, as mentioned in reference [21]
6. A localization process, based on a received signal strength indicator (RSSI), can be used along with wearable sensor measurements in medical diagnosis, as mentioned in reference [22]

3.16 SUMMARY

In this chapter, implantable electronics and the integration of bio-interfaces, wearable sensors and devices have been discussed. Different types of wearable sensors, which can be used in healthcare monitoring along with the activity tracking sensors, have also been discussed. Fast and rapidly evolving sensors have been presented and the preferred placement of sensors has also been discussed. A practical BAN, which includes wearable sensors and provides real-time monitoring, has been presented and the results have been discussed. Finally, the challenges and future scope have also been presented and discussed.

REFERENCES

1. Mukhopadhyay, S.C., Wearable sensors for human activity monitoring: A review. *IEEE Sensors Journal*, 2015. 15(3): p. 1321–1330.
2. Sen, K. and S. Anand, Design of microstrip sensor for non invasive blood glucose monitoring. *In 2017 International Conference on Emerging Trends & Innovation in ICT (ICEI)*, 2017, Pune India. IEEE.
3. Rachim, V.P. and W.-Y. Chung, Wearable noncontact armband for mobile ECG monitoring system. *IEEE Transactions On Biomedical Circuits And Systems*, Volume 10 Issue 6, 2016. 0(6): p. 1112–1118.
4. De Rossi, D. et al., Electroactive fabrics and wearable biomonitoring devices. *AUTEX Research Journal*, 2003. 3(4): p. 180–185.
5. Pacelli, M. et al., Sensing fabrics for monitoring physiological and biomechanical variables: E-textile solutions. In *3rd IEEE/EMBS International Summer School on Medical Devices and Biosensors*, 2006, Cambridge, MA, USA. IEEE.
6. Oehler, M. et al., A multichannel portable ECG system with capacitive sensors. *Physiological Measurement*, 2008. 29(7): p. 783.
7. Lee, B.-G. and W.-Y. Chung, A smartphone-based driver safety monitoring system using data fusion. *Sensors*, 2012. 12(12): p. 17536–17552.
8. Park, C. et al., An ultra-wearable, wireless, low power ECG monitoring system. In *IEEE Biomedical Circuits and Systems Conference, 2006. BioCAS*, 2006, London,UK. IEEE.
9. Lee, K., S. Lee, and K. Park, Belt-type wireless and non-contact electrocardiogram monitoring system using *flexible active electrode. Int J Bioelectromagn*, 2010. 12: p. 153–157.
10. Instruments, N. *Build Your Own Heart Rate Monitor.* 2012. http://www.ni.com/example/31560/en/
11. Alzahrani, A. et al., A multi-channel opto-electronic sensor to accurately monitor heart rate against motion artefact during exercise. *Sensors*, 2015. 15(10): p. 25681–25702.
12. Sato, H. et al., Blood pressure monitor with a position sensor for wrist placement to eliminate hydrostatic pressure effect on blood pressure measurement. In *2013 35th Annual International Conference of the IEEE Engineering in Medicine and Biology Society (EMBC)*, 2013, Osaka Japan IEEE.
13. Lee, Y.-D. and W.-Y. Chung, Wireless sensor network based wearable smart shirt for ubiquitous health and activity monitoring. *Sensors And Actuators B: Chemical*, 2009. 140(2): p. 390–395.
14. Dey, N., A.S. Ashour, and S. Borra, *Classification in BioApps: Automation of Decision Making.* Nilanjan Dey, Amira Ashour, Surekha Borra (Eds.). Vol. 26. 2017, Springer.
15. Abate, N. et al., Relationships of generalized and regional adiposity to insulin sensitivity in men. *The Journal of Clinical Investigation*, 1995. 96(1): p. 88–98.

16. Dey, N. et al., Developing residential wireless sensor networks for ECG healthcare monitoring. *IEEE Transactions on Consumer Electronics*, 2017. 63(4): p. 442–449.
17. Dey, N. and A.S. Ashour, Computing in Medical Image Analysis, In Nilanjan Dey Amira Ashour Fuquian Shi Valentina E. Balas (Eds.). *Soft Computing Based Medical Image Analysis*, 2018, Elsevier. p. 3–11.
18. Dey, N. et al., *Internet of Things and Big Data Analytics Toward Next-Generation Intelligence*. N. Dey, A.E. Hassanien, C. Bhatt, A. Ashour, S.C. Satapathy (Eds.) 2018, Springer.
19. Elhayatmy, G., N. Dey, and A.S. Ashour, Internet of Things Based Wireless Body Area Network in Healthcare, In N. Dey, A.E. Hassanien, C. Bhatt, A. Ashour, S.C. Satapathy (Eds.). *Internet of Things and Big Data Analytics Toward Next-Generation Intelligence*. 2018, Springer. p. 3–20.
20. Moraru, L. et al., Quantitative diffusion tensor magnetic resonance imaging signal characteristics in the human brain: A hemispheres analysis. *IEEE Sensors Journal*, 2017. 17(15): p. 4886–4893.
21. Kamal, M.S., N. Dey, and A.S. Ashour, Large scale medical data mining for accurate diagnosis: A blueprint. In Samee U. Khan, Albert Y. Zomaya, Assad Abbas (Eds.). *Handbook of Large-Scale Distributed Computing in Smart Healthcare*. 2017, Springer. p. 157–176.
22. Dey, N. et al., Wireless capsule gastrointestinal endoscopy: Direction-of-arrival estimation based localization survey. *IEEE Reviews In Biomedical Engineering*, 2017. 10: p. 2–11.

4 Challenges in Designing Software Architectures for Web-Based Biomedical Signal Analysis

Alan Jovic, Kresimir Jozic, Davor Kukolja, Kresimir Friganovic, and Mario Cifrek

CONTENTS

4.1 INTRODUCTION: BACKGROUND AND DRIVING FORCES

There has been significant progress made recently in scientific, computerized exploration of biomedical data where novel software platforms are developed which offer remote access and data mining capability. Traditionally, biomedical time series (signal) analysis tasks include, among many others: detection of disorders by processing of electrocardiogram (ECG) recordings [1], analysis of sleep in electroencephalogram (EEG) recordings [2], analysis of gait and muscle dynamics using surface electromyogram (EMG) [3], calm/distress classification by skin conductivity [4], and so on. All of these tasks are considered difficult to accomplish, since mathematical methods for accurate analysis are computationally demanding and the difference between a disorder and a healthy state may not be totally reflected in the analysed signal. There are many approaches developed offering solutions to such analysis problems, but the mainstay approach is still considered to be offline analysis (after all the data has been collected) and off-the-web analysis, using either MATLAB® (or Octave) based scripts or frameworks [5], or other specialized software – for example for heart rate variability analysis [6].

The primary reason for such a mainstay, aside from biomedical engineering tradition, lies in the complex algorithmic procedures for signal processing and analysis which take a lot of computer time to complete, visualize and interpret [7].

In Figure 4.1, we depict one of the usual workflows for a complex biomedical signal analysis task. The primary steps included in such a workflow are:

1. The acquisition of signal data, either from an existing web repository, local database, file storage or directly from a measuring instrument [8,9];
2. Signal pre-processing, including noise filtering, signal transformations, characteristic points detection [10];
3. Feature extraction, usually recommended by medical experts or medical guidelines for a particular type of signal and analysis goal [11];
4. Feature selection and/or dimensionality reduction, usually based on several statistical and machine learning techniques (e.g. principal component analysis, filter based selection, wrappers, etc.) [12–14];
5. Model construction, usually involving multivariate inductive statistics, classification, regression or unsupervised learning machine learning methods [15,16]; and
6. Results visualization and interpretation, using a variety of statistical and machine learning evaluation tools and methods [17].

Although some steps may be skipped, depending on the analysis goal, input signal type and methods used, the entire process may be automated only partially and is thus cumbersome on the hardware and software resources of the platform on which it is performed. Also, special expertise in biomedical engineering research is needed in order to intervene in the process.

In order to ameliorate the issues with the hardware and software resources, the majority of novel research focuses on remote healthcare, where services operate in a way as to offer remote collection of data, transfer of data via network, medical

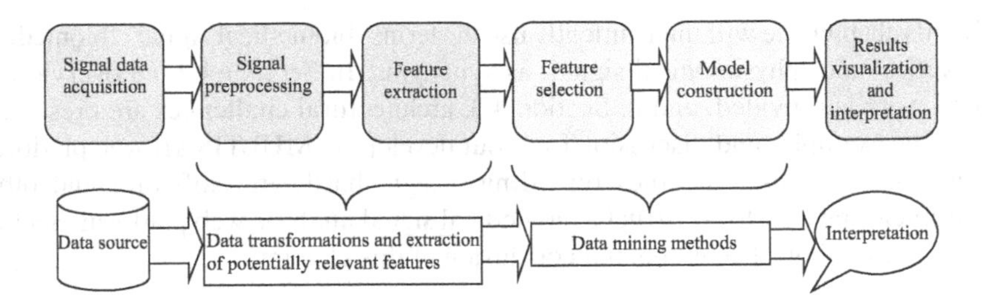

FIGURE 4.1 Complex biomedical signal analysis scenario.

centre-based analysis on a server or cloud architecture using a variety of complex signal processing tools [18]. Still, the main issue with such a process is that both signal analysis and signal inspection are considered to be reserved only for medical and biomedical engineering experts, acting locally in hospitals or research centres.

There has been little effort involved in complete web-based solutions for biomedical signal analysis that would enable on-the-web analysis of a patient's data. The advantages of web-based solutions for biomedical signal analysis are plenty: remote and ubiquitous access, browser-only software installation requirements, reliance on fast server solutions for calculations, independent client and server side development and so on.

In this chapter, we mostly consider web-based systems which enable biomedical signal analysis. We focus on software architecture aspects in designing such systems, from the perspective and experience of an ongoing research project named MULTISAB. It was aimed at developing both a research and application-based web platform for parallel, heterogeneous biomedical time series analysis intended for medical disorder diagnostics and classification [19–21], which would support complex analysis scenarios such as the one depicted in Figure 4.1. For the sake of completeness we also examine typical applications of web-based systems for home healthcare that have very little or no options for the end user [22]; medical educational software platforms offering training to medical staff regarding biomedical signal analysis [23]; biomedical signal repositories offering both signal recordings and signal visualization [8] as well as electronic and personal health record systems related to the current topic [24].

The chapter does not consider offline and off-the-web biomedical time series analysis software developed in specific medical domains, although there are many examples of such software [5,6,25,26]. We also do not discuss numerous biomedical image applications and software in this work, given either as a web or as an offline solution [27–30].

Challenges in developing web-based biomedical signal analysis software are considered, such as: data privacy, security and user roles, frontend workflow organization, frontend and backend interactions, changes in implementation languages and libraries, database design, integration of existing data analysis and reporting libraries and workload amelioration. We would like to emphasize that not much work was published regarding these practical aspects of biomedical time series analysis software design and implementation in a web setting. Therefore, through an in-depth exploration of these challenges, we highlight the importance of further development of web-based solutions to end-users.

In this chapter, we will intermittently use the terms 'biomedical signal', 'biomedical time series' and 'physiological signal' as synonyms. In Section 4.2, an overview of related work is provided, and in Section 4.3, architectural challenges are presented, with some examples and discussion from our developing MULTISAB web platform. Section 4.4 discusses specifics regarding design, hardware, software and other requirements needed to construct a biomedical signal analysis web platform. Section 4.5 provides a short discussion and conclusion of this chapter.

4.2 OVERVIEW OF WEB-BASED SYSTEMS FOR BIOMEDICAL TIME SERIES ANALYSIS

Web-based systems for biomedical time series analysis have evolved significantly since the late 1990s (the advent of major internetization). The evolution of such systems has had several directions:

- Web physiological data repositories, highlighted mostly by the well known PhysioNet website [8];
- Remote healthcare systems for online patient monitoring, based on improvements in wireless sensor technologies [31–33], with different storage capabilities, such as cloud infrastructure [34], and governed today by a multitude of companies covering various aspects of online service, see e.g. [35];
- Medical educational software, intended for better understanding of complex physiological processes, elucidated by biomedical time series visualization [36];
- Electronic health record systems, primarily intended to replace traditional paper health records and to allow information sharing among medical specialists and primary care physicians for the purpose of improving patient care [37], but also intended for clinical decision support, usually taking into account multiple information sources available in the record [38] which still have many architectural challenges [39]; and
- Research and application-oriented platforms/environments enabling decision support and data mining of time series data, with a major focus in bioinformatics [39] and on recent development in biomedical time series data analysis [19,40].

4.2.1 Web Physiological Data Repositories

Web physiological data repositories, such as the renowned openly available PhysioNet platform [8], or such as in reference [41] were developed in the last two decades as an ongoing effort to promote better understanding of biomedical time series data through scientific exploration of their mechanisms and behaviours. Although a web repository may provide a user with somewhat limited analytical capabilities, its primary use is to expose reliable and available anonymised patient data to interested researchers and other users. Contributors to databases in web repositories are usually hospitals and established medical research centres. The web repository allows better comparison of developed feature extraction and data mining algorithms among researchers, usually with standard physiological signal databases available as reference points (e.g. MIT-BIH Arrhythmia Database on PhysioNet web portal).

Web-based biomedical signal visualization and off-the-web use of biomedical signal analysis software (such as WFDB toolkit available from PhysioNet [8]) are also important aspects contributing to widespread influence of the web signal repositories. Aside from open-web physiological data repositories, there also exist commercial or membership-oriented repositories, such as the THEW project [42] and Ann Arbor Electrogram Libraries [43], which offer similar or even improved (better annotated, higher sampling quality) data records, compared to the openly available repositories – but at a price.

4.2.2 Remote Healthcare Systems

In aging societies, the importance of remote healthcare systems cannot be overstated. Indeed, immediate assistance in the case of medical emergency is imperative, and organization of technological environment so that it best suits patient's needs and limitations has become a largely investigated issue. Specialized scientific conferences (e.g. HealthyIoT, AmIHEALTH) and journals (e.g. *Journal of Healthcare Engineering*, *Journal of Biomedical and Health Informatics*, *Sensors*) cover this research area. Scientific literature usually focuses on descriptions of wireless sensors and connectivity protocols [21,44], with the majority of applications in home environments [45,46]. Particular effort is usually attributed to server side data processing [47], where many different architectures and technologies may be used, for example cloud-based architecture [27,48], standard server with relational database architecture [49] and standard server with online (script-based) processing [47].

Data mining in remote healthcare systems may be related to simple outlier detection (whether sensor anomaly or patient emergency) [50], fuzzy rule-based diagnostic systems that alert physicians in the case of emergency [51], prediction of disorder onset (as is the case with blood glucose in diabetes) [52] or a more complex expert system allowing multiple diagnoses based on a variety of measured parameters [47], to name just a few. Some of the analyses are more suitable for online processing, such as anomaly detection and rule-based alerting, while others are more appropriate for offline analysis, such as prediction of disorders. There are multiple challenges involved in remote health monitoring systems, such as data acquisition and pre-processing (in particular, a lack of standardized wireless sensor solutions), medical equipment pricing, data privacy preserving, network coverage and bandwidth allocation, data modelling methodology, results evaluation options and so on. [27,53,54].

Web solutions in remote healthcare system usually offer unrestricted access only to qualified physicians and medical personnel [22], with the home (patient) side of the web interface being either non-existent [55] or limited to simple actions [54,56,57]. For example, a user may see some measured health parameters through a web-based interface and request a feedback from a medical expert [57].

Remote monitoring by smart phones and tablets may offer price and access related improvements [58], although mobile applications may present an obstacle for some patients if the usability is not in focus [59]. Web interface on the server side differs from one system to another, but usually supports patient vital signs (emergency) monitoring, recorded signal visualization for inspection and in some occasions, data mining [47,60].

4.2.3 Medical Educational Software

Although there are many instances of web-based medical software intended for physicians' education, such as software for medical 2D [61] and 3D imaging [62,63], there are only a handful of online systems available for biomedical time series analysis education, most of them ECG related [23,36,64].

The purpose of online learning system is to provide medical students with easier access to a large number of interesting examples of physiological recordings, as well as to accelerate students marking. Also, automated computer-based detection and annotation of morphological signal features (such as waves in ECG) and comparison with expert-based annotations is another application of the on-the-web educational approach [23]. Aside from specialized educational software, medical personnel also have the option to use web physiological data repositories, as the ones described in Section 4.2.1, for educational purposes.

4.2.4 Electronic and Personal Health Record Systems

Contemporary electronic health record (EHR) system may be considered as a big data system, particularly in the sense of large data volume and considerable data variety (structured text, unstructured text, signals, images, 3D data) [65]. This is clearly reflected in various categories of medical knowledge required to describe the state of a patient through time. The knowledge may include standard medical history, demographical information, medications taken, various laboratory tests results, undergone treatments, and many forms of recorded signals and images (e.g. ECGs, echocardiography, radiology, MRI [66]). To support this kind of information diversity and communication between medical institutions, and not only within institutions, as is the case with standard electronic medical records (EMR), a set of standards and regulations were developed such as HL7, ISO 13606 (and its underlying standards), DICOM and ISO/TC 215's set of health informatics standards.

The experience of using EHRs in hospitals is mostly positive, with reported improvements in hospitals efficiency [67]. However, focusing specifically on the information coming from biomedical signals in EHRs, there are very few proposed and implemented solutions for storing ECG, EEG and other important biomedical signals directly in an EHR [68]. The main issue lies in integrating various heterogeneous formats of the signal recordings with the existing architectural standards in EHRs [69]. As a commonly used alternative, the signals may be stored (and also transformed to images) in a separate record system unavailable to the EHR. In such cases, usually, only a final report is summarized and presented in the EHR [70]. This certainly prohibits the interoperability between medical centres and therefore lowers the quality of service to patients.

There are some efforts to standardize various formats, such as the one in ECG analysis domain, by conversion of various ECG formats to DICOM-PACS image format [24] and integrating ECG-as-image in EHR stored in the cloud. Such images may be used later for learning disease models based on image mining techniques, involving automatic detection and classification [71]. Nevertheless, aside from individual uses in medical centres on a pay-per-use basis, the full service of EHR, especially in the context of biomedical signals analysis has not yet been established [72]. The hospitals and medical centres are sometimes unwilling to use the available cloud-based solutions for EHRs, and this is not without a cause. Some of the major challenges influencing hospitals in having such a stance, aside from the incompleteness of the structure of EHRs due to lack of standards, are data safety and security issues, limited storage capacity, network unreliability and low transfer capacity [73]. Also, in the context of use of EHR in mobile devices (mobile health), which is considered as especially convenient for medical personnel and patients in remote areas, limitations such as short battery life, small storage capacity and limited processing ability of the mobile devices represent additional challenges [69]. However, using big data architecture and resources can lead to many opportunities in healthcare: data quality improvement, improved population management and health, early detection of diseases, accessibility, improved decision-making and cost reduction [73].

Personal health records (PHR) are a more recent development in the field of electronic health records. Essentially, the main difference between EHR and PHR is access privileges and user roles. Whereas EHR may be looked upon and updated only by medical professionals, PHR may be read and, sometimes, modified by the patient since it is considered as the patient's property. Alyami et al. [74] mention two types of PHR systems: untethered and tethered. In untethered PHR, patients have full control over their PHR, where they can collect, manage and share data in their records. Tethered PHR is linked to specific healthcare provider's EHR (or hospital EHR), and can be used remotely by patients to view their data. Patients sharing untethered PHR with healthcare providers may be one way in supporting interoperability between medical centres. In this respect, blockchain technology developed for data privacy, secure access and scalability [75] may play an important role in developing applications for serving PHR to interested and authorized users [76]. We note that we are unaware of any current research investigating individualized web-based biomedical signal analysis and processing based on PHR information.

4.2.5 Research and Application Oriented Web Platforms/Environments for Biomedical Signal Analysis

Decision support systems (DSS) may be an important part of EHR systems, with the intent of helping the medical practitioner reach an accurate diagnosis, treatment or prognosis. DSS are of particular importance to less experienced physicians and general physicians in rural areas, where absence of experts in a particular domain (e.g. cardiology) as well as medical emergency requires immediate and educated action [77]. Web-based solutions in the form of web applications accessing EHRs in order to provide decision support should allow easy access to relevant data and conclusions to doctors in a well-defined clinical domain [78].

Research focused web platforms that offer some analysis and decision support based on biomedical signals are mostly developed for some specific domain, for example stress level in virtual reality environments [40] or ECG diagnostic interpretation [79]. As far as we are aware, there are currently no complete research and application-oriented web platforms available allowing biomedical signal analysis and data mining of multiple heterogeneous signals, whether in real time, or offline. Still, at least one such platform is currently in development [20,21].

4.3 ARCHITECTURAL CHALLENGES IN WEB-BASED BIOMEDICAL SIGNAL ANALYSIS SOFTWARE

4.3.1 Data Privacy, Security and User Roles

Data privacy is a very important topic in biomedical data analysis. In recent times, a number of software exploits have arisen. Most security issues are related to bugs in software and weak server configurations. Often, programmers don't keep security and possible consequences of software malfunctions in mind. Common errors are related to poor or absent verification of user input and to an absence of verification of sizes of data buffers. Possible consequences are crashes caused by invalid user input or leakage of confidential data [80]. The solution for this kind of problems is to use defensive programming techniques [81].

A weak server configuration allows an attacker to trick the server to use an older version of SSL or TLS security protocols that have many exploits and then to use one or more exploits to get confidential data. The solution is to pay more attention while writing server configuration files. Namely, all unneeded services should be disabled, all non-user ports should be blocked by a firewall, and only the latest version of TLS cryptographic protocol should be used [82].

The most common attacks fall in the following groups: (1) weak configuration of servers, (2) protocol design flaws, and (3) protocol implementation flaws. We provide examples for each group of attacks, as follows.

Group 1 example: Downgrade attack. A protocol downgrade attack tricks a server to use one of the older versions of SSL/TLS protocols that have design flaws and then employ one of the many known exploits to get confidential data. An attacker performs this attack in the handshake phase of the connection. The client initiates a handshake by sending the list of supported TLS and SSL versions. An attacker intercepts the traffic, posing as a server (Man-in-The-Middle attack) and persuades the client to accept one of the older versions of the TLS or SSL protocols. Once the connection between the client and the server is established on an older protocol version, the attacker can perform one of many known attacks exploiting protocol design or implementation flaws.

Group 2 example: SSL 2.0 design flaws. SSL version 2.0 has many known flaws. One of the flaws is to use the same cryptographic keys both for message integrity and encryption. This is a problem if a chosen cryptographic algorithm is weak. If an attacker successfully breaks the encryption he/she can change the content of a message, and also change the message integrity part used to verify its content. Another flaw is an unprotected handshake, which leads to Man-in-The-Middle attacks that can possibly go undetected.

Group 3 example: Heartbleed attack. Heartbleed attack exploits the bug in the OpenSSL library [83]. The client sends a 'heartbeat' message to the server, which contains data and data size. The server responds with the same data and the size of the data received from the client. The problem is that if a client sent the false data size (bigger than the real size), the server responds with the data received from the client + random data from the server's RAM to fill the response to the required size. That random data can be, for example, password or encryption keys.

We can safely conclude that using only cryptographic protocols is not enough. For additional protection, authentication should be used. There are two major approaches: cookies-based and token-based [84]. Most websites use a strategy that stores a cookie in the browser. After a user logs in, he receives a cookie with the session identifier, which is used in a later request to the server. Cookie-based authentication is stateful. This means that a session must be kept both on a server and on a client. Token-based authentication is similar to cookies, but the major difference is that token-based authentication is stateless. The server does not keep a record of which users are logged in. Every request from a client to the server contains a token, which the server uses to verify the request.

Arguably, the most popular token-based authentication technology nowadays is JWT – JSON Web Tokens [85]. JWT are used for securely representing claims between two parties. Representation is in the form HEADER.DATA.SIGNATURE. The header describes the token type and encryption algorithm. Data is user data that is protected by JWT. The signature contains header and user data signed by the encryption algorithm, for example:

Header:
```
{
  "alg": "HS256",
  "typ": "JWT"
}
```

Data:
```
{
  "name": "Alan",
  "admin": true
}
```

Password:
```
"secret" - without quotes
```

JWT is constructed in the following form:
```
HMACSHA256(
 base64UrlEncode(header) + "." +
 base64UrlEncode(payload),
 secret
).
```

The JWT token looks like this:
```
eyJhbGciOiJIUzI1NiIsInR5cCI6IkpXVCJ9.
eyJuYW1lIjoiQWxhbiIsImFkbWluIjp0cnVlfQ.
fTHiKsW8gH_Bp5AUzqoOTx7FVTL0PZZrnVBnto05He0
```

In a later communication, the JWT token is used in this way:
Authorization: Bearer

```
eyJhbGciOiJIUzI1NiIsInR5cCI6IkpXVCJ9.
eyJuYW1lIjoiQWxhbiIsImFkbWluIjp0cnVlfQ.fT
HiKsW8gH_Bp5AUzqoOTx7FVTL0PZZrnVBnto05He0
```

JWT token is generated on the backend after the successful login of a user. It is used in the authorization header of each HTTP request, which goes from frontend to backend afterwards. When the backend receives the JWT token, it verifies it, extracts the user credentials and generates a response to the HTTP request. If TLS encryption is compromised, an attacker can inject or modify the JWT data, but the attack will be unsuccessful because the JWT token will not pass verification. It is a recommended practice to limit the lifetime of a JWT token, which is done using the 'exp' field.

One additional layer of security can be implemented on a server, even before transmission of data. That layer is the storage of files in an encrypted form. By implementing this layer, security is increased substantially because the only point left open to attack is the RAM. Encryption can be achieved in two ways, depending on the file systems used. On older file systems (e.g. FAT, ext2, ext3), which do not support encryption, a file is run through some encryption software or library and then stored as a regular file. On newer file systems, it is only necessary to enable encryption as the files are then encrypted on the fly before being stored. Most notable examples of the newest file systems are ZFS and APFS, which contain many other improvements in addition to file encryption [86]. The approach to use the newest file systems is the preferable method for encryption and it is very easy to implement, because a server administrator just needs to create a file system that supports encryptions and enable it.

Most recently, the exploitation of hardware vulnerabilities has begun to emerge. Most notable are Meltdown and Spectre [87]. These vulnerabilities are related to CPU parts responsible for speculative execution and caching (used for speeding up programme execution), but they follow different routes. Meltdown is used to attack a kernel (the core of and operating systems) and Spectre is used to attack another user programme. Meltdown is simpler and easier to protect from. The solution is to isolate the kernel and user programme page tables. Spectre is more complex and more dangerous, because it can get confidential data from a programme that doesn't have bugs. Also, spectre patches are complex and significantly slow down the performance of a CPU.

Data anonymisation is important in medicine in order to avoid identification of patients by unauthorized users. As part of data security for a biomedical signal analysis platform, the users should be made aware that if possible, only anonymised data should be sent to the analysis platform. Thus, data stripped of any possibility for patient identification should be used to reach a diagnosis, or to model a disorder through the use of these platforms. Since data anonymisation procedures may be complex and have been investigated in detail in the literature, they are considered beyond the scope of this work, but we nevertheless point the interested reader to some of the existing work [88,89].

User roles are the usual mechanism used to prevent unauthorized users to access confidential data and to prevent damage to a database [90]. The administrator role does not have restrictions, and a regular user is restricted to only access data for which the administrator has granted access. It is also possible to define multiple levels of user roles, which grant access to more or less data. As an example, the administrator of a hospital EHR system can change everything: user data (user is a doctor), patient data, password, system information and so on. Superuser of a hospital EHR system can change user data, patient data and password, but cannot change system information. A regular user (doctor) can change only patient data and his/her own password.

4.3.2 Frontend Workflow Organization

The workflow organization of a frontend solution for complex signal analysis poses a problem because many interconnected analysis steps and actions are not directly related to signal analysis. There are several problems that must be solved: how to organize the display, how to switch between the steps of analysis, how to pause and continue the analysis session and so on.

The MULTISAB frontend is built using an Angular web framework [91]. The central concept of Angular is a component. A component consists of HTML, CSS and TypeScript (or JavaScript) code. The MULTISAB consists of a large number of components. Some components are unrelated to biomedical signal analysis (e.g. user login, main windows, change the password, used data editor), but most of them are. Navigation between components is done by using routes. An example of routes in a MULTISAB frontend is shown in the following code segment:

```
export const routes: Routes = [
    {path: "", redirectTo: "/login", pathMatch: "full"},
    {path: "login", component: UserLogin},
    {
       path: "panel", component: PanelComponent,
        children: [
           {path: "", redirectTo: "/panel/multisab", pathMatch:
            "full"},
           {path: "multisab", component: MultisabComponent},
           {path: "type_selection", component:
            TypeSelectionComponent},
           {path: "scenario_selection", component:
            ScenarioSelectionComponent},
           {path: "file_upload", component: FileUploadComponent},
           {path: "record_inspection", component:
            RecordsInspectionComponent},
           {path: "records_preprocessing", component:
            RecordsPreprocessingComponent},
           {path: "features_extraction", component:
            FeaturesExtractionComponent},
           {path: "features_selection", component:
            FeaturesSelectionComponent},
           {path: "model_construction", component:
            ModelConstructionComponent},
           {path: "reporting", component: ReportingComponent},
           {path: "manage_users", component: ManageUsersComponent},
           {path: "change_password", component:
            ChangePasswordComponent},
           {path: "edit_self_data", component: EditSelfDataComponent}
       ]    }
];
```

Navigation is orchestrated by the main component, which is also responsible for the layout of components on the web page. Navigation between routes not related to biomedical signal analysis is not restricted, except for limitations imposed by user roles. For example, an ordinary user can only do signal analysis and change his password. Navigation between routes related to biomedical signal analysis is regulated by a finite state machine. This allows navigation to components that are related to current state and to actions a user wants to perform. For example, when a user creates a new analysis, he is permitted only to close the analysis or to select the type of analysis. This is important in order to support the usual process of biomedical signal analysis, in which one usually moves forward, starting from a state in which the analysis goal is set, and ending in a state where reporting the analysis results is performed. A transition between the states is resolved on the frontend and sent to the backend, which stores it in the database. The backend does not have any semantics regarding the states, except when it filters open and closed analysis sessions. We show all the possible state transitions in Table 4.1. For a selected state in a row, X represents the allowed state for transition (in a column).

4.3.3 Frontend and Backend Interactions

Because of data confidentiality, all communication in MULTISAB goes over a secure connection. The contemporary HTTP/2 protocol was chosen for communication,

TABLE 4.1
State Transitions in MULTISAB Frontend Finite State Machine

State	S1	S2	S3	S4	S5	S6	S7	S8	S9	S10	S11	S12
S0	X	X										
S1			X	X								
S2												
S3	X											
S4			X	X	X							
S5			X	X	X	X						
S6			X	X	X	X						
S7			X			X	X	X	X			
S8			X			X	X	X	X			
S9			X			X	X	X	X			
S10			X							X	X	X
S11			X							X	X	X
S12			X									X

Abbreviations: S0: Start state; S1: New analysis session; S2: Continue analysis session; S3: Close analysis session; S4: Select analysis type; S5: Scenario selection; S6: Input data selection; S7: Records inspection; S8: Records preprocessing; S9: Features extraction; S10: Features selection; S11: Model construction; S12: Reporting.

because it supports TLS and many other features [92]. Although HTTP/2 does not require encryption, the majority of web browsers only support HTTP/2 over TLS. HTTP/2 has several strong points, of which the most prominent are:

- It is a binary protocol, so it is much faster to parse than the previous versions of the protocol, which were textual; and
- It is multiplexed, so it significantly reduces round trip times without any additional optimisations. In previous versions of the protocol, every request was followed by the response. In version 2 of the protocol, many requests can be sent at once followed by one (or more) response.

After a successful login, the backend generates a random number, which is used as an authorization token, and sends it to the frontend. The token represents a session maintained on the backend. The frontend sends the token to the backend in every request. The backend verifies the token and, according to the session data, determines if the user is logged in and if the user is allowed to access a resource, with respect to his role. The frontend often sends the request to the backend to refresh the token. If the token expires, the backend automatically logs out the user for safety reasons.

All communication between the backend and the frontend is done over RESTful API. However, contrary to common practice of using GET, PUT, POST and DELETE HTTP requests, we have chosen to use only POST request. Consequently, we need to use one element or URL path more than in the usual approach. For example, for deletion of user data:

Common practice:

```
DELETE /api/users
{
     userid: 123456
}
```

MULTISAB implementation:

```
POST /api/users/delete-user
{
    userid: 123456
}
```

We consider that our approach is better, because the API structure looks more uniform (although a bit more complex) and it reduces the number of errors during programming, since we can copy-paste functions that create the HTTP request and change only the URL. Although it is generally believed that RESTful architectural style should mirror CRUD (create, read, update, delete) to HTTP methods POST, GET, PUT and DELETE, according to the author of the RESTful protocol, this is not mandated [93].

4.3.4 Changes in Implementation Languages and Libraries

Programming languages evolve over time. Some changes improve performance, other simplify or improve syntax and thus make writing programmes easier. Changes in programming languages, no matter how small, often break working programmes. Changes in languages also cause changes in its libraries [94]. If a programme uses more libraries, then there is a higher probability it will not work correctly when a programming language change occurs. It is often necessary to create wrappers or workarounds for the libraries that cause problems due to language changes. These generally work only temporarily. For some changes, there is no easy solution and parts of a programme must be rewritten. From our experience, changes occur faster in frontend technologies compared to backend technologies and it requires more time to modify a programme and to ensure that security issues are not introduced.

On the MULTISAB backend, we had an occasion where we had to modify several dozen lines of code, because Java 9 changed legal identifier names [95]. Apparently, Java 9 removed the symbol "_" (underscore) from the set of legal identifier names. This problem was easy to solve, because the compiler issued an error for each occurrence of "_" as an identifier name, so we only had to follow the reported errors and rename the variables.

Additionally, Java 9 introduced the module system. We soon recognized the importance of the module system, because it adds an additional layer of security to the usual private/protected/public visibility scope. By using the module system, it is possible to completely hide the classes that the programmer does not want to export. For the exported classes, the programmer continues to use the usual visibility scope. When we tried to implement the module system in the MULTISAB backend, we ran into a problem, because some of the external libraries that we use had issues with the module system and stopped working. We had to revert our code to the previous version and wait for the libraries to be converted to Java 9.

On the MULTISAB frontend, we also had issues with the RxJs library a few times. The Angular library changes often (rarely causing a problem), but the RxJs library, which we use for reactive programming, changes somewhat slower. On several occasions, we had a problem with the inclusion of the RxJs operators. The initial strategy was to wait several days for RxJs to resolve these problems. The strategy was soon abandoned, because we wasted time waiting. Therefore, we decided to create a workaround for each problem as a temporary solution. When the next version of RxJs solved some problems, the workarounds were removed.

Not all changes are negative or require a lot of work. For instance, when Angular introduced HttpClient service as a replacement for the old Http service, we saw this as an opportunity to simplify our programme. HttpClient response returns JSON object by default, so it became unnecessary to explicitly parse textual response. Changes were trivial to implement and the end result was a smaller code base.

On the other hand, the use of contemporary frontend development frameworks such as Angular for developing the platform also poses a challenge. For example, in order to allow the upload of multiple files in the platform in an elegant, drag-and-drop manner, as shown in Figure 4.2, we needed to write our own component with

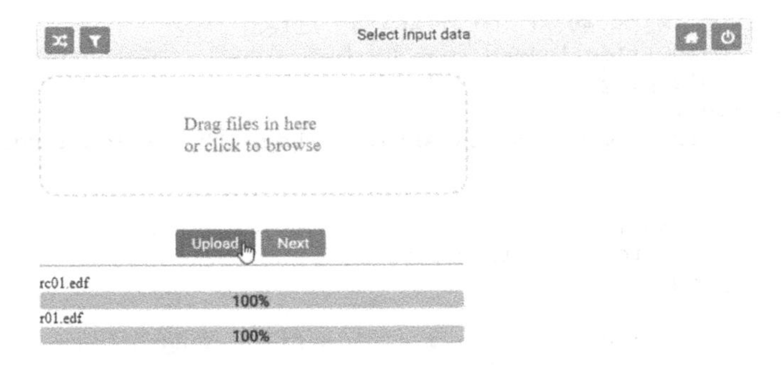

FIGURE 4.2 Multiple file upload page on frontend, as an example of a complicated Angular component, with some example files uploaded onto the MULTISAB platform.

a significant number of lines of code (as there was no readily available one), as is shown in the example:

```
import {Component} from "@angular/core";
import {Observable} from "rxjs";
import {AnalysisUrls} from "../../services/url/AnalysisUrls";
import {PrintMessages} from "../../services/PrintMessages";
import {Router} from "@angular/router";
import {HttpErrorResponse} from "@angular/common/http";
import {Language, LocaleService, TranslationService} from
"angular-l10n";

@Component({
    templateUrl: "/app/components/analysis/FileUploadComponent.html"
     })
export class FileUploadComponent extends PrintMessages {
    @Language() lang: string;
    files: FileList;
    progress: Array<number> = [];

    constructor(private analysisUrls: AnalysisUrls, private router:
     Router,private locale: LocaleService, private translation:
     TranslationService) {
        super(translation);
    }
    fileChangeEvent(event: Event): void {
        this.files = (event.target.valueOf() as
              HTMLInputElement).files;
        if (this.files.length === 0) {
            this.progress = [];
        } else {
            this.progress = new Array<number>(this.files.length);
        }
    }
      dragHandler(event: DragEvent): void {
        event.stopPropagation();
        event.preventDefault();
    }
    dropHandler(event: DragEvent): void {
        event.stopPropagation();
        event.preventDefault();
```

```
        this.files = event.dataTransfer.files;
        if (this.files.length === 0) {
            this.progress = [];
        } else {
            this.progress = new Array<number>(this.files.length);
        }
    }
    submit(): void {
        for (let prog of this.progress) {
            prog = 0;
        }
        for (let i: number = 0; i < this.files.length; i++) {
             this.uploadFile(i);
        }
    }
...
}
```

We also need to write an HTML template for the component (FileUploadComponent.html). In this case, the code is not long and is given here in full:

```
<p-growl [value]="msgs"></P-growl>
<div style="position: absolute; width: 450px; top: 10px; left: 10px;">
    <form>
        <div id="drop_zone" (drop)="dropHandler($event)"
            (dragenter)="dragHandler($event)"
            (dragover)="dragHandler($event)" (click)="inputFiles.
            click()"
            style="border: dashed 2px lightgray; border-radius: 20px;
            height:
          125px; font-size: 20px; color: gray; width: 100%; display:
             table;">
            <span style="text-align: center; display: table-cell;
        vertical-align: middle;">
                 {{'FileUpload.dragFiles' | translate:lang}}
                 {{'FileUpload.dragFilesOrClick' | translate:lang}}
            </span>
        </div>
        <div hidden>
            <input type="file" id="file" (change)="fileChangeEvent
            ($event)"
            multiple #inputFiles>
        </div>
        <div style="text-align: center;">
            <button type="submit" pButton (click)="submit()" label="
            {{'FileUpload.buttonSend' | translate:lang}}"
       [disabled]="files == null"></button>
            <button type="submit" pButton (click)="nextPage()"
            label="{{'FileUpload.buttonNext' | translate:lang}}">
            </button>
        </div>
        <hr *ngIf="files != null">
        <div style="width: 100%" *ngFor="let file of files; let i =
        index;">
            <label>{{files[i].name}}</label>
```

```
            <p-progressBar [value]="progress[i]"></P-progressBar>
        </div>
    </form>
</div>
```

Hence, in our experience, there are significant challenges in keeping up with frontend development languages and technologies, which slows down the overall platform development. We suspect that fast language and libraries evolution, in particular related to frontend, may be one of the reasons for having such a small number of relevant online biomedical signal analysis platforms. Another reason may be because the traditional biomedical signal analysis community did not need to concern itself much with frontend development technologies, and due to the complexity of its programming, the implementation of such a platform remained out of the community's reach.

4.3.5 Database Design

Changes in database inevitably lead to changes on the backend and likely to changes in the frontend. The opposite direction is true as well, as changes in backend and frontend cause changes in database. In the period of planning and implementation of a database, it is hard to predict all the possible use cases, and therefore the database design should evolve as implementation progresses, much as in any other programme. An important advice is to try avoiding changes in the database, because then, the backend and frontend suffer. Changes made to one of these three segments (database, backend or frontend) should trigger changes in the other two segments. Hence, it is necessary to have good code organization in all three segments.

In MULTISAB, we have only one central place for data prototypes in each of the three segments. In the database, we have the file 'Database.ddl' that holds data definitions of the database tables. All changes in the file are propagated to the database implementation. In the example below, we provide the definition of a class Phase that is a part of implementation of our database.

```
CREATE TABLE Phase (
  Id  BIGINT NOT NULL,
  Name VARCHAR(256)  NOT NULL,
  PRIMARY KEY (Id),
  CHECK (LENGTH(TRIM(Name)) > 0),
  UNIQUE (Name)
);
```

On the backend, we use Java Persistence API–JPA. This is a mechanism that maps Java classes to database tables by using annotations. JPA defines Java Persistence Query Language, which is a simplified version of SQL language and is adapted to the object-oriented way of programming. Changes in the database reflect easily to the backend. One only needs to change the variable definitions and annotations. In the following example, we provide the definition of the class Phase in backend.

```
@Entity
@Table(name = "PHASE")
public class Phase implements Serializable {
    private static final long serialVersionUID = 1L;
    @Id
    @Basic(optional = false)
    @Column(name = "ID")
    private Long id;
    @Basic(optional = false)
    @Column(name = "NAME")
    private String name;
    public Long getId() { return id; }
    public void setId(Long id) { this.id = id; }
    public String getName() { return name; }
    public void setName(String name) { this.name = name; }
}
```

On the frontend, we keep the TypeScript class prototypes in services, which are responsible for their part of the RESTful API (e.g. LoginService, SessionService). Changes in frontend are not 1:1 related to changes in databases. Instead, they are related to changes in the RESTful API.

To simplify the documentation of MULTISAB RESTful API, we use OpenAPI. OpenAPI was originally known as the Swagger Specification. Description of a user RESTful API is contained in a YAML file [96]. For editing the YAML file, the Swagger Editor tool can be used, while for viewing, the Swagger UI tool is available [97]. Both Swagger Editor and Swagger UI translate YAML file to HTML and visually display the users' RESTful API; see our example in Figure 4.3.

Our experience has shown that the RESTful API should be edited first. Afterwards, the backend and frontend can be programmed according to the changes, and the database can also be changed if needed. If the RESTful API needs to be changed due to changes in database, backend or frontend, we recommend to change the documentation first and then propagate the changes accordingly. Thus, one can always have an up-to-date documentation and develop different parts of the programme independently, using the documentation to synchronize.

In the MULTISAB project development, we use a simplified approach, skipping the use of the Swagger Editor. We first write the backend code and annotate it with comments. In this case, we merge two steps into one – writing of documentation and implementation in the backend. Annotations in the code are transformed by the SpringFox library [98] into documentation. When the backend is launched, documentation can be accessed over the Swagger UI. The Swagger UI is packaged into the SpringFox so the end-user just needs to visit the URL at http://localhost:8081/swagger-ui.html to view the documentation. As an example, we show the "closeSession()" function annotated with documentation comments. The end result for the code can be seen in Figure 4.3.

```
@RequestMapping(value = "/close",
      method = RequestMethod.POST,
      consumes = MediaType.APPLICATION_FORM_URLENCODED_VALUE,
      produces = MediaType.APPLICATION_JSON_UTF8_VALUE)
 @ApiOperation(value = "Zatvaranje sjednice", notes = "Zatvara
  sjednicu analize")
```

```
  @ApiResponses(value = {
      @ApiResponse(code = 200, message = "Uspješno je zatvorena
        sjednica"),
      @ApiResponse(code = 401, message = "Korisniku nedostaju
        ovlasti za zatvaranje sjednice"),
      @ApiResponse(code = 500, message = "Dogodila se interna
        greška na poslužitelju")
})
synchronized ResponseEntity<Void> closeSession(@RequestParam
   long connectionToken) {
  Optional<UserData> data =
      loggedInUsers.findUserByToken(connectionToken);
      if (!data.isPresent()) {
          return new ResponseEntity<>(HttpStatus.UNAUTHORIZED);
      }
      if (closeSessionInternal(connectionToken, data.get())) {
          return new ResponseEntity<>(HttpStatus.OK);
      } else {
          return new
              ResponseEntity<>(HttpStatus.INTERNAL_SERVER_ERROR);
      }
}
```

In the beginning of the MULTISAB backend project, we wanted a solution easily portable between different computers and that could be copy-pasted from a portable medium without installation. Therefore, we chose Java as the programming language. For the main development framework on the backend, we chose Spring Boot, due

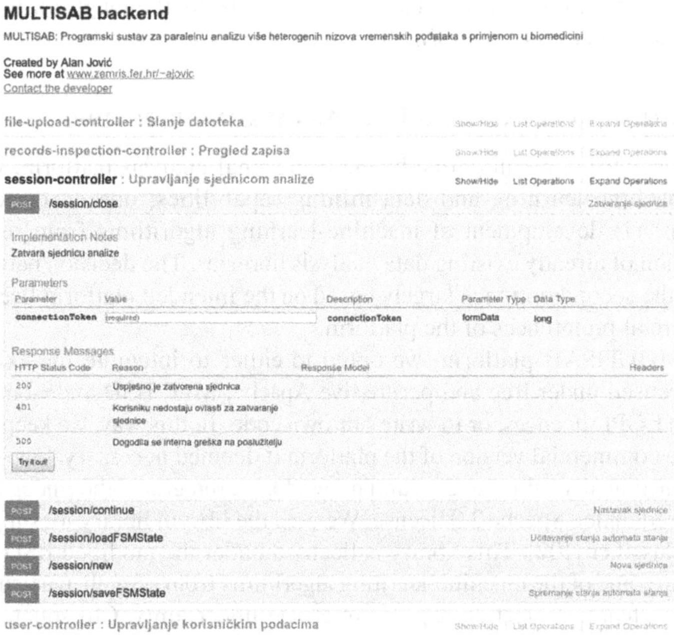

FIGURE 4.3 Screenshot of Swagger UI displaying part of the MULTISAB RESTful API.

to its high efficiency in creating web applications [99]. Database candidates were Apache Derby and H2 [100].

Apache Derby had the advantage of being a part of the Java distribution (Java DB). Its main disadvantage was that it stores database files in many files and folders within the main folder. We wanted as few files as possible, so H2 database was chosen instead, because it stores all data in a single file. The disadvantage was that we had to pull the library from the separate repository. As we use Maven in our project, this disadvantage is insignificant because Maven pulls the H2 database library files while pulling other libraries we use. H2 is a very compact DMBS (only 1.5 MB) and it lacks some features larger DBMSs have. For example, it does not have a language for stored procedures and it does not support multi-threaded processing.

Our current database model may be found in Friganovic et al. [20]. The database is designed to support the workflow of the biomedical signal analysis platform. It is not designed to store signal records, but rather to contain user session data, including the scenario that is conducted through the platform, phase of the analysis in which the user is currently found (so the continuation of the session last discontinued is possible), the pre-processing methods and the features the user selected for the analysis. Also note that, aside from the analysis-related data, we only store user personal data. The MULTISAB backend is not demanding regarding data storage, so the H2 database is just enough for this purpose. This is because the files containing biomedical signals, extracted feature vectors, models and reports are all stored in the server file system and only their names are stored in the database as part of an analysis session. Although we may modify some parts of the database design in the future, it should be noted that the initial design proposed was found to be satisfactory thus far. We contend that such a database solution is good enough for a web-based biomedical signal analysis platform.

4.3.6 Integration of Existing Data Analysis and Reporting Libraries

When one is developing an online biomedical signal analysis platform, which also includes machine learning and data mining capabilities, one needs to consider two options: (1) development of machine-learning algorithms from scratch, and (2) integration of already existing data analysis libraries. The decision between using the first or the second option is largely based on the intended platform licensing, and the commercial preferences of the platform.

In the MULTISAB platform, we opted to either to integrate the existing data libraries licensed under free and permissive Apache, MIT, BSD and (somewhat less permissive) LGPL licences, or to write our own code. In this way, we keep the option to develop a commercial version of the platform if deemed necessary. For some of the machine learning algorithms (e.g. neural network), we integrated the Encog framework [101], licensed under Apache 2.0 license. We used the libsvm library for support vector machines classifier [102]. This library is licensed under the modified BSD license.

We wrote some of the machine-learning algorithms from scratch, particularly those pertaining to feature selection (e.g. symmetrical uncertainty, Chi square, ReliefF), as we did not find any implementation under permissive licenses. For reporting the results of the analysis process (final statistics and evaluation measures of results), we opted to use the JasperReports (JR) library [103]. JR is licensed under LGPL, briefly meaning that in order to keep the possibility of commercializing our platform, we can only use its API, not change its source code. The use of a separate reporting library should be considered suitable for a web platform in the case of biomedical signal analysis applications, as many different types of statistical results may be obtained from the analysis scenario (e.g. for classification: class distribution, confusion matrix, total classification accuracy, sensitivity, specificity, F1 measure, etc.) [104]. Some of the results may be presented in a tabular form or as a list of evaluation measures, while others may use pie charts, histograms and so on.

Achieving a successful and uniform data connection between several different libraries and our implementations is done through classes that adapt the data for particular API requirements of certain libraries. As the platform evolves, we expect to add more machine-learning algorithms in time, either through our own implementations or by integrating permissive licensed implementations. We consider the approach we propose here reasonable and viable for similar purpose online analysis software.

4.3.7 Workload Amelioration

During the backend planning phase for the MULTISAB project, we decided to separate it into two parts. The first part is the core of the backend, which is responsible for database access, RESTful API and user rights management. The second part is a sub-project called *processing*. *Processing* implements all signal analysis algorithms, including machine learning, which are used in the MULTISAB project. We made this decision because we wanted to have better control over the intellectual property protection and have the possibility to improve the performance of the algorithms without touching the core backend functionality.

We started with the implementation of algorithms in a sequential way. After that, we conducted experiments with the OpenCL library. We did not use OpenCL directly, but over the Aparapi library [105]. Aparapi translates native Java bytecode in OpenCL kernels dynamically at runtime. We initially did synthetic benchmark tests on GPU and they looked promising. After that, we did the implementation of several concrete feature extraction algorithms in Aparapi. Results were far below the results achieved on synthetic tests. The conclusion was that the overhead of real-time compilation of Java bytecode to OpenCL kernels, as well as data transfer between the CPU and GPU, took too much time. Another cause was that the data analysis algorithms were either too simple, or if they were complex, too little data was used to mask the latency of transfer to the GPU. We also tried to use the CPU as an OpenCL computing device, but the results were similar.

Finally, we made a decision to rely on the classical approach with Java threads. The parallelization procedure we implemented for the feature extraction step of the platform is as follows:

- If there are multiple segments present in a signal, then these are resolved (all the features are extracted) in parallel;
- Otherwise, if there are no multiple segments, but there are signals of the same type (e.g. EEG), then these are resolved (all the features are extracted) in parallel;

- If there are no multiple segments nor signals of the same type, but there are multiple patient files, then these are resolved (all the features are extracted) in parallel; and
- Lastly, if there is only a single file with a single signal of a particular type and a single segment, then it is treated as sequential and single threaded.

Currently, we use parallelization over primitive data arrays; however, in the future, we plan to use Java streams (introduced in Java 8). The concept of streams is simple; instead of viewing data as a collection (sets, maps), the user sees data as a 'stream of elements'. A stream can be created from existing collections using functions 'stream()' and 'parallelStream()'. By using 'parallelStream()', elements are automatically parallelized. The user only needs to ensure that algorithms are suitable for parallel execution, which may be difficult in signal analysis.

Another optimization is possible, which is the usage of several computers. In the future, we will have to implement a kind of message-passing interface to enable communication of the core of the backend with the *processing* instances running on several physical machines.

4.4 REQUIREMENTS FOR CONSTRUCTING A BIOMEDICAL SIGNAL ANALYSIS WEB PLATFORM

4.4.1 Hardware and Software Requirements

Architectural requirements for a web platform designed to perform biomedical signal analysis may vary depending on the amount of expected data processing and the analysis scenarios supported by the platform. For example, if one wants only to record some physiological signals in a home environment for a single (or a few) persons and provide a few health markers, which may be sent to a medical expert acting remotely for further (manual) evaluation, then such a system does not need to have significant hardware requirements. A usual personal computer, or even a handheld device, may suffice for the client side provided it works as a gateway to a remote server in a health institution [57,106]. In such a setting, a remote server is usually used for storing collected patient data for visual inspection by the medical professional, and perhaps for a future offline analysis.

However, supporting many users at the same time and performing complex analysis scenarios (as depicted in Figure 4.1) may require more resources for a general and expandable solution. A typical minimum solution would include a single computer, acting as a server for data analysis with fast multi-core processor capabilities (4 or more logical cores), as well as large hard drive and RAM capacities. The computer would need to have a web server installed to support the web application (e.g. Apache Tomcat), H2 or similar in-memory relational DBMS, and would need to provide software support for the whole web development technological stack in order to accommodate for potential software improvements. It would hence include (1) frontend technologies, such as HTML, CSS, Bootstrap, JavaScript/TypeScript, Angular or similar frontend development frameworks; and (2) backend technologies, such as Java 9, Spring Boot and JPA (or related backend Microsoft, PHP, or Python technologies). Additionally, permissive license libraries used to cover the various steps in biomedical signal analysis would be a welcome – but not a necessary – requirement for the construction of the web platform, as some of the required methods may be efficiently implemented from scratch.

Any expansion of the proposed minimal solution would be primarily concerned with workload improvements, as we already elaborated in Section 4.3.7. Although we focused mostly on single-computer parallelism in our work, distributed backend processing, through the use of OpenMP [107] bindings or through service-oriented architecture based on remote method invocation [108] should be considered.

A step further might be the integration of the web-based biomedical signal analysis platform within a cloud infrastructure. Although some cloud-based solutions for medical big data analysis were already proposed in the literature [109,110], we are only aware of one solution dealing in any way with biomedical signal analysis in such a setting [111]. This solution, called Cloudware, focuses on ECG pre-processing and visualization through the use of Hadoop big data technology in a cloud. Nevertheless, the Cloudware platform still lacks the complete complex analysis scenarios for several different types of biomedical signals that we proposed in the MULTISAB platform and which we depicted in Figure 4.1, as it is primarily intended for better visualization capabilities of ECG recordings.

4.4.2 Other Requirements

Designing use cases for the requirements of the web platform proved to be a difficult task in our case, due to the complexity of possible analysis scenarios. Nevertheless, the use of UML use case diagrams to specify possible user behaviour in communication with the system proved to be a valuable asset in the development of the platform [112]. Therefore, we would recommend the use of UML tools, especially in specifying platform requirements through the use of scenarios.

Development of this web analytical platform requires that the team members be well-versed in a variety of frontend, backend and database technologies, aside from an expertise in signal analysis. This is something not often encountered or required in practice, and this may be one of the reasons such a platform has not been proposed earlier. As we have elaborated in Section 4.3.4, numerous challenges with implementation languages and libraries exist, especially in frontend technologies. A well-versed competent team requirement might seem trivial at first, but in our experience – and this is also corroborated by other researchers [113] – it is very important that all of the research team members be at least well acquainted with all aspects of the platform. However, we also agree with the conclusions of the Software Sustainability Institute that it may be counterproductive to instill all the software development details to all team members, as some of them might be better in, for example, signal analysis than in frontend development [114].

4.5 DISCUSSION AND CONCLUSION

Designing a web platform intended for biomedical signal analysis is challenging, as we showed in this work. Indeed, there are many contributing factors to design and implementation complexity. In Section 4.3, we presented some of the most pervasive challenges we encountered in developing the MULTISAB platform. Not all the challenges were reported in detail here, though. For example, the issue of biomedical signal data itself is problematic, because analysing multiple heterogeneous time series poses difficulties with respect to input data formats, feature extraction and parallelization. We reported progress on some of these issues in an earlier work [21] and we also plan additional publications to cover these issues in more detail.

We would like to stress that one of the biggest problems in designing the web platform are the web development technologies, which are still mostly not standardized (apart from HTML) and are in continuous development. Thus, designing an offline biomedical signal analysis platform that would accommodate all the complex scenarios related to multiple heterogeneous signal analysis may be difficult, but designing the same platform for the web is even more difficult. Aside from the development technology, we consider that the only other serious issue in web platform construction is security. In biomedical applications, software safety and security need to be of the highest degree possible. As we have elaborated in this work, this issue has not been solved in a general sense. Although employing the most recent security protocols, message encryption, and other advanced techniques should suffice in most cases [115], there are really no theoretical guarantees with the safety of web systems.

Integrating all of the technologies to have a working platform is something we are currently working on. Despite the challenges discussed in this chapter, we still consider that having such an integrated analysis platform would greatly benefit both researchers and medical professionals. Overcoming the challenges would allow distant access to platforms with advanced diagnostical and analytical capabilities. In the future, we plan to investigate how the platform could be applied in medical practice, including:

- The modelling of body state, based on multiple patient records and signal types, for example for stress detection [116], pregnancy characterization [117], etc.;
- Possible smooth integration with remote patient monitoring technologies to enable online analysis through its decision support capabilities [21]; and
- The integration with medical centre computer infrastructures, in order to enable EHR information processing in the platform to achieve better medical diagnostics.

ACKNOWLEDGEMENTS

This work has been fully supported by the Croatian Science Foundation under the project number UIP-2014-09-6889.

REFERENCES

1. Clifford, G.D., Azuaje, F., McSharry, P.E. 2006. *Advanced Methods and Tools for ECG Data Analysis*. Artech House, Norwood MA, USA.
2. Campbell, I.G. 2009. EEG recording and analysis for sleep research. *Curr Protoc Neurosci*. chapter: Unit 10.2.; doi:10.1002/0471142301.ns1002s49.

3. Cifrek, M., Medved, V., Tonkovic, S., Ostojic, S. 2009. Surface EMG based muscle fatigue evaluation in biomechanics. *Clinical Biomechanics*. 24(4), pp. 327–340.
4. Zangroniz, R., Martinez-Rodrigo, A., Pastor, J.M., Lopez, M.T., Fernandez-Caballero, A. 2017. Electrodermal activity sensor for classification of calm/distress condition. *Sensors*. 17(10), p. 2324.
5. Robbins, K.A. 2012. EEGVIS: A MATLAB® toolbox for browsing, exploring, and viewing large datasets. *Front Neuroinform*. 6, p. 17.
6. Tarvainen, M.P., Niskanen, J.P., Lipponen, J.K., Ranta-aho, P.O., Karjalainen, P.A. 2014. Kubios HRV – heart rate variability analysis software. *Comput Methods Programs Biomed*. 113(1), pp. 210–220.
7. Srhoj-Egekher, V., Cifrek, M., Medved, V. 2011. The application of Hilbert-Huang transform in the analysis of muscle fatigue during cyclic dynamic contractions. *Med & Biol Eng & Comput*. 49(6), pp. 659–669.
8. Goldberger, A.L., Amaral, L.A.N., Glass, L. et al. 2000. PhysioBank, PhysioToolkit, and PhysioNet: Components of a new research resource for complex physiologic signals. *Circulation*. 101(23), pp. e215–e220.
9. Baumert, M., Porta, A., Cichocki, A. 2016. Biomedical signal processing: From a conceptual framework to clinical applications [scanning the issue]. *Proceedings of the IEEE*. 104(2), pp. 220–222.
10. Friganovic, K., Jovic, A., Kukolja, D., Cifrek, M., Krstacic, G. 2017. Optimizing the detection of characteristic waves in ECG based on exploration of processing steps combinations. *Joint Conf of the European Medical and Biological Engineering Conference (EMBEC'17) and the Nordic-Baltic Conf on Biomedical Engineering and Medical Physics (NBC'17), IFMBE Proceedings*, vol. 65, Tampere, Springer Nature Singapore, pp. 928–931.
11. Sassi, R., Cerutti, S., Lombardi, F., Malik, M., Huikuri, H.V., Peng, C.-K., Schmidt, D., Yamamoto, Y. 2015. Advances in heart rate variability signal analysis: Joint position statement by the e-cardiology ESC working group and the European Heart Rhythm Association co-endorsed by the Asia Pacific Heart Rhythm Society. *Europace*. 17, pp. 1341–1353.
12. Mladenic, D. 2006. Feature Selection for Dimensionality Reduction. In: Saunders, C., Grobelnik, M., Gunn, S., Shawe-Taylor, J. (Eds.), *Subspace, Latent Structure and Feature Selection. Lecture Notes in Computer Science* 3940. Springer, Berlin, Heidelberg, pp. 84–102.
13. Nandi, D., Ashour, A.S., Samanta, S., Chakraborty, S., Salem, M.A.M., Dey, N. 2015. Principal component analysis in medical image processing: A study. *Int. J. Image Mining*. 1(1), pp. 65–86.
14. Dey, N., Ashour, A.S., Borra, S. (Eds.). 2017. *Classification in BioApps: Automation of Decision Making. Lecture Notes in Computational Vision and Biomechanics* Vol. 26. Springer; doi:10.1007/978-3-319-65981-7.
15. Kamal, M.S., Dey, N., Ashour, A.S. 2017. Large scale medical data mining for accurate diagnosis: A blueprint. In: *Handbook of Large-Scale Distributed Computing in Smart Healthcare*, Springer, Cham; pp. 157–176, doi:10.1007/978-3-319-58280-1_7.
16. Rangayyan, R.M. 2015. *Pattern Classification and Diagnostic Decision. In: Biomedical Signal Analysis*. John Wiley & Sons, Inc., pp. 571–632; doi:10.1002/9781119068129.ch9.
17. Hall, M., Frank, E., Holmes, G., Pfahringer, B., Reutemann, P., Witten, I.H. 2009. The WEKA data mining software: An update. *Sigkdd Explor Newsl*. 11(1), pp. 10–18.
18. Bezerra, V.L., Leal, L.B., Lemos, M.V., Carvalho, C.G., Filho, J.B., Agoulmine, N. 2013. A pervasive energy-efficient ECG monitoring approach for detecting abnormal cardiac situations. *IEEE 15th Int. Conf. on e-Health Networking, Applications and Services (Healthcom 2013)*, Lisbon, Portugal, pp. 340–345.

19. Jovic, A., Kukolja, D., Jozic, K., Horvat, M. 2016. A web platform for analysis of multivariate heterogeneous biomedical rime-series–a preliminary report. *Proc 23rd Int. Conf. on Systems, Signals and Image Processing (IWSSIP 2016)*, Bratislava, Slovakia, p. 70.
20. Friganovic, K., Jovic, A., Jozic, K., Kukolja, D., Cifrek, M. 2017. MULTISAB project: A web platform based on specialized frameworks for heterogeneous biomedical time series analysis–an architectural overview. *Proc. Int. Conf. on Med & Biol Eng (CMBEBiH 2017)*, Sarajevo, Bosnia and Herzegovina, Springer Nature, pp. 9–15.
21. Jovic, A., Kukolja, D., Friganovic, K., Jozic, K., Car, S. 2017. Biomedical time series preprocessing and expert-system based feature extraction in MULTISAB platform. *Proc. Int. Conf. MIPRO 2017*, Opatija, Croatia, pp. 349–354.
22. Chen, C.M. 2011. Web-based remote human pulse monitoring system with intelligent data analysis for home health care. *Expert Syst Appl.* 38(3), pp. 2011–2019.
23. Zunic, E., Djedovic, A., Boskovic, D. 2016. Web-based and mobile system for training and improving in the field of electrocardiogram (ECG). *5th Mediterranean Conference on Embedded Computing (MECO 2016)*, Bar, Montenegro, pp. 441–445.
24. Hsieh, J.-C., Hsu, M.-W. 2012. A Cloud Computing Based 12–Lead ECG Telemedicine Service. *Bmc Med Inform Decis Mak.* 12, p. 77.
25. Jovic, A., Bogunovic, N., Cupic, M. 2013. Extension and detailed overview of the hrvframe framework for heart rate variability analysis. *Proc. Int. Conf. Eurocon 2013*, Zagreb, Croatia, IEEE Press, pp. 1757–1763.
26. Bao, F.S., Liu, X., Zhang, C. 2011. PyEEG: An open source python Module for EEG/MEG Feature Extraction. *Comput. Intell. Neurosci.* 2011, p. 406391; doi:10.1155/2011/406391.
27. Moraru, L., Moldovanu, S., Dimitrievici, L.T., Shi, F., Ashour, A.S., Dey, N. 2017. Quantitative diffusion tensor magnetic resonance imaging signal characteristics in the human brain: A hemispheres analysis. *IEEE Sensors J.* 17(15), pp. 4886–4893.
28. Wang, D., Li, Z., Cao, L. et al. 2017. Image fusion incorporating parameter estimation optimized Gaussian mixture model and fuzzy weighted evaluation system: A case study in time-series plantar pressure data set. *Ieee Sensors Journal.* 17(5), pp. 1407–1420.
29. Chakraborty, S., Chatterjee, S., Ashour, A.S., Mali, K., Dey, N. 2017. Intelligent computing in medical imaging: A study. In: Dey, N., Ashour, A.S., Shi, F., Balas, V.E. (Eds.), *Advancements in Applied Metaheuristic Computing*, Academic Press, London, p. 143; doi:10.4018/978-1-5225-4151-6.ch006.
30. Dey, N., Ashour, A.S. 2018. Computing in medical image analysis. In: *Soft Computing Based Medical Image Analysis*, pp. 3–11.
31. Elhayatmy, G., Dey, N., Ashour, A.S. 2018. Internet of things based wireless body area network in healthcare. In: Dey, N., Hassanien, A.E., Bhatt, C., Ashour, A.S., Satapathy, S.C. (Eds.), *Internet of Things and Big Data Analytics Toward Next-Generation Intelligence*, Springer, Cham, pp. 3–20.
32. Dey, N., Hassanien, A.E., Bhatt, C., Ashour, A., Satapathy, S.C. (Eds.). 2018. *Internet of Things and Big Data Analytics Toward Next-Generation Intelligence. Studies in Big Data* Vol. 30. Springer; doi:10.1007/978-3-319-60435-0.
33. Banaee, H., Ahmed, M.U., Loutfi, A. 2013. Data mining for wearable sensors in health monitoring systems: A review of recent trends and challenges. *Sensors.* 13(12), pp. 17472–17500.
34. Kaur, P.D., Chana, I. 2014. Cloud based intelligent system for delivering health care as a service. *Comput Methods Programs Biomed.* 113(1), pp. 346–359.
35. Xtelligent Media, LLC. 2018. Top 10 Remote Patient Monitoring Companies for Hospitals. https://mhealthintelligence.com/news/top-10-remote-patient-monitoring-solutions-for-hospitals (accessed: 2018-01-07)
36. Lessard, Y., Sinteff, J.P., Siregar, P. et al. 2009. An ECG analysis interactive training system for understanding arrhythmias. *Stud Health Technol Inform.* 150, pp. 931–935.

37. Ross, M.K., Wei, W., Ohno-Machado, L. 2014. "Big data" and the electronic health record. *Yearb Med Inform.* 9(1), pp. 97–104.
38. Murdoch, T.B., Detsky, A.S. 2013. The inevitable application of big data to health care. *Jama.* 309(13), pp. 1351–1352.
39. Hsu, C.-L., Wang, J.-K., Lu, P.-C., Huang, H.-C., Juan, H.-F. 2017. DynaPho: A web platform for inferring the dynamics of time-series phosphoproteomics. *Bioinformatics.* 33(12), pp. 3664–3666.
40. Gaggioli, A., Cipresso, P., Serino, S. et al. 2014. A decision support system for real-time stress detection during virtual reality exposure. *Stud Health Technol Inform.* 196, pp. 114–120.
41. National Sleep Research Resource. 2018. https://sleepdata.org/ (accessed: 2018-01-07).
42. University of Rochester Medical Center. 2018. *Telemetric and Holter ECG Warehouse (THEW).* http://thew-project.org/ (accessed 2018-01-07).
43. University of Michigan in Ann Arbor. 2018. *Ann Arbor Electrogram Libraries.* http://electrogram.com/ (accessed 2018-01-07).
44. Gardasevic, G., Fotouhi, H., Tomasic, I., Vahabi, M., Björkman, M., Linden, M. 2017. A heterogeneous IoT-based architecture for remote monitoring of physiological and environmental parameters. *4th EAI Int. Conf. on IoT Technologies for HealthCare (HealthyIoT 2017)*, Angers, France, p. 4869.
45. Guan, K., Shao, M., Wu, S. 2017. A remote health monitoring system for the elderly based on smart home gateway. *Journal Of Healthcare Engineering.* 2017, p. 5843504.
46. Dey, N., Ashour, A.S., Shi, F., Fong, S.J., Sherratt, R.S. 2017. Developing residential wireless sensor networks for ECG healthcare monitoring. *Ieee Trans. Consum. Electron.* 63(4), pp. 442–449.
47. Granulo, E., Becar, L., Gurbeta, L., Badnjevic, A. 2016. Telemetry system for diagnosis of asthma and chronical obstructive pulmonary disease (COPD). *3rd EAI Int. Conf. IoT Technologies for HealthCare (HealthyIoT 2016)*, Vasteras, Sweden, pp. 113–118.
48. Matallah, H., Belalem, G., Bouamrane, K. 2017. Towards a new model of storage and access to data in big data and cloud computing. *Int. J. Ambient Comput. Intell. (Ijaci).* 8(4), pp. 1–14.
49. Tomasic, I., Petrovic, N., Fotouhi, H., Linden, M., Bjorkman, M. 2017. Relational database to a web interface in real time. *Eur Med Biol Eng Conf & Nordic-Baltic Conf. on Biomed Eng & Med Physics (EMBEC & NBC 2017)*, Tampere, Finnland, pp. 89–92; doi:10.1007/978-981-10-5122-7_23.
50. Salem, O., Guerassimov, A., Mehaoua, A., Marcus, A., Furht, B. 2013. Sensor fault and patient anomaly detection and classification in medical wireless sensor networks. *IEEE Int. Conf. on Communications (ICC 2013)*, Budapest, Hungary, pp. 4373–4378; doi: 10.1109/ICC.2013.6655254.
51. Ivascu, T., Aritoni, O. 2015. Real-time health status monitoring system based on a fuzzy agent model. *E-Health and Bioengineering Conference (EHB 2015)*, Iasi, Romania, pp. 1–4; doi:10.1109/EHB.2015.7391502.
52. Chatterjee, S., Dutta, K., Xie, H.Q., Byun, J., Pottathil, A., Moore, M. 2013. Persuasive and pervasive sensing: A new frontier to monitor, track and assist older adults suffering from type-2 diabetes. *Proc. 46th Hawaii Int. Conf. on System Sciences*, Grand Wailea, HW, USA, pp. 2636–2645.
53. Nangalia, V., Prytherch, D.R., Smith, G.B. 2010. Health technology assessment review: Remote monitoring of vital signs--current status and future challenges. *Crit Care.* 14(5), p. 233.
54. Villarrubia, G., Bajo, J., De Paz, J.F., Corchado, J.M. 2014. Monitoring and detection platform to prevent anomalous situations in home care. *Sensors.* 14(6), pp. 9900–9921.

55. Mainetti, L., Patrono, L., Secco, A., Sergi, I. 2016. An IoT-aware AAL system for elderly people. *Int. Multidisc. Conf. on Computer and Energy Science (SpliTech 2016)*, Split, Croatia, pp. 1–6; doi:10.1109/SpliTech.2016.7555929.
56. Lin, S.-S., Hung, M.-H., Tsai, C.-L., Chou, L.-P. 2012. Development of an ease-of-use remote healthcare system architecture using RFID and networking technologies. *J Med Syst*. 36(6), pp. 3605–3619.
57. Tsujimura, S., Shiraishi, N., Saito et al. 2009. Design and implementation of web-based healthcare management system for home healthcare. *13th Int. Conf. on Biomedical Engineering, IFMBE Proceedings*, vol. 23. Berlin, Heidelberg, Springer, pp. 1098–1101; doi:10.1007/978-3-540-92841-6_270.
58. Secerbegovic, A., Suljanovic, N., Nurkic, M., Mujcic, A. 2015. The usage of smartphones in remote ECG monitoring scenarios. *6th Eur. Conf. Int. Federation for Med & Biol Eng (MBEC 2014)*, IFMBE vol. 45, Dubrovnik, Croatia, Springer, pp. 666–669.
59. Zapata, B.C., Fernandez-Aleman, J.L., Idri, A., Toval, A. 2015. Empirical studies on usability of mhealth apps: A systematic literature review. *J Med Systems*. 39(2), p. 1.
60. Fong, E.-M., Chung, W.-Y. 2013. Mobile cloud-computing-based healthcare service by noncontact ecg monitoring. *Sensors*. 13(12), pp. 16451–16473.
61. Huang, Q., Huang, X., Liu, L., Lin, Y., Long, X., Li, X. 2018. A case-oriented web-based training system for breast cancer diagnosis. *Comput Methods Programs Biomed*. 156, pp. 73–83.
62. Qualter, J., Sculli, F., Oliker, A. et al. 2012. The biodigital human: A web-based 3D platform for medical visualization and education. *Stud Health Technol Inform*. 173, pp. 359–361.
63. Hackett, M., Proctor, M. 2016. Three-dimensional display technologies for anatomical education: A literature review. *J. Sci. Educ. Technol*. 25(4), pp. 641–654.
64. Porras, L., Drezner, J., Dotson, A. et al. 2016. Novice interpretation of screening electrocardiograms and impact of online training. *J. Electrocardiol*. 49(3), pp. 462–466.
65. Hilbert, M. 2016. Big data for development: A review of promises and challenges. *Dev. Policy Re*. 34(1), pp. 135–174.
66. Dey, N., Ashour, A.S., Shi, F., Balas, V.E. 2018. *Soft Computing Based Medical Image Analysis*. Academic Press. ISBN:978-0-12-813087-2.
67. van Poelgeest, R., van Groningen, J.T., Daniels, J.H. et al. 2017. Level of digitization in Dutch Hospitals and the lengths of stay of patients with colorectal cancer. *J Med Syst*. 41(5), p. 84.
68. Rubio, O.J., Alesanco, A., Garcia, J. 2013. Secure information embedding into 1D biomedical signals based on SPIHT. *J Biomed Inform*. 46(4), pp. 653–664.
69. Hsieh, J.-C., Li, A.-H., Yang, C.-C. 2013. Mobile, cloud, and big data computing: Contributions, challenges, and new directions in telecardiology. *Int J Environ Res Public Health*. 10(11), pp. 6131–6153.
70. McCarthy, L.H., Longhurst, C.A., Hahn, J.S. 2015. Special requirements for electronic medical records in neurology. *Neurol Clin Pract*. 5(1), pp. 67–73.
71. Karaa, W.B.A., Dey, N. 2015. *Biomedical Image Analysis and Mining Techniques for Improved Health Outcomes*. IGI Global, Hershey, PA, USA. ISBN:1466688114.
72. Griebel, L., Prokosch, H.-U., Kopcke, F. et al. 2015. A scoping review of cloud computing in healthcare. *BMC Med Inform Decis Mak*. 15, p. 17.
73. Kruse, C.S., Goswamy, R., Raval, Y., Marawi, S. 2016. Challenges and opportunities of big data in health care: A systematic review. *Jmir Med Inform*. 4(4), p. e38.
74. Alyami, M.A., Almotairi, M., Aikins, L., Yataco, A.R., Song, Y.-T. 2017. Managing personal health records using meta-data and cloud storage. *IEEE/ACIS 16th Int. Conf. on Computer and Information Science (ICIS 2017)*, Wuhan, China, pp. 265–271; doi:10.1109/ICIS.2017.7960004.

75. Rifi, N., Rachkidi, E., Agoulmine, N., Taher, N.C. 2017. Towards using blockchain technology for eHealth data access management. *4th Int. Conf. on Advances in Biomed Eng (ICABME 2017)*, Beirut, Lebanon, pp. 1–4; doi:10.1109/ICABME.2017.8167555.
76. Azaria, A., Ekblaw, A., Vieira, T., Lippman, A. 2016. MedRec: Using blockchain for medical data access and permission management. *2nd Int. Conf. on Open and Big Data (OBD 2016)*, Vienna, Austria, pp. 25–30; doi:10.1109/OBD.2016.11.
77. Mitchell, J., Probst, J., Brock-Martin, A., Bennett, K., Glover, S., Hardin, J. 2014. Association between clinical decision support system use and rural quality disparities in the treatment of pneumonia. *J Rural Health.* 30(2), pp. 186–195. doi:10.1111/jrh.12043.
78. Miller, P., Phipps, M., Chatterjee, S. et al. 2014. Exploring a clinically friendly web-based approach to clinical decision support linked to the electronic health record: Design philosophy, prototype implementation, and framework for assessment. *Jmir Med Inform.* 2(2), p. e20.
79. Cloughley, R.G., Bond, R.R., Finlay, D.D., Guldenring, D., McLaughlin, J. 2016. An interactive clinician-friendly query builder for decision support during ECG interpretation. *Comput Cardiol Conf. (CinC 2016)*, Vancouver, BC, Canada, pp. 381–384.
80. Kell, S. 2016. Dynamically diagnosing type errors in unsafe code. *Splash OOPSLA 2016 Conf.*, Amsterdam, Netherlands, pp. 800–819; doi:10.1145/2983990.2983998.
81. McConnell, S. 2004. *Code Complete.* 2nd ed. Microsoft Press. ISBN: 0735619670.
82. Dierks, T., Rescorla, E. 2008. The Transport Layer Security (TLS) Protocol, Version 1.2. Network Working Group Request for Comments 5246. https://tools.ietf.org/html/rfc5246 (accessed: 2018-02-20).
83. Prokhorenko, V., Choo, K.-K.R., Ashman, H. 2016. Web application protection techniques: A taxonomy. *Journal Of Network And Computer Applications.* 60, pp. 95–112.
84. Huang, X.-W., Hsieh, C.-Y., Wu, C.-H., Cheng, Y.-C. 2015. A token-based user authentication mechanism for data exchange in RESTful API. *18th Int. Conf. on Network-Based Information Systems (NBiS 2015)*, Taipei, Taiwan, pp. 601–606; 10.1109/NBiS.2015.89.
85. Jones, M., Bradley, J., Sakimura, N. 2015. *JSON Web Token (JWT).* Internet Engineering Task Force. Request for Comments 7519. https://tools.ietf.org/html/rfc7519.
86. Shen, Q., Yang, Y., Wu, Z., Wang, D., Long, M. 2013. Securing data services: A security architecture design for private storage cloud based on HDFS. *Int. J. Grid Util. Comput.* 4(4), pp. 242–254; doi:10.1504/IJGUC.2013.057118.
87. Graz University of Technology. 2018. *Meltdown and Spectre.* https://meltdownattack.com (accessed: 2018-01-21).
88. Gkoulalas-Divanis, A., Loukides, G. 2013. *Anonymization of Electronic Medical Records to Support Clinical Analysis.* SpringerBriefs in Electrical and Computer Engineering. doi:10.1007/978-1-4614-5668-1_2.
89. Emam, K.E. Rodgers, S., Malin, B. 2015. Anonymising and sharing individual patient data. *BMJ.* 350, p. h1139.
90. Fernandez-Aleman, J.L., Senor, I.C., Lozoya, P.A.O., Toval, A. 2013. Security and privacy in electronic health records: A systematic literature review. *J Biomed Inform.* 46(3), pp. 541–562.
91. Google. 2018. *Angular.* https://angular.io/ (accessed: 2018-01-21).
92. Belshe, M., Peon, R., Thomson, M. 2015. Hypertext Transfer Protocol Version 2 (HTTP/2). Internet Engineering Task Force (IETF). RFC 7540. http://httpwg.org/specs/rfc7540.html (accessed: 2018-01-19).
93. Fielding, R.T. 2009. *It is Okay to Use POST.* Untangled, Blog. http://roy.gbiv.com/untangled/2009/it-is-okay-to-use-post (accessed 2018-01-19).

94. Orchard, D. 2016. *Programming Language Evolution and Sustainable Software*. Blog. The Software Sustainability Institute, UK, https://www.software.ac.uk/blog/2016-09-12-programming-language-evolution-and-sustainable-software (accessed: 2018-02-20).
95. Oracle. *Java Platform, Standard Edition What's New in Oracle JDK 9*. https://docs.oracle.com/javase/9/whatsnew/toc.htm (accessed: 2018-01-19).
96. Ben-Kiki, O., Evans, C., Net, I.D. 2009. *YAML Ain't Markup Language (YAML™) Version 1.2*. http://www.yaml.org/spec/1.2/spec.html (accessed: 2018-01-20).
97. SmartBear Software. 2018. *Swagger Editor and Swagger UI*. https://swagger.io/ (accessed: 2018-01-20).
98. Krishnan, D., Kelly, A. 2018. *Springfox Reference Documentation*. https://springfox.github.io/springfox/docs/current/ (accessed: 2018-01-20).
99. Pivotal Software, Inc. 2018. *Spring Boot*. https://projects.spring.io/spring-boot/ (accessed: 2018-01-20).
100. H2 database. 2018. http://www.h2database.com/html/features.html (accessed: 2018-01-20).
101. Heaton, J. 2015. Encog: Library of interchangeable machine learning models for Java and C#. *Jmlr*. 16, pp. 1243–1247.
102. Chang, C.-C., Lin, C.-J. 2011. LIBSVM: A library for support vector machines. *Acm Tist*. 2(27), pp. 1–27.
103. Tibco Software, Inc. 2018. *JasperReports® Library, Open Source Java Reporting Library*. https://community.jaspersoft.com/project/jasperreports-library (accessed: 2018-01-21).
104. Witten, I.H., Frank, E., Hall, M.A. 2011. *Data Mining: Practical Machine Learning Tools and Techniques*. 3rd ed. Morgan Kaufmann. ISBN:0123814790.
105. Syncleus. 2018. *Aparapi*. http://aparapi.com (accessed: 2018-01-21).
106. Majumder, S., Mondal, T., Deen, M.J. 2017. Wearable sensors for remote health monitoring. *Sensors*. 17(1), p. 130.
107. Chapman, B., Jost, G., van der Pas, R. 2007. *Using OpenMP: Portable Shared Memory Parallel Programming*. Scientific and Engineering ed., The MIT Press, Boston, MA, USA, ISBN:0262533022.
108. Kalin, M. 2013. *Java Web Services: Up and Running: A Quick, Practical, and Thorough Introduction*. 2nd ed., O'Reilly Media, Sebastopol, CA, USA, ISBN:1449365116.
109. Li, W.-S., Yan, J., Yan, Y., Zhang, J. 2010. Xbase: Cloud-enabled information appliance for healthcare. *Proc 13th Int. Conf. on Extending Database Technology (EDBT '10)*, New York, NY, USA, ACM, pp. 675–680. doi:10.1145/1739041.1739125.
110. Vukicevic, M., Radovanovic, S., Milovanovic, M., Minovic, M. 2014. Cloud based metalearning system for predictive modeling of biomedical data. *The Sci. World J.* 2014, p. 859279.
111. Sahoo, S.S., Jayapandian, C., Garg, G., Kaffashi, F., Chung, S., Bozorgi, A., Chen, C.-H., Loparo, K., Lhatoo, S.-D., Zhang, G.-Q. 2014. Heart beats in the cloud: Distributed analysis of electrophysiological 'Big Data' using cloud computing for epilepsy clinical research. *Jamia*. 21(2), pp. 263–271, 2014.
112. Jovic, A., Kukolja, D., Jozic, K., Cifrek, M. 2016. Use case diagram based scenarios design for a biomedical time-series analysis web platform. *Proc. Int. Conf. MIPRO 2016*. Opatija, Croatia, pp. 326–331.
113. Brett, A., Croucher, M., Haines, R., Hettrick, S., Hetherington, J., Stillwell, M., Wyatt, C. 2017. Research Software Engineers: State of the Nation Report 2017. University of Southampton, UK: The Research Software Engineer Network (RSEN); doi:10.5281/zenodo.495360.

114. Grieve, S., Mueller, E., Morley, A., Upson, M., Adams, R., Clerx, M. 2018. *Bridging the Gap: Convincing Researchers with Different Backgrounds to Adopt Good (Enough) Software Development Practices.* Blog. The Software Sustainability Institute, UK, https://software.ac.uk/blog/2018-02-09-bridging-gap-convincing-researchers-different-backgrounds-adopt-good-enough (accessed: 2018-02-20).
115. Cairns, C., Somerfield, D. 2017. *The Basics of Web Application Security.* MartinFowler.com blog, https://martinfowler.com/articles/web-security-basics.html (accessed: 2018-02-20).
116. Healey, J.A., Picard, R.W. 2005. Detecting stress during real-world driving tasks using physiological sensors. *Ieee Tits.* 6(2), pp. 156–166.
117. Chudacek, V., Spilka, J., Bursa, M., Janku, P., Hruban, L., Huptych, M., Lhotska, L. 2014. Open access intrapartum CTG database. *Bmc Pregnancy And Childbirth.* 14, p. 16.

5 Handling of Medical Imbalanced Big Data Sets for Improved Classification Using Adjacent_Extreme Mix Neighbours Oversampling Technique (AEMNOST)

Sachin Patil and Shefali Sonavane

CONTENTS

FOCUS OF CHAPTER TITLE IN RELATION TO THE SCOPE AND TITLE OF THE BOOK

The emerging digital world has outdrawn data scarcity. Zettabytes of data are churned out every year. Digital data will reach Zettabytes by the year 2025, which is predicted to be 30 times more than presently. This gigantic data flow has led to today's catchword 'Big Data'. To assimilate, exploit and further analyse this data has become a focus of research. The issues unaddressed by conventional data analytics need to be handled proficiently. The capabilities of ecosystem have evolved down the line for management of Big Data.

The huge availability of medical data sets has provided an opportunity to construct refined models with supporting expertise within several medical domains. These diagnostic models are suffering from numerous issues such as missing values, multi-class domains and overlapping instances, including class imbalance scenarios.

Exhibiting and assimilating the medical data using advanced front-end apps assists in the improved prediction of diseases, thus allowing for their timely treatment. The use of huge in-hand data sets for artificial machine training helps to build precise prediction models. It affects various domains such as security, life sciences and preventive disclosures. The proposed technique (AEMNOST) in this chapter benefits and enhances classification outcomes of imbalanced medical data sets, leveraging traditional classifiers. It also considers the intrinsic medical data characteristics affecting classification improvement. The results demonstrate the superiority of planned technique over other benchmarking techniques.

5.1 MEDICAL IMBALANCED BIG DATA SETS: OUTLINE AND IMPRINT CHALLENGES

The exploration on the field of digital data has led to enormous data while also enriching the field of advanced analytics [1,2]. Furthermore, the available medical data contains a wide range of variability and volume [6,36,37]. The review of such data with exceptional investigation has set a new dimension for advanced analytics. Additionally, there is a need for an innovative model responding to the recent necessities of Big Data management which have evolved over the last decade [2–5,7,8].

Numerous real-world applications representing the extreme imbalance between class sizes such as web author identification [9], medical judgment [41,42], financial frauds, threat supervision, network invasion, software defect detection [10] have diverted attention towards its analysis. Moreover, in certain medical domains [38–40,43,44] a count of samples in one class is measurably less compared to others classes within a data set. It ultimately leads to a situation called as a class imbalance problem [11–14].

In recent studies of machine learning, precisely classifying such class disparities has become the foremost interest of the research community [15]. Traditional classifiers fail to predict precise classification of minority instances in imbalanced data sets, ignoring their laxity in forming rule sets. Moreover, they fail to discover the well-known result boundaries present in imbalanced data sets. Dealing with data sets characterizing a skewed division is a vital issues across different learning algorithms different learning algorithms.

The class disparity problem is addressed by numerous available techniques. These techniques work basically at three levels, namely, algorithmic level, data level [15] and cost-sensitive level [16]. At the data level, the focus is based on altering the volume of the original set for further analysis. At the algorithmic level, the approach is to revise a prevailing algorithm so as to promote the processes dealing with imbalanced data. A cost-sensitive approach is a hybridization of the data level technique with the algorithmic level technique. It aids in attaining enriched accuracy while reducing the misclassification cost. The literature survey including the experimental results has revealed the supremacy of data level techniques. The new technique discussed in this chapter primarily works at the data level.

The data level techniques are further distributed into three categories namely under sampling, oversampling and a hybrid tactic of the earlier two assemblies [15,16]. Every other technique does have its own advantages and disadvantages. The techniques underlying the oversampling approaches mechanizes the required density, however, they may be inclined to duplicate noisy data. Likewise, undersampling techniques assists in reducing imbalance concentration but may miss out the significant data. The cost-sensitive techniques implicitly inherit the merits and demerits of oversampling as well as undersampling approaches. The outcomes of random oversampling supports to emphasize its dominance compared to the other two techniques. The proposed technique in this chapter basically works with the oversampling method.

The Synthetic Minority Oversampling TEchnique (SMOTE) is an elementary oversampling technique used to deal with imbalanced data set problems by randomly synthesizing the minority class examples [17]. It achieves simple and improved oversampling with less variance within the data set. Correspondingly, the classifiers based on variances along with class-specific mean values are less influenced by SMOTE-based synthetic samples. SMOTE fails while dealing with various shortcomings like addressing disjuncts and random synthetization. Additionally, it is applicable only for the binary-class data sets with explicitly considering its limited usage over the minority class samples itself. The process of attaining synthetization for the required balance association is discussed as follows. The 'K' Nearest Neighbors (KNN) of the sample under consideration is selected on a random basis.

The 'K' Nearest Neighbors (KNN) of the sample under consideration is selected. The necessary subset of KNN's is further chosen for basic over sampling process to fulfil the required imbalance ratio among classes. To overcome these drawbacks, various SMOTE-based flavours such as Borderline SMOTE [18], Safe-Level SMOTE [19] and Adaptive Synthetic Sampling (ADASYN) [20] were evolved over time. The Borderline SMOTE technique helps to oversample only the minority examples near the borderline. Similarly, the Safe-Level SMOTE technique specifically over samples the minority instances which exhibits higher safe level whereas improving classification accuracy. The ADASYN technique progresses to learn and analyse data distribution by dropping the partiality. It adaptively changes the classification border near the hard examples to deeply diagnose the features. An evolutionary algorithm uses the method which fits to a generalized model considering objects in Euclidean space, in order to resolve the problems of imbalanced data sets [21]. Neighborhood Rough Set-M based, SMOTE+GLMBoost [22] and NRBoundary-SMOTE are specially working for boundary based oversampling. The ensemble methods functioning in alignment to SMOTE namely SMOTEBoost [23], RUSBoost and AdaBoost [24], work for the issues in imbalanced data sets. Almost the above-discussed techniques primarily focus on the binary-class imbalanced data set problems. Alberto et al. proposed a fuzzy rule-based classification model along with pre-processing and integrating the pairwise learning to address multi-class deficiencies [25]. Ordinal classification of imbalanced data sets approximating the class probability distribution is the focus of discussion in [26].

In this chapter, a better oversampling technique, AEMNOST, which deals efficiently with imbalanced data sets, is elaborated in detail. Additionally, it addresses various data characteristics like small disjuncts, lack of density, borderline and overlapping instances to further improve classification performance. The classifiers such as Multilayer Perceptron, AdaBoostM1 and Random Forest (RF) are used to perform classification [27,28,29]. The experiments are performed over the experimental skeleton of the MapReduce environment [16,30]. The worthiness of these techniques is fundamentally evaluated using two measures: Geometric Mean (GM) and Area Under Curve (AUC) [27,28].

5.2 EXPERIMENTAL APPROACH

The work involves experimentation on imbalanced big medical data sets using projected oversampling techniques. AEMNOST technique proposes to acquire data and appropriately manipulate it for improving classification outcomes.

5.2.1 Architecture and Experimental Drift

The overall architecture illustrated in Figure 5.1. depicts the distinctive stages involved in investigating big medical imbalanced data [31]. The varied streaming input data is recorded and processed to deliver preliminary conception. Furthermore, the intermediate data is treated with an oversampling technique, producing a balanced data set for the required examination. A prerequisite oversampling rate, including the value of 'K', is provided to maintain an imbalance ratio (IR) and to

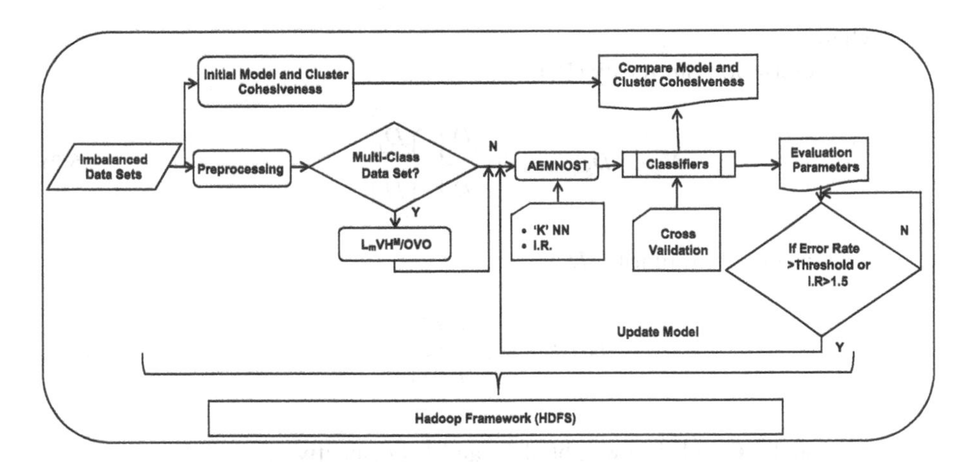

FIGURE 5.1 Overall system architecture.

discover the required number of nearest neighbours respectively. The distributed implementation architecture is based on the Hadoop framework which capably addresses heterogeneity and dynamic scaling.

Figure 5.1 clarifies the notion of experimental workflow for acquiring and processing of medical imbalanced Big Data sets. At first, the data sets are realized for binarization prerequisite as per their respective categories (binary/multi-class data sets). Further, the data sets are oversampled using AEMNOST or benchmarking techniques. Lastly, the results are analysed including the foreseen updates to the model.

The theoretical flow of examination is as follows:

1. Acquire input data in HDFS format
2. Mark initial cluster cohesiveness and assess the initial classification model
3. Recognize the category of data sets (binary/multi-class) and process the binarization stages
4. Accomplish oversampling using AEMNOST or benchmarking techniques
5. Create a new model of oversampled balanced data set and analyse its precision
6. Modify the model based on observations of IR and error rate

5.2.2 Preprocessing

Clustering of unstructured input data is carried out based on characteristic similarities using the following modules [32].

1. Pre-processing input data to find '*n*' semantic terms
2. Calculate the Characteristic Similarity (D_C):

$$D_C(d_i, d_j) = \alpha * D_S(d_i, d_j) + \beta * D_F(d_i, d_j) \tag{5.1}$$

where:

- Description Similarity (D_S):

$$D_S(d_i,d_j)=\frac{|D_i \cap D_j|}{|D_i \cup D_j|} \tag{5.2}$$

- Functional Similarity (D_F):

$$D_F(d_i,d_j)=\frac{|F_i \cap F_j|}{|F_i \cup F_j|} \tag{5.3}$$

- α and $\beta \in (0,1)$ is a weight of D_S and D_F, respectively

$$\alpha+\beta=1$$

3. Cluster construction

5.2.3 Lowest Minority Versus Highest Majority Method ($L_M VH^M$): Addressing Multi-Class Data Sets

Traditional methods such as One versus Al (OVA) and One versus One (OVO) incur numerous drawbacks for managing multi-class imbalanced data sets. The conceived disadvantages of these two methods are as:

1. Increase in computation time
2. Non-compliance of oversampling rate adhering the realistic need of classification
3. Excess oversampling of the minority classes may sometimes surpass the majority sub-classes
4. Rise in time for building a classification model

The suggested method such as the Lowest Minority versus Highest Majority ($L_m VH^M$) technique [31] overcomes most of the disadvantages of the OVO and OVA method. In the $L_m VH^M$ method, the process of oversampling considers only the highest majority class to be compared with the individual minority class satisfying IR > 1.5.

The formulation for this method is more effective for treating multi-class domains. It produces focused over sampling instances along with improving classification performance. The advantages of $L_m VH^M$ method are as:

1. Reduces computation time
2. Avoids superseding the majority sub-classes
3. Fulfills implicitly the IR in comparison to the residual majority classes

The $L_m VH^M$ method is exemplified by a scenario where a data set is considered to have five classes P, Q, a, b and c. The classes P and Q are assumed to be majority classes and the remaining others are minority classes. Class *P* is designated as the highest majority class and class '*a*' as the lowest minority class. At first, the L_mVH^M method leads to balancing the lowest minority class (*a*) with respect to the highest majority class (*P*), while complying the oversampling rate. Likewise, the remaining minority classes, classes '*b*' and '*c*', are further seen for complying the oversampling rate with respect to class '*P*'.

5.3 ENHANCED OVERSAMPLING TECHNIQUE

5.3.1 Adjacent Extreme Mix Neighbours Oversampling Technique (AEMNOST)

The AEMNOST technique generates oversampling instances by studying the influence of an equal mix of near and far neighbours of the instance under consideration. It mainly assists in overcoming the issues of overlapping and replication by considering a wide spectrum of inputs, which helps in improving classification. The technique is discussed in detail as:

- D_i – 'N' instances
- Minority samples s_n (n = 1,2,….n)//binary-class data set or the lowest minority class in the multi-class data set

Compute safe levels of all cases [33].//based on number of minority samples present in KNN of each individual instance
Input: a set of all instances (D_i)
Output: a set of all synthetic minority instances (D_o)

1. $D_o = \varnothing$
2. *Repeat*
3. *Search* (I) set of instances for each sample in s_n such that
   i. I/2 adjacent neighbours
   ii. I/2 extreme neighbours and
   iii. Mid neighbour (except for the even value of 'I')
   where N > 1 and I <= N
4. *Check* set (I)
   if (I contains all minority instances)
      follow step 5 to 7
   else if (I contains all majority instances)
      follow step 8 to 12
   else
      follow step 13 to 16
5. Interpolated instances = *Safe-Level based Synthetic Sample creation* (SSS) of s_n.[i] with each sample in set (I) individually
6. New synthetic instance (S_Y) = average (interpolated instances)
7. *if* (S_Y = duplicate instance)
   delete the neighbour having a lowest safe level from the set (I)
      delete the respective interpolated instance generated from that neighbour
      goto step 6

else
$D_o = S_Y$
goto step 17
8. *Random instance* R_i(I)
9. Nearest minority instance $R_{in}(R_i)$
10. Interpolated instances = SSS (s_n.[i] and R_i) and SSS (s_n.[i] and R_{in})
11. New synthetic instance (S_Y) = average (interpolated instances)
12. *if* (S_Y = duplicate instance)
goto step 10
else
$D_o = S_Y$
goto step 17
13. *Random instance* R_i(I)
14. *if* (R_i = minority)
SSS (s_n.[i] and R_{in})
goto step16
15. *else*
Maximum safe level minority instance R_{is}(I or D_i)
Interpolated instances SSS (s_n.[i] and R_i) and SSS (s_n.[i] and R_{is})
New synthetic instance (S_Y) = average (interpolated instances)
goto step16
16. *if* (S_Y = duplicate instance)
Next nearest minority instance $R_{isn}(R_i)$
goto step 14 or 15
else
$D_o = S_Y$
17. *Until* (s_n)//for remaining minority classes

The oversampling process applies to rest of the minority classes in the multi-class domain fulfilling IR > 1.5. This proposed technique delivers a better quality of synthetic instances, while additionally improving the classification outcomes.

The executional flow of AEMNOST is stated in Figure 5.2. It helps to visualize the rationale behind the idea to process medical imbalanced Big Data sets.

5.3.2 Safe Level Centred Synthetic Samples (SSS) Creation

The purpose of the proposed method (SSS) is to arithmetically authorize the displacement factor in the oversampling process. The foundation behind the concept helps to improve the pre-eminence of interpolated instances with implicitly refining classification performance. In addition, it helps overcoming the issues of overlapping and outliers. The details of SSS are as follows:

1. Find *Safe levels*
   a. Minority instance under consideration (SL_m) and
   b. All its KNN instances (SL_{KNNi})

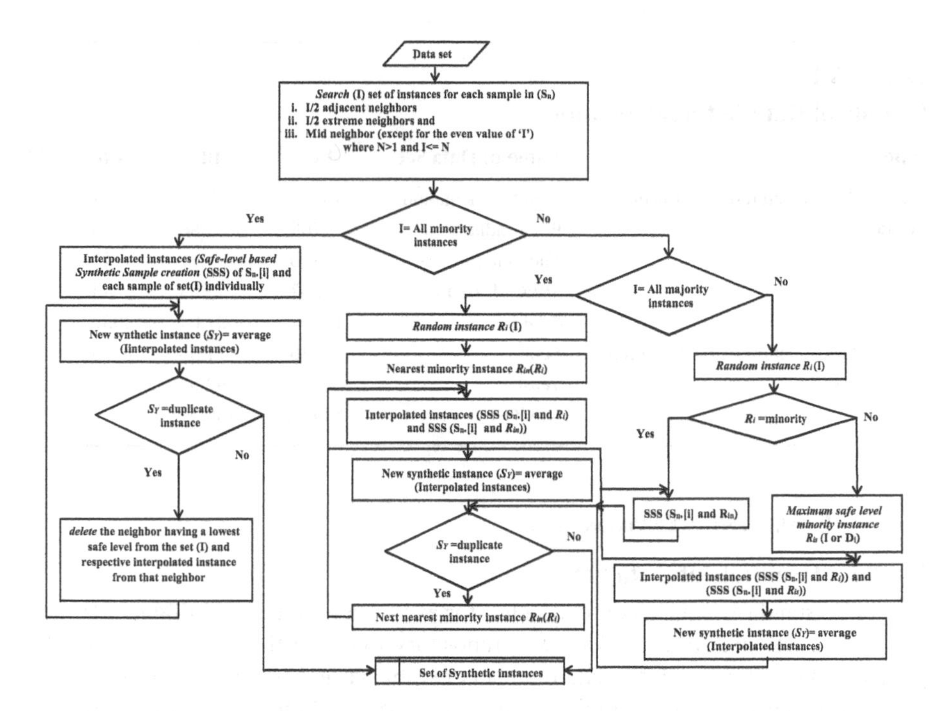

FIGURE 5.2 AEMNOST Flow Chart.

2. Summation of safe levels of all instances
   $T = sum\ (SL_m + SL_{KNNi})$
3. Normalized Value (NV_i) of safe levels (SL_m and SL_{KNNi}):
   NV_i = individual's safe level value/T
   - NV_i is between 0 to 1
   - $\Sigma\ NV_i$ equals to 1
4. Denoting the value of the Displacement Factor (DF) in oversampling process:
   if (NV_m is safer compared to NV_{KNNi})
   $DF = NV_{KNNi}$
   else if ($NV_m = NV_{KNNi}$)
   $DF = = 0.5$
   else
   if ($NV_{KNNi} < 0.5$)
   $DF = 1 - NV_{KNNi}$
   else
   $DF = NV_{KNNi}$

5.4 INVESTIGATIONAL FACTS

The investigational facts presented here will establish the efficacy of the proposed technique compared to diverse benchmarking techniques. The experimental setup deals with the issue of class imbalance in medical Big Data sets.

TABLE 5.1
Details of Data Set Specifications

Type	Name of Data Set	Ex	IR	Attr	Cl
Binary-Class Structured/Un-Structured Data Sets	Skin Segmentation	245057	3.81	4	2
	Pima Indians	768	1.86	9	2
	Thoracic Surgery	470	5.71	17	2
	SPECTF Heart	267	3.85	45	2
	SIDO	12678	27.04	4932	2
Multi-Class Structured/Un-Structured Data Sets	Ecoli	336	71.5	8	8
	Yeast	1484	92.6	9	10
	KEGG-D	53413	13156.5	23	13

5.4.1 Experimental Environment

5.4.1.1 Essentials of Data Set

The eight standard data sets are selected for experimental analysis (Table 5.1). The seven data sets are from the UCI ML repository [34] and the residual is from the pharmacology domain [35]. The data sets exhibit varied features such as sample counts ranging from low to high, and a huge set of attributes and a comprehensive IR. Moreover, they are categorized into two groups, namely binary and multi-class data sets.

The binary-class data sets encompasses a diverse range of IR (1.86 to 27.04), a varied number of attributes (4 to 4932) and an extreme number of instances (245,057). The multi-class data sets contain an easily manageable and practical number of attributes and IR. The two multi-class data sets, Ecoli and Yeast, are of a structured form and the remaining data set (SIDO) is of an unstructured type. The details of the data sets are specified in Table 5.1.

Table 5.1 reviews the features of these medical imbalanced data sets including their type, name, samples count (Ex), IR, number of attributes (Attr) and classes (Cl). The IR stated for multi-class data sets is in proportion to the ratio of the uppermost majority class versus the lowermost minority class. The data sets listed for analysis in Table 5.1 are integrated for storage in HBase.

5.4.1.2 Assessment Factors

The assessment of techniques is performed using two standard evaluation factors such as the GM and AUC.

- *GM*: This measure provides a balanced accuracy. It indirectly assists in judging the importance of the estimated technique over the given classifier.

$$GM = \sqrt{\text{sensitivity} * \text{specificity}} = \sqrt[k]{\pi \sum_{i=1}^{k} R_i} \tag{5.4}$$

- k - number of classes
- R_i - recall of ith call
- *AUC*: It is recorded by calculating the true positive rate versus the false positive rate, while indirectly demonstrating the performance of the classification system. Similarly, it further benefits to choose an optimal model and analyze its cost.

5.4.1.3 Environmental Pre-settings

1. Mappers are reused for the next iteration until they are explicitly terminated
2. RAID/LVM storage processes are avoided on the TaskTracker/DataNode systems
3. Compression methods are used for intermediate data
4. Writable objects are reused

5.4.1.4 Assumptions

1. Data sets are contextually transformed into required forms
2. The mapper slots are effectively utilized by maintaining its ratio to the number of mapper tasks

5.4.1.5 Symbolizations

The notations used for algorithms during the experimental analysis are given in Table 5.2.

5.4.2 Experimental Explorations

The proposed technique (AEMNOST) is applicable to binary-class, as well as multi-class data sets. Furthermore, the comparison of L_mVH^M over OVA method is highlighted. The experimental performance is evaluated in terms of GM and AUC values while performing a tenfold cross-validation. The value of 'K' = 5 is set as constant for finding the nearest neighbours. Three classifiers (Multilayer Perceptron, AdaBoostM1 and Random Forest) are used in this experimental work. The performance of the proposed technique is compared to the outcomes of the three pre-selected benchmarking techniques (SMOTE, Safe-level SMOTE and Borderline SMOTE).

The experimental work is conducted on four nodes of Hadoop clusters. Each node has an Intel i7@3.4 GHz with 4 GB RAM. Moreover, the clusters are built on the

TABLE 5.2
Symbols of Technique

Symbol	Technique
A	Unprocessed data set
B	SMOTE
C	Safe-Level SMOTE
D	Borderline SMOTE
E	AEMNOST

TABLE 5.3
GM Values (Binary-Class Data Sets)

Classifier	Data Set	Oversampling Techniques A	B	C	D	E
Multilayer Perceptron	Skin Segmentation	0.80	0.81	0.83	0.84	0.89
	Pima Indians Diabetes	0.74	0.75	0.76	0.78	0.80
	Thoracic Surgery	0.65	0.66	0.66	0.68	0.70
	SPECTF Heart	0.76	0.77	0.78	0.78	0.82
	SIDO	0.85	0.88	0.88	0.89	0.92
AdaBoostM1	Skin Segmentation	0.82	0.84	0.85	0.85	0.90
	Pima Indians Diabetes	0.75	0.76	0.77	0.80	0.81
	Thoracic Surgery	0.66	0.67	0.68	0.69	0.72
	SPECTF Heart	0.77	0.78	0.79	0.80	0.82
	SIDO	0.86	0.89	0.90	0.90	0.93
Random Forest	Skin Segmentation	0.82	0.85	0.87	0.88	0.92
	Pima Indians Diabetes	0.76	0.78	0.79	0.81	0.83
	Thoracic Surgery	0.67	0.68	0.69	0.71	0.74
	SPECTF Heart	0.80	0.82	0.82	0.83	0.84
	SIDO	0.87	0.91	0.91	0.92	0.95
Average		0.77	0.79	0.80	0.81	**0.84**

top of Hadoop 2.7.4 encompassing Ubuntu 16.04. Easy distribution of data in the MapReduce computing cluster helps in the effective processing of data-intensive tasks and to attain higher throughputs.

The results obtained aid to demonstrate a measurable rise in GM and AUC values for the proposed technique (AMENOST), which ultimately represents an enhanced classification. Random forest use leads to encouraging results for almost all techniques compared to the other two classifiers.

5.4.2.1 Binary-Class Structured/Unstructured Data Sets

The results comprising GM and AUC values for the five data sets are shown in Tables 5.3 and 5.4, respectively. The results of the initial data set, along with the benchmarking and proposed techniques, are considered for further analysis. There is a marginal growth of classification improvement in benchmarking techniques. It is seen from the results that the AEMNOST technique represents leads to improved classification ratios over the benchmarking techniques.

The graphs in Figures 5.3 and 5.4 illustrate the average GM and AUC values respectively for all techniques under consideration. The binary-class data sets under experimental analysis are investigated using three classifiers.

In Figures 5.3 and 5.4, the x-axis represents the oversampling techniques (B–E), including initial data set results (A), and the y-axis represents the values for GM and AUC, respectively. Analysing the graphs, the AEMNOST technique achieves the higher values of GM and AUC compared to all other techniques for almost all classifiers.

TABLE 5.4
AUC Values (Binary-Class Data Sets)

		Oversampling Techniques				
Classifier	**Data Set**	**A**	**B**	**C**	**D**	**E**
Multilayer Perceptron	Skin Segmentation	0.84	0.85	0.87	0.87	0.93
	Pima Indians Diabetes	0.80	0.81	0.82	0.82	0.84
	Thoracic Surgery	0.66	0.68	0.69	0.70	0.73
	SPECTF Heart	0.83	0.84	0.85	0.85	0.87
	SIDO	0.88	0.89	0.91	0.92	0.96
AdaBoostM1	Skin Segmentation	0.85	0.87	0.87	0.88	0.94
	Pima Indians Diabetes	0.81	0.82	0.82	0.83	0.86
	Thoracic Surgery	0.67	0.70	0.70	0.72	0.75
	SPECTF Heart	0.84	0.85	0.85	0.86	0.89
	SIDO	0.89	0.91	0.92	0.93	0.97
Random Forest	Skin Segmentation	0.86	0.88	0.89	0.89	0.96
	Pima Indians Diabetes	0.82	0.83	0.84	0.84	0.88
	Thoracic Surgery	0.69	0.71	0.72	0.74	0.78
	SPECTF Heart	0.85	0.86	0.87	0.87	0.91
	SIDO	0.90	0.92	0.93	0.94	0.98
Average		0.81	0.83	0.84	0.84	**0.88**

5.4.2.2 Multi-Class Structured/Unstructured Data Sets

Tables 5.5 and 5.6 showcase the outcomes of GM values underlying the effect of L_mVH^M versus OVA, respectively, on handling the multi-class data sets. The results of all techniques are showcased considering all three classifiers.

The average GM scores in Table 5.5 shows the dominance of the proposed technique over the other benchmarking techniques. AEMNOST outperforms all of the benchmarking techniques. The handling of multi-class data sets is effectively

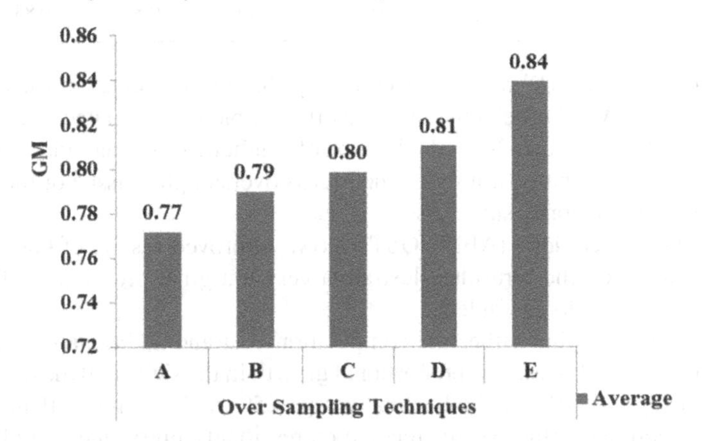

FIGURE 5.3 Average GM values (binary-class data sets).

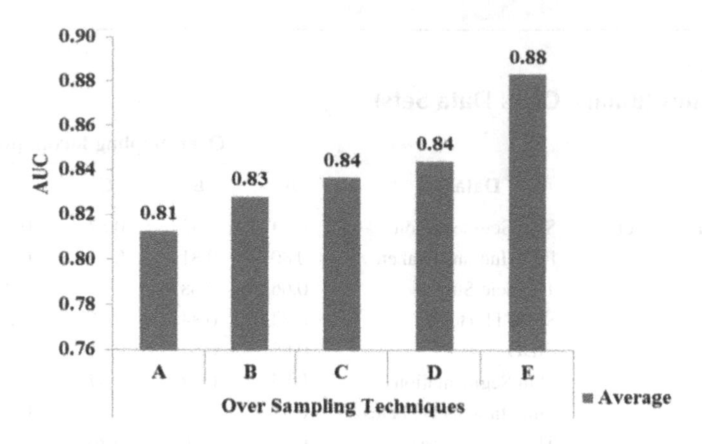

FIGURE 5.4 Average AUC values (binary-class data sets).

TABLE 5.5
GM Values (Multi-Class Data Sets using L_mVH^M)

		Oversampling Techniques				
Classifier	Data Set	A	B	C	D	E
Multilayer Perceptron	Ecoli	0.87	0.88	0.89	0.90	0.94
	Yeast	0.58	0.61	0.62	0.63	0.68
	KEGG-D	0.88	0.90	0.90	0.92	0.94
AdaBoostM1	Ecoli	0.88	0.89	0.90	0.92	0.95
	Yeast	0.59	0.62	0.63	0.65	0.69
	KEGG-D	0.89	0.90	0.91	0.93	0.95
Random Forest	Ecoli	0.89	0.91	0.92	0.94	0.97
	Yeast	0.60	0.64	0.65	0.67	0.71
	KEGG-D	0.91	0.92	0.92	0.94	0.96
Average		0.79	0.81	0.82	0.83	**0.87**

carried out using L_mVH^M method, overcoming the disadvantages of the outmoded OVA method. L_mVH^M efficiently addresses the problems associated with handling multi-class data sets. It avoids the duplication of synthetic instances, and involves less computation. Additionally, it justifies the excess oversampling issue of the minority classes above the majority sub-classes.

The planned technique (AEMNOST) shows improved results of GM values in Table 5.6. However, these results illustrate a very marginal progress in GM scores compared to the results in Table 5.5.

The OVA method intensifies the computation cost and additionally may exceed some majority sub-classes. The proportional growth in the size of all the oversampled data sets, using the OVA method, is on average 25% to 35% greater than using the L_mVH^M method. Nevertheless, the relative regressional improvement of GM values using OVA method in comparison to L_mVH^M method is very less (approximately near

TABLE 5.6
GM Values (Multi-Class Data Sets Using OVA)

		Oversampling Techniques				
Classifier	**Data Set**	**A**	**B**	**C**	**D**	**E**
Multilayer Perceptron	Ecoli	0.88	0.89	0.90	0.91	0.92
	Yeast	0.57	0.62	0.62	0.64	0.69
	KEGG-D	0.89	0.91	0.92	0.93	0.95
AdaBoostM1	Ecoli	0.89	0.90	0.91	0.93	0.96
	Yeast	0.60	0.63	0.64	0.67	0.70
	KEGG-D	0.91	0.91	0.93	0.94	0.96
Random Forest	Ecoli	0.90	0.92	0.94	0.95	0.98
	Yeast	0.61	0.64	0.66	0.68	0.72
	KEGG-D	0.92	0.93	0.93	0.95	0.98
Average		0.80	0.82	0.83	0.84	**0.87**

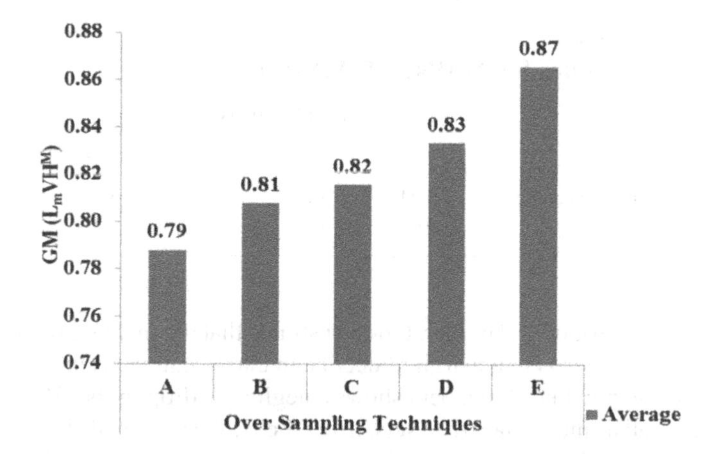

FIGURE 5.5 Average GM values (multi-class data sets using L_mVH^M).

to 1 to 2%). It demonstrates the efficiency of applying the L_mVH^M method to handle multi-class data sets, all the while providing better performance.

The average values of GM using L_mVH^M and OVA are depicted in Figures 5.5 and 5.6. The x-axis characterizes the techniques, and the y-axis denotes the average GM values. The Multilayer Perceptron and Random Forest classifiers focus mainly on addressing overfitting problems. Likewise, the AdaBoostM1 classifier avoids the impact of noisy data and outliers on classification outcomes.

5.4.2.3 The Influence of the number of Mappers on Outcomes

The results from Table 5.7 provides an insight of classification results encompassing distinct numbers of mappers. These results of GM values, using the AEMNOST technique in Table 5.7, are for two data sets, namely the Skin segmentation and

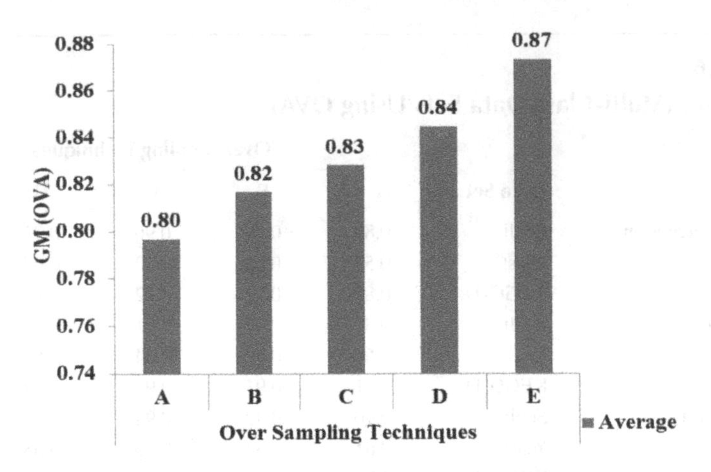

FIGURE 5.6 Average GM values (multi-class data sets using OVA).

TABLE 5.7
GM Values for AEMNOST (L_mVH^M)

	Number of Mappers			
Data Set	8	16	32	64
Skin Segmentation	0.94	0.93	0.91	0.90
KEGG-D	0.95	0.94	0.92	0.91

KEGG-D using AdaBooost classifier. It demonstrates that the increase in the number of mappers (8 to 64) leads to a marginal decline in classification accuracy.

The increasing number of mappers shows a negligible drop in the GM values for the binary, as well as multi-class data sets, using the L_mVH^M method. This behaviour appears to be due to the lessened availability of the oversampled minority instances per mapper.

It is evident from Figure 5.7 that such an increase in the number of mappers explicitly affects more the data sets containing a high number of classes and extreme IR (KEGG-D). These results openly indicate the relevance of the number of instances existing in the mapper for the construction of the ultimate model.

5.5 AEMNOST: RESOLVING THE ISSUES OF DATA CHARACTERISTICS AND IR

5.5.1 Data Characteristics

- Duplicate oversampling
  - The algorithm AEMNOST checks each of the generated synthetic samples for evading duplication before further consideration

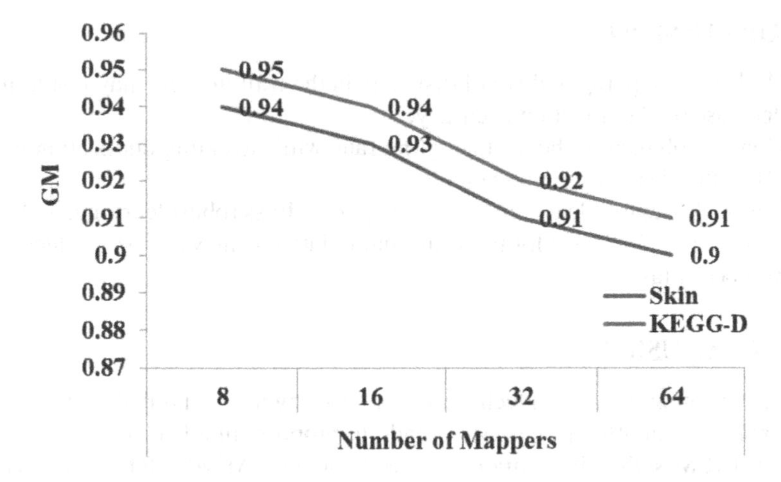

FIGURE 5.7 GM values of Skin and KEGG-D for discrete mappers using AEMNOST.

- Overlapping instances
  - The consideration of the nearest and farthest neighbours in the oversampling process, along with averaging of interpolated instances, aids in assisting the issues related to the overlapping instances
- Borderline instances
  - The technique addresses a wide spectrum of neighbours, which represents the valid borderline instances
- Small disjuncts
  - The consideration of nearest as well as farthest neighbours in the oversampling process relieves the concern for small disjuncts while delivering precise synthetic samples

5.5.2 Imbalance Ratio

The imbalance ratio is a parameter to measure the proportion of between-class imbalances. The earlier studies show the feasibility of maintaining the IR threshold of data sets near 1.5. The achievement of IR is accomplished with an oversampling process generating the synthetic samples based on a first-come-first-serve (FCFS) basis. It certainly justifies the balanced state of the classes within the data sets. However, the inferred oversampling set may not be the best set for building precise models.

The excess oversampling is to be planned based on FCFS with aligning the IR near '1'. Furthermore, the better learning samples can be selected from this FCFS set based on certain criteria's such as safe levels, ease of learning and other data characteristics, while achieving an IR near the required rate of '1.5'. This sorted set of better learning samples, filtered from the excess oversampled set, may provide support for an enhanced classification.

Furthermore, the set of redundant samples should be eliminated in the pre-processing stage, as it impacts on achieving the oversampling rate while additionally producing extra minority samples.

5.6 DISCUSSION

- Higher overlapping and small disjuncts in the training sets may result in a decrease of classification accuracy;
- Non-compliance of the oversampling rate, while avoiding duplication may affect the classification outcomes;
- Ensemble-centred learning models help to address robust learning; and
- Planning of block size for big voluminous data sets may assist to reduce the number of tasks.

5.7 CONCLUSION

This chapter compares several techniques for handling medical imbalanced data sets. More explicitly, an attempt has been made to propose an advanced oversampling technique (AEMNOST) in addition to the SSS method. AEMNOST and L_mVH^M are able to efficiently deal with binary as well as multi-class medical Big Data sets while improving the classification results by an average of 3% to 4%. This helps to reduce bias and deals with the drawbacks of traditional techniques. The SSS technique successfully addresses the issues raised due to fundamental data characteristics such as overlapping, influence of borderline instances and small disjuncts. Experimentations are projected mostly on the UCI ML repository with a additional data set from causality workbench domain set. Data sets under consideration exhibit a wide-range of IR, massive instance count and a varied number of attributes.

Well-known standard classifiers (Multilayer Perceptron, AdaBoostM1 and Random Forest) were used for prototypical analysis. The ensemble-based classifier (Random Forest) indicates a promising improvement in performance of 2%–3%, in comparison to the other two classifiers. The Hadoop framework underlying the MapReduce environment is used to experiment the necessities of Big Data management. The increase in the count of mappers (Figure 5.7) practically illustrates the static value for GM.

The projected AEMNOST technique improves the precision and recall, while achieving better GM and AUC values (Tables 5.3 to 5.6). The work presented here validates the efficacy of proposed technique for learning from medical imbalanced Big Data sets.

5.8 FUTURE WORK

The work carried out can further be extended:

1. To examine the consequence of mislabelling noise samples with respect to classification; and
2. To study and inspect the balance between the number of mappers and the size of a minority class.

REFERENCES

1. A. Gandomi, M. Haider, "Beyond the hype: Big data concepts, methods, and analytics," *International Journal of Information Management*, vol. 35, no. 2, pp. 137–144, 2015.

2. D. Agrawal et al., "Challenges and opportunity with big data," *Community White Paper*, pp. 01–16, 2012.
3. W. Zhao, H. Ma, Q. He., "Parallel k-means clustering based on mapreduce," *CloudCom*, pp. 674–679, 2009.
4. X.-W. Chen et al., "Big Data deep learning: Challenges and perspectives," *IEEE Access Practical Innovations: Open Solutions*, vol. 2, pp. 514–525, 2014.
5. "Big Data: Challenges and Opportunities, Infosys Labs Briefings - Infosys Labs," http://www.infosys.com/infosys-labs/publications/Documents/bigdata-challenges-opportunities.pdf.
6. M. Kamal, N. Dey, A. Ashour, "Large Scale Medical Data Mining for Accurate Diagnosis: A Blueprint," In *Handbook of Large-Scale Distributed Computing in Smart Healthcare*, Springer, Cham., pp. 157–176, 2017.
7. N. Dey, A. Hassanien, C. Bhatt, A. Ashour, S. Satapathy, "*Internet of Things and Big Data Analytics Toward Next-Generation Intelligence*," Springer, 2018.
8. G. Elhayatmy, N. Dey, A. Ashour, "Internet of Things Based Wireless Body Area Network in Healthcare," In *Internet of Things and Big Data Analytics Toward Next-Generation Intelligence*, Springer, Cham., pp. 3–20, 2018.
9. Vorobeva, "Examining the performance of classification algorithms for imbalanced data sets in web author identification," *Proceeding of the 18th Conference of FRUCT-ISPIT Association*, pp. 385–390, 2016.
10. P. Byoung-Jun, S. Oh, W. Pedrycz, "The design of polynomial function-based neural network predictors for detection of software defects," *Elsevier: Journal of Information Sciences*, pp. 40–57, 2013.
11. N. Japkowicz, S. Stephen, "The class imbalance problem: A systematic study," *ACM Intelli. Data Analysis Journal*, vol. 6, no. 5, pp. 429–449, 2002.
12. H. He, E. Garcia, "Learning from imbalanced data," *IEEE Transactions on Knowledge and Data Engineering*, vol. 21, no. 9, pp. 1263–1284, 2009.
13. Y. Sun, A. Wong, M. Kamel, "Classification of imbalanced data: A review," *Int Journal of Pattern Recognition Artificial Intelligence*, vol. 23, no. 4, pp. 687–719, 2009.
14. H. Guo et al., "Learning from class-imbalanced data: Review of methods and applications," *Elsevier Expert Systems With Applications*, vol. 73, pp. 220–239, 2017.
15. V. López et al., "An insight into classification with imbalanced data: Empirical results and current trends on using data intrinsic characteristics," *Elsevier: Journal of Information Sciences*, vol. 250, pp. 113–141, 2013.
16. R. Sara, V. Lopez, J. Benitez, F. Herrera, "On the use of MapReduce for imbalanced big data using Random Forest," *Elsevier: Journal of Information Sciences*, pp. 112–137, 2014.
17. N. Chawla, K. W. Bowyer, L. O. Hall, W. P. Kegelmeyer, "SMOTE: Synthetic minority over-sampling technique," *Journal of Artificial Intelligence Research*, vol. 16, pp. 321–357, 2002.
18. H. Han, W. Wang, B. Mao, "Borderline-SMOTE: A new over-sampling method in imbalanced data sets learning," *Proceedings of the 2005 International Conference on Intelligent Computing*, vol. 3644 of Lecture Notes in Computer Science, pp. 878–887, 2005.
19. C. Bunkhumpornpat, K. Sinapiromsaran, C. Lursinsap, "Safe-level-smote: Safelevel-synthetic minority over-sampling technique for handling the class imbalanced problem," *PAKDD Springer Berlin Heidelberg*, pp. 475–482, 2009.
20. H. He et al., "ADASYN: Adaptive synthetic sampling approach for imbalanced learning," *IEEE International Joint Conference on Neural Networks*, pp. 1322–1328, 2008.
21. S. Garcia et al., "Evolutionary-based selection of generalized instances for imbalanced classification," *Elsevier: Journal of Knowledge-Based Systems*, pp. 3–12, 2012.

22. H. Feng, L. Hang, "A novel boundary oversampling algorithm based on neighborhood rough set model: NRSBoundary-SMOTE," *Hindawi Mathematical Problems in Engineering*, 2013.
23. N. Chawla, L. Aleksandar, L. Hall, K. Bowyer, "SMOTEBoost: Improving prediction of the minority class in boosting," *PKDD Springer Berlin Heidelberg*, pp. 107–119, 2003.
24. H. Xiang, Y. Yang, S. Zhao, "Local clustering ensemble learning method based on improved AdaBoost for rare class analysis," *Journal of Computational Information Systems*, vol. 8, no. 4, pp. 1783–1790, 2012.
25. F. Alberto, M. Jesus, F. Herrera, "Multi-class imbalanced data-sets with linguistic fuzzy rule based classification systems based on pairwise learning," *Springer IPMU*, pp. 89–98, 2010.
26. S. Kim, H. Kim, Y. Namkoong, "Ordinal classification of imbalanced data with application in emergency and disaster information services," *IEEE Intelligent Systems*, vol. 31, no. 5, pp. 50–56, 2016.
27. M. A. Nadaf, S. S. Patil, "Performance evaluation of categorizing technical support requests using advanced K-means algorithm," *IEEE International Advance Computing Conference*, pp. 409–414, 2015.
28. R. C. Bhagat, S. S. Patil, "Enhanced SMOTE algorithm for classification of imbalanced bigdata using Random Forest," *IEEE International Advance Computing Conference*, pp. 403–408, 2015.
29. J. Hanl, Y. Liul, X. Sunl, "A scalable random forest algorithm based on MapReduce," *IEEE*, pp. 849–852, 2013.
30. H. Jiang, Y. Chen, Z. Qiao, "Scaling up MapReduce-based big data processing on multi-GPU systems," *SpingerLink Cluster Computing*, vol. 18, no. 1, pp. 369–383, 2015.
31. S. Patil, S. Sonavane, "Enhanced over_sampling techniques for imbalanced big data set classification," in *Data Science and Big Data: An Environment of Computational Intelligence: Studies in Big Data*, Springer International Publishing AG, 2017, ch. 3, vol. 24, pp. 49–81.
32. H. Rong, D. Wanchun, L. Jianxun, "ClubCF: A Clustering-Based Collaborative Filtering Approach for Big Data Application," *IEEE Transactions on Emerging Topics in Computing*, vol. 2, no. 3, pp. 302–313, 2014.
33. W. A. Rivera, O. Asparouhov, "Safe Level OUPS for Improving Target Concept Learning in Imbalanced Data Sets," *Proceedings of the IEEE Southeast Con.*, pp. 1–8, 2015.
34. https://archive.ics.uci.edu/ml/datasets.html
35. SImple Drug Operation mechanisms (SIDO). http://www.causality.inf.ethz.ch/data/SIDO.html.
36. W. Karra, N. Dey, "Biomedical image analysis and mining techniques for improved health outcomes," *IGI Global*, 2015.
37. N. Dey, A. Ashour, F. Shi, V. Balas, "*Soft Computing Based Medical Image Analysis*," Elsevier, vol. 1, 2018.
38. S. Fadlallah, A. Ashour, N. Dey, "Advanced Titanium Surfaces and Its Alloys for Orthopedic and Dental Applications Based on Digital SEM Imaging Analysis," In *Advanced Surface Engineering Materials*, Scivener Publishing, Wiley, Canada, 2016.
39. L. Moraru, S. Moldovanu, A. Culea-Florescu, D. Bibicu, A. Ashour, N. Dey, "Texture analysis of parasitological liver fibrosis images," *Microscopy Research and Technique*, 2017.
40. D. Wang, Z. Li, L. Cao, V. Balas, N. Dey, A. Ashour, P. McCauley, S. Dimitra, F. Shi, "Image fusion incorporating parameter estimation optimized Gaussian mixture model and Fuzzy weighted evaluation system: A case study in time-series plantar pressure data set," *IEEE Sensors Journal*, vol. 17, no. 5, pp. 1407–1420, 2017.

41. N. Dey, A. Ashour, S. Borra, "Classification in BioApps: Automation of Decision Making," Springer, vol. 26, 2017.
42. N. Dey, A. Ashour, "Computing in medical image analysis," *In Soft Computing Based Medical Image Analysis*, pp. 3–11, 2018.
43. L. Moraru, S. Moldovanu, L. Dimitrievici, F. Shi, A. Ashour, N. Dey, "Quantitative diffusion tensor magnetic resonance imaging signal characteristics in the human brain: A hemispheres analysis," *IEEE Sensors Journal*, pp. 4886–4893, 2017.
44. S. Chakraborty, S. Chatterjee, A. Ashour, K. Mali, N. Dey, "Intelligent computing in medical imaging: A study," *Advancements in Applied Metaheuristic Computing*, vol. 143, 2017.

6 A Big Data Framework for Removing Misclassified Instances Based on Fuzzy Rough

Mai Abdrabo, Mohammed Elmogy, Ghada Eltaweel, and Sherif Barakat

CONTENTS

6.1 INTRODUCTION

Every day, data is becoming larger and larger, as described by the concept of big data. Qualities and variables put together for forming data which united a special sense, till data measure want reliably been stretched [1,4]. Until 2003, 5 exabytes (EB) (10^{18} bytes) of data have been devoted eventually by way of human, but nowadays human is finishing this entirety of cash over simply days. In 2012, the size of records has reached 2.72 zettabytes (ZB) (10^{21} bytes). It is going to would have preferred should repeat at fashionable durations, arriving at over 8 ZB of data by 2015 [2,4]. Data volumes will become 35.2 ZB by 2020, that is, 37.6 billion hard drives of 1TB required to store this data [3,4]. Therefore, this enormous sum of data leads to a new concept termed big data. Big data is a period for important statistics sets which might be got to be larger, greater extended, and confounded shape with the obstacles of storing, analyzing, and visualizing for methods [2,4]. Big data is not just about volume, variety, or velocity of data, but it is also an important focus in both academic and industrial communities.

Developments in biomedical and health furnish phenomenal data with illuminating a health awareness framework that should take from its normal fill in. The rapid increase in the size and variety of biomedical data was one of the most important factors for the emergence of big data in this field. Will successfully utilize this data, there will make a compelling reason for masters that would prepare clinched alongside biomedical and health informatics to perform knowledge discovery techniques to big data [5].

Therefore, the pre-processing stage of big data systems is still a fundamental issue, which needs techniques to contrast estimations crosswise over sources and with recognizing notable features and instances quickly. The primary objective of managing biomedical big data is to accomplish the best classification accuracy because it could have an impact on individuals' lives. To attain this aim, we need to solve the problems of biomedical data, such as missing, unsure, inconsistent, and actual-valued data types. Data inconsistency is an issue, but it is a target for all big data analysts. Many solutions have been utilized to handle this issue. Every one of them is utilized as part of the pre-processing stage of data. Features elimination is used for removing features with missing data. It also helps for reducing the volume of data, particularly when we manage a big dataset.

On the other hand, data inconsistency is considered as a significant issue causing misclassification of instances. There are some critical factors that must be resolved in order to manage misclassified instances in data. Accordingly, it is first recommended to reduce attributes with missing values that are repeated and not fundamental [6].

If the output data (image) turned into utilized in a fuzzy weighted evaluation system, that must include the subsequent assessment indices: imply, preferred deviation, entropy, common gradient, and spatial frequency; the distinction with the reference image, which include the basis suggest rectangular errors, sign to noise ratio (SNR), and the height SNR; and the difference with supply photo such as the move entropy, joint entropy, mutual facts, deviation index, correlation coefficient, and the degree of distortion. Those parameters were used to evaluate the consequences of the comprehensive evaluation cost for the synthesized image [34].

Our goal in this research is to pre-process big data sets to reduce features and instances to reach a high classification accuracy. Anomalies and mislabelled instances are deleted before performing a learning algorithm. For handling outlier and mislabelled instances, learning algorithms are used to delete an instance that is misclassified. The techniques mainly used are genetic rough set attribute reduction (GenRSAR) for features selection and fuzzy-rough nearest neighbour for removing misclassified instances. This research aims to produce data in high quality with a minimum number of misclassified data to help us to perform a classification process with high accuracy. By applying the previous pre-processing steps, it helps to increase classification technique performance, especially for a classification tree. Our model helps to remove instances which cause problems in the classification process. The proposed model raised classification performance by 20%, reaching 89.24% for a decision tree.

This paper is organized as follows. Section 6.2 reviews previous modern-day research on eliminating misclassified instances for classification techniques. In Section 6.3, the proposed framework is defined in detail. The experimental results of the proposed framework are presented in Section 6.4. The conclusion and future outlooks are discussed in Section 6.5.

6.2 RELATED WORK

This section introduces an overview of some current studies for removing misclassified instances based on classification techniques. Sebban and Nock [7] studied different reduction techniques. They focused on two reduction strategies, which consist of removing irrelevant features and instances for building better trees. Many reduction methods were proposed, but their prototypes proved that removing misclassified instances based on KNN helped to noise filtering pass and avoided over-fitting.

The problem of over-fitting and mislabelled examples of data during the classification steps was discussed by Verbaeten and Van Assche [8]. Removing outliers was proposed as a good solution to enhance classification performance. The proposed solution improves performance and the size of trees based on filter techniques like a validated cross, voting and begging. So noisy instances are deleted by using C4.5 decision tree algorithm.

In order to filter noise in datasets, Zeng and Martinez [9] presented a neural network-based approach ANR. Each instance attaches a class probability vector to involve in neural network training. A large fraction of noise is removed by ANR. A small error in misidentifying of non-noise data is shown by the results. Removing misclassified instances increases accuracy by 24.5% while training neural networks. The neural network calculates the low probability of correctly labelled instances based on its output.

A versatile commotion decrease strategy called FaLKNR. The predictions of a situated from claiming nearby SVM models manufactured on the training set help Segata et al. [10] for presenting FaLKNR. To attaining a hypothetical complexity bound of (Onlog (n)) for non-high-dimensional data FaLKNR included some optimizations. More than 500,000 samples are possible to apply this technique to datasets. FaLKNR is the fastest and permits the highest NN accuracy improvements. Remove the too close instances to the wrong side that produced by support vector machine.

To achieve better results in most classification problems, the basic step is data cleaning. Data cleaning is used for deleting noise, inconsistent data and errors in the training data. So, Jeatrakul et al. [11] emphasized that data and classification models can be represented in better form with the help of data cleaning. The misclassified analysis used for data cleaning. They concluded that results showed that artificial neural network support by construing a better classification model. In order to handle label noise, pre-processing has an important role. Pre-processing can be in a different form like noisy instances removing, identifying the weight of the instances, or correcting incorrect on recognizing which instances would loud Toward Different criteria. For increasing classification accuracy, filtering noisy instances is needed to gain a significant part consideration and largely brought. Particular cases every now and again utilized the method of filtering any mislabelled instance using learning algorithms.

Jensen and Cornelis [12] developed Sakar's fuzzy-rough approach. They proposed two methods for fuzzy rough classification based on the traditional definition of fuzzy rough classification. The two techniques are 'FRNN-FRS (Fuzzy-Rough Nearest Neighbour classifier–Fuzzy Rough Sets)' and 'FRNN-VQRS (Fuzzy-Rough Nearest Neighbour classifier–Vaguely Quantified Rough Sets)'. They developed techniques are based on the use of lower and upper approximations and t-norm. They proved that these methods are effective and are competitive with existing classification techniques. They faced a problem in implementation, where is the similarity between a pair of objects is zero and causes a problem in feature selection. However, the authors suggested an alternative technique to combine the relationships.

Ougiaroglou and Evangelidis [13] explained how mislabelled instances caused other instances in the dataset to be mislabelled. So, removing mislabelled instances can be effective for noise reduction. They suggested instances based on algorithms like the nearest neighbour. KNN is less noise tolerant than other learning algorithms. They aimed to eliminate internal instances in a cluster of an instance in the same class to remove noise or corrupt instances.

Smith and Martinez [14,26] proposed PRISM to delete misclassification. It helps for enhancing classification performance over outlier identification strategies. PRISM helps for increasing the classification accuracy from 78.5% with 79.8% on a set of 53 data sets, which may be statistically critical. In addition, the accuracy on the non-outlier instances increments increased from 82.8% to 84.7%. They showed how machine-learning algorithms handle noise and outlier for generating better models.

In order to organize hyperglycemia in the hospitalized inpatient, the authors of reference [15] conducted a project. It helps for most non-ICU (intensive care unit) patients; they say that inpatient administered economy is optional What's all the more regularly every one of the prompts perhaps no solution on the other hand absolutely differences already, glucose at standard managed economy philosophies need help used. Consequently, traditions are proposed. Convincing expectation investigating readmissions engages recuperating offices to recognize Furthermore target patients amid a high peril. Accordingly, the targets are discovering genuine elements helping on mending focuses readmissions and also finding that capable framework for envisioning the kind of readmissions. The last precision using a boosting tree classifier is around 70%. That best execution is around 70% to 80% precision.

Arunkumar and Ramakrishnan [16] proposed a novel strategy that makes use of connection based clear out for dimensionality reduction took after with the aid of rough brief reduct for function selection on a particle swarm optimization seek space.

The primary stage evacuated the repetitive traits making use of connection coefficient channel on a molecule swarm enhancement look to space. The second stage created a fuzzy tough short reduct that could be applied so as. The characteristics acquired after aspect determination are subjected to grouping using customary classifiers. It has been found that the proposed approach provides to lessening in the combination wide variety of qualities and alternate within the classifier exactness contrasted with first-rate desire and order making use of relationship coefficient and fuzzy difficult short reduct set of rules. This methodology likewise lessens the number of misclassifications which can occur in these kinds of methodologies.

Chaudhuri [17] stated that category undertaking is completed through FRSVM. It is a variation of FSVM and MFSVM. The fuzzy rough set offers affectability of boisterous examples and handles inaccuracy. The participation capacity is created because of the capacity of centre and span of each elegance in highlight space. It assumes a critical part of checking out the selection surface. The instruction tests are either direct or non-linear distinct. In non-linear getting ready exams, enter the area is mapped into excessive dimensional detail space to system isolating floor. The numerous information publications determine one of the type commitments in the direction of the selection floor. The execution of the classifier is surveyed as a long way as the quantity of bolster vectors. The impact of variability in expectation and speculation of FRSVM is analyzed as for estimations of C. It viably determines lopsidedness and covering class issues, standardizes to concealed information and unwinds reliance amongst components and marks. The test comes about on both manufactured and genuine datasets support that FRSVM accomplishes predominant execution in diminishing anomalies' belongings than existing SVMs.

Amiri and Jensen [18] discussed missing values exist in numerous produced datasets in science. Subsequently, using missing information attribution strategies is a typical and vital practice. These techniques are a sort of treatment for instability and unclearness existing in datasets. Then again, strategies in view of fuzzy-rough sets give phenomenal instruments to managing vulnerability, having exceptionally alluring properties, for example, vigor and clamor resistance. Besides, they can discover negligible representations of data and don't require conceivably incorrect client inputs. Accordingly, using fuzzy-rough sets for ascription ought to be a successful methodology, it is proposed three missing worth ascription strategies in view of fuzzy-rough sets and its late augmentations.

6.3 METHODOLOGY

6.3.1 Rough Set

Pawlak [27] proposed the rough sets theory to cope with uncertain and fuzzy materials and to simplify knowledge. Within the rough sets theory, people utilize their trendy mastering to order their trendy surroundings as unique or cement. The attributes are the principle general for arranging the whole thing. The attributes may be categorized by accumulating similar ones. This is known as indiscernible relation Ind, denoted as and is the basis of rough sets theory.

One of the principle focal points of the rough set theory is that it needn't bother with any preparatory or extra data about information. The main important problems that moved toward utilizing noisy sets hypothesis like combine information decrease, detection the conditions of information, a rating of information essentialities, a period of choice calculation from data, suppose data characteristics, detection patterns in data and reveal effect relations.

6.3.1.1 Upper and Lower Approximations

The upper and lower approximation are the essential ideas for analysis data based on rough sets. The idea of lower and upper approximation is to identify which element in the set surely have a place or potentially have a place. The definition is appeared as takes after:

Let X denotes the subset of elements of the universe U, the lower approximation of BX, denoted as, is described because of the union of these kinds of fundamental units, which might be contained in X, greater formally:

$$\underline{BX} = \{x_i \in U \big| [x_i]_{Ind(B)} \subset X\} \tag{6.1}$$

The lower approximation of the set X for objects x_i become proven at the above Equation 6.3, which belong to the simple units contained in X (within the space B), $\underline{BX}$ is known as the lower approximation of the set X in B. The upper approximation of the set X, denoted as $\overline{BX}$ computed as the union of these simple units, which have a non-empty intersection with X:

$$\overline{BX} = \{x_i \in U \big| [x_i]_{Ind(B)} \cap X \neq 0\} \tag{6.2}$$

The above statement is to be read as: the upper approximation of the set X is a set of objects x_i, the difference between upper approximation and the lower approximation is called a boundary of X in U.

$$BNX = \overline{BX} - \underline{BX} \tag{6.3}$$

6.3.1.2 Core and Redact of Attributes

The ideas of center and reduct are two critical ideas of rough sets theory. On the off chance that the arrangement of characteristics is reliant, one can be changed on discovering all conceivable negligible subsets of traits. These prompt an indistinguishable number of rudimentary sets from the entire arrangement of properties (reducts) in finding the arrangement of every single essential characteristic (core).

Disentanglement of the information system can be utilized to perceive a few estimations of traits, which are redundant. For instance, a few traits, which are excess, can be erased or be sifted by methods for the rearrangements strategies. If, $Ind(A) = Ind(A - a_i)$, then the attribute a_i is dispensable, otherwise, a_i is indispensable in A. In other words, if after deleting the attribute a_i, the number of elementary sets in the information system is the same, then it concludes that attribute a_i is dispensable.

6.3.2 Fuzzy-Rough Sets

To enhance the attributes selection, the previous rough set strategies ought to be extended to fuzzy rough sets. Sincerely due to the fact, most datasets incorporate actually valued attributes, and the rough set cannot cope with noisy data. Fuzzy equivalence classes are the central concept of fuzzy rough sets. It is necessary to apply discretization for performing fuzzification step. The process to apply membership degrees of feature values to fuzzy set helps in dimensionality reduction process. FuzzyRoughset is a better guide for selection. The Fuzzy-Rough set is a generalization of the Rough set, derived from the approximation of a fuzzy set in crisp approximation [28].

6.3.3 Genetic Rough Set

Attribute reduction is the approach of deleting a subset of attributes from the dataset. One of the most famous tools used for solving the attribute reduction hassle is a rough set theory. The attribute reduction strategies in rough set idea are failed for finding the maximum efficient reduction because of no ideal heuristic can ensure optimality. So it's far taken into consideration a singular rough set approach to feature reduction primarily based on a heuristic genetic set of regulations. A genetic algorithm is proposed as attribute reduction. The proposed approach uses new suitable crossover and mutation operators that in shape the considered trouble. Genetic plays seek technique for the greatest discount. Proposed method finished on 13 of famous datasets from UCI machine learning repository. The consequences prove that the algorithm is extra effective, it has stepped forward the global searching for the capability to keep away from falling into a close by the gold standard, and it may get relative minimal attribute reduction for better performance [29].

The benefits of the proposed answer method embody: (a) fixing issues that may be decomposed into beneficial requirements and redact the scale of statistics, and (b) difficult set improves the overall performance of the GA with the aid of reducing the region form of the initial population and limited crossover using rough set idea.

6.4 THE PROPOSED FRAMEWORK

Our framework is proposed to enhance classification performance that consists of four basic stages, as shown in Figure 6.1. the most important stages in the framework are removing irrelevant features and instances. The Genetic Rough Set Attribute Reduct (GenRSAR) with FuzzyRoughNN are integrated to improve classification performance. The first phase is to prepare data for transforming data, handling messy attributes, and removing non-predictive ones. The second phase is to reduce attributes for saving training time. The third phase is to reduce instances by removing misclassification based on FuzzyRoughNN to increase classification accuracy and save training time. The last phase is classification phase that contains three main algorithms Naïve Bayes, SVM, and decision tree. Our framework has clearly impact on decision tree accuracy. Inside the following subsections, we are able to discuss the primary stages of our proposed framework in greater info.

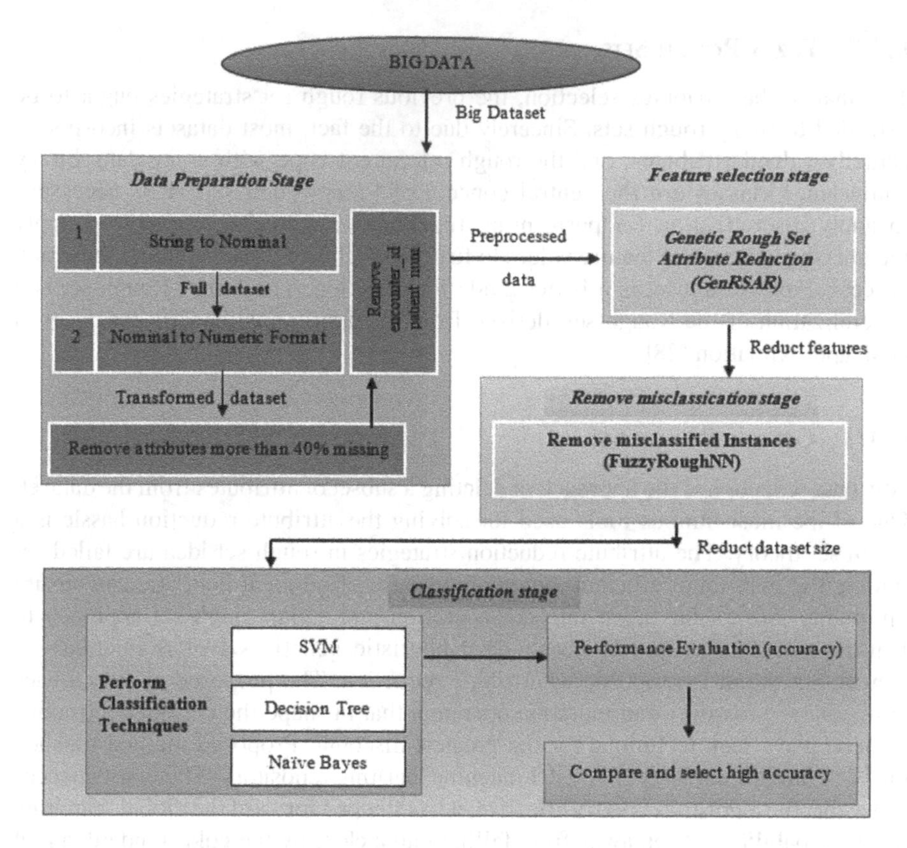

FIGURE 6.1 The block diagram of proposed framework.

6.4.1 Data Preparation Stage

Data preparation stage helps to enhance the quality of classification process. It is a critical stage to supply data in a suitable format for processing stage. We did not use discretization techniques to convert continuous data into discrete format because discretization causes the loss of large amount of information [21]. Discretization has a big problem with missing data that can be interpreted as attribute value and causes a problem in classification results.

Many attributes might not be treated straightforwardly because of a high percentage of missing values. Weight attribute might have been acknowledged on be excessively sparse. It might have been not incorporated for further examination. Payer code might have been deleted since it required identification of absent values [22].

In this stage, we prepare the tested dataset with the help of previous to two studies. We prefer transforming data type into numerical type by assigning a numerical value to each nominal attribute's value to represent it. SVM can be used when the data has exactly two classes. We also removed attributes that contain more than 40% missing data. As well as removing previous three attributes, patient and encounter numbers are removed because they are not predictive attributes.

6.4.2 Feature Selection Stage

Feature selection alludes all of the issues from claiming to select applicable attributes, which handle a majority predictive result. Datasets are holding an enormous amount from claiming attributes, so feature selection is used. The rough set theory is one of the most important techniques for selecting attributes. The principle objective of selecting attributes is to reduce the number of features and eliminate problem domain with a great performance of representing attributes [23].

Our framework proposes GenRSAR be the main filtering technique. It helps to extract the most relevant attributes for machine learning algorithm. Since genetic algorithms are powerful to fast search for large data, nonlinear also poorly seen spaces. Dissimilar to traditional selection of feature methodologies where the result will be optimized, some results could be altered at the same time. It can result in several optimal feature subsets as output.

6.4.3 Handling Misclassified Instances Stage

We proposed a stage named instance reduction for removing corrupted and mislabeled instances. It helps to reach high classification accuracy. As shown in Figure 6.2, received data from feature reduction stage passes to partition it into small datasets. After partitioning it into small datasets, we individually performed fuzzy-rough nearest neighbor in each one. We aggregated the resulting small datasets from

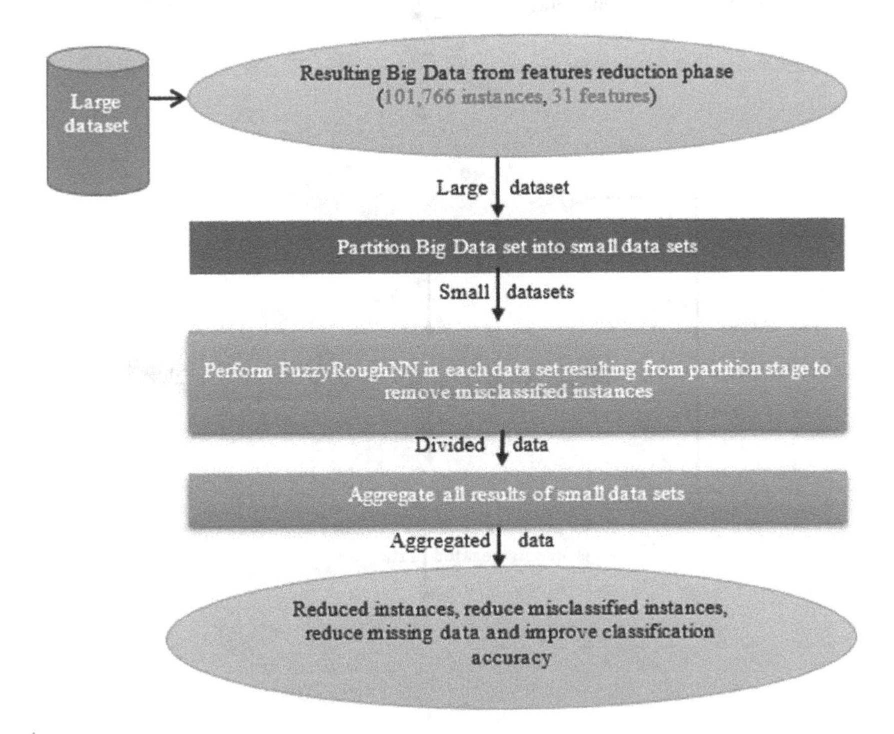

FIGURE 6.2 The block diagram of the handling of the misclassified instances.

removing misclassified based on FuzzyRoughNN in one dataset. We realized that the instance is reduced, the misclassified instances are reduced, the missing data is reduced, and classification accuracy is increased.

FuzzyRoughNN for handling misclassified instances pseudo code as shown in Figure 6.3.

```
Step 1: input reduct big dataset.
        Reduct big dataset ← features reduction using GenRSAR
Step 2: if dataset size is small then fuzzyRoughNN for instances
        Else
        Divide dataset into small ones with ratio 4%
        Small dataset ← 4% * resulting dataset
Step 3: each small dataset performs fuzzyRoughNN for instances
Step 4: aggregate all resulting results from step 3 in one dataset
```

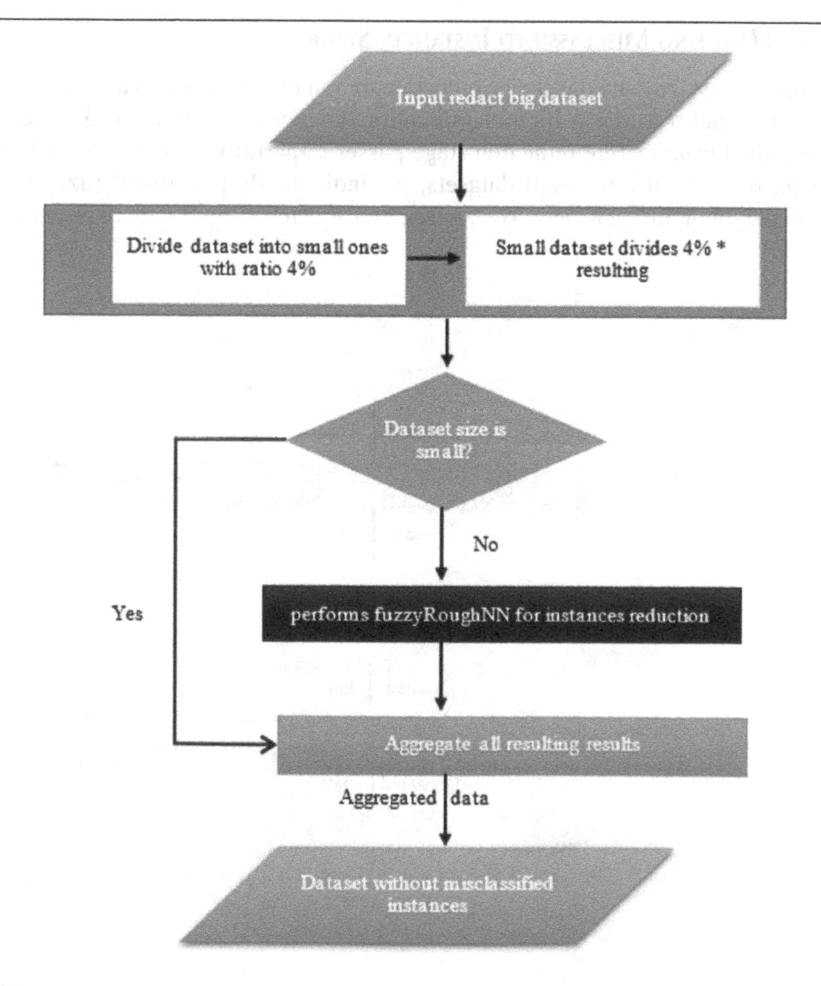

FIGURE 6.3 FuzzyRoughNN for handling misclassified instances.

6.4.4 Classification Stage

Classification is a critical data mining method with expansive applications to classify different sorts of data utilized for each field in our existence. Classification is used to classify data based on attributes of a dataset of the predefined classes. We computed classification performance dependent on correct and incorrect instances of data. The most important classification techniques are Naïve Bayes, J48, SVM and ANN is considered one of the famous, powerful nonlinear information processing systems that designed from interconnected basic processing modules, specifically neurons. Some of the basic benefits of the ANN consist of fault tolerance, adaptive learning, parallelism, and self-adjustment. ANN has several packages in knowledge extraction and category [37].

The Equations 6.4 to 6.9 illustrate how to measure classification performance [25]. Three parameters evaluate the proposed model: accuracy of the classifier, sensitivity, specificity, precision, recall, and f-measure. Accuracy means what is the percentage of instances correctly classified. So we can calculate error rate by 1-accuracy. Precision means a measure of correctness in instances that positive prediction. The recall is the measure of actual instances that correct prediction. F-measure is a ratio to measure the importance of either recall or precision. Specificity and sensitivity are the measures to evaluate receiver operating characteristics ROC on any distribution for accuracy.

$$\text{Accuracy} = \frac{\text{TP} + \text{TN}}{\text{TP} + \text{TN} + \text{FP} + \text{FN}} \tag{6.4}$$

$$\text{Specificity} = \frac{\text{TN}}{\text{FP} + \text{TN}} \tag{6.5}$$

$$\text{Sensitivity} = \frac{\text{TP}}{\text{TP} + \text{FN}} \tag{6.6}$$

$$\text{Precision} = \frac{\text{TP}}{\text{TP} + \text{FP}} \tag{6.7}$$

$$\text{Recall} = \frac{\text{TP}}{\text{TP} + \text{FN}} \tag{6.8}$$

$$\text{F-measure} = \frac{2 * \text{Percision} * \text{Recall}}{\text{Percision} + \text{Recall}} \tag{6.9}$$

Wherein TP (true positive) is the quantity of instances efficiently labeled to that magnificence. The FP (false positive) is the number of instances incorrectly rejected from that class. Ultimately, the FN (false negative) is the wide variety of instances incorrectly classified to that class. The TN (true negative) is the quantity of instances successfully rejected from that magnificence.

6.5 EXPERIMENTAL RESULTS

We tried to improve the performance of classification techniques. We applied our framework to two different benchmark datasets. The first one is diabetes data [15,19]. The second one is Electroencephalography test results [19]. We collected data from UCI data repository. We describe the two studies in the following subsections in detail. The 'Waikato Environment for Knowledge Analysis (Weka) version 3.7.10' had been used to perform model. It required windows 7 64bit with 3 GB RAM.

6.5.1 Data Description

IoT is associated with healthcare for joining the physicians and sufferers thru smart devices while each individual is roaming disadvantaged of any boundaries. So one can upload the patient's data, cloud offerings can be employed the usage of the big data era after which, the transferred statistics may be analysed [31,32,36].

6.5.1.1 DATASET 1: Diabetes Dataset

It is data for 130 US hospitals that collected delivery networks. Dataset covered the full decade of time between 1999 and 2008 as shown in Table 6.1. It represents the patient's basic data, data log, procedures conducted by the hospital during his/her residence time. It consists of 50 features and 101.766 instances facts was extracted from the database for diabetic encounters. The dataset is too large, which has many features with missing values for example race (2% missing), weight (97% missing), payer code (40% missing), medical especially (50% missing) and diagnosis 3 (1% missing). Therefore, this data needs a lot of preprocessing. The dataset contained some feature to describe patients like race, gender, age, and patient number. There are many features to indicate multiple inpatient visits and their observations. There are also 24 features to describe some types of medicine proceed to patients. The dataset contains 80% features are nominal, and remaining ones are numeric [15,19].

6.5.1.2 DATASET 2: EEG Dataset

It contains results of EEG (electroencephalography) test, which used to identify the electrical activity of human brain and disorders. ECG is beneficial technique for monitoring electrical activity the usage of an electrode array placed at specific positions at the body. ECG tracking systems support a medical doctor for expertise the patient's situation. Currently, low power [30]. It includes 15 features and 14,980 instances as shown in Table 6.2. It does not have a missing value. All attributes are in real data type. EEG measurement plays with the Emotiv EEG Neuroheadset. The period of the dimension changed into 117 seconds. The eye country changed into detected through a digital camera at a few levels inside the EEG length and manually added later to the report after reading the video frames. '1' indicates the attention-closed and '0' the attention-open nation. All values are in chronological order with the primary measured fee at the pinnacle of the facts [20,33]. Data obtained from UCI. We pre-processed it according to previously showed steps in the framework. We perform classification. 10-fold cross-validation might have been utilized within training dataset [24]. We divide our dataset into 33% testing and 67% training. Decision tree regards

TABLE 6.1
Listing of Features and Their Descriptions Inside the Preliminary Dataset

Feature Name	Type	Feature's Description and Values	Missing (%)	Distinct (%)
encounter_id	Numeric	Unique identifier of an encounter	0	100
Patient _nbr	Numeric	Unique identifier of a patient	0	100
Race	Nominal	Values: Caucasian, Asian, African American, Hispanic, and other	2	0
Gender	Nominal	Values: male, female, and unknown/invalid	0	0
Age	Nominal	Grouped in 10-year intervals: [0, 10), [10, 20), …, [90, 100)	0	0
Weight	Nominal	Weight in pounds	97	0
admission _type_id	Nominal	Integer identifier corresponding to 8 distinct values, for example, emergency, urgent, elective, newborn, and not available	0	0
discharge _disposition_id	Nominal	Integer identifier corresponding to 26 distinct values, for example, discharged to home, expired, and not available	0	0
Admission _source_id	Nominal	Integer identifier corresponding to 17 distinct values, for example, physician referral, emergency room, and transfer from a hospital	0	0
Time_in _hospital	Numeric	Integer number of days between admission and discharge 0% corresponding to14 distinct values	0	0
Payer_code	Nominal	Integer identifier corresponding to17 distinct values, for example, Blue Cross\Blue Shield, Medicare, and self-pay	40	0
Medical _specialty	Nominal	values, for example, cardiology, internal medicine, family\general practice, and surgeon	50	0
Num_lab_procedures	Numeric	Number of lab tests performed during the encounter	0	0
Num_procedures	Numeric	Number of procedures (other than lab tests) performed during the encounter	0	0
Num_medications	Numeric	Number of distinct generic names administered during the encounter	0	0
Number_outpatient	Numeric	Number of outpatient visits of the patient in the year preceding the encounter	0	0
Number_emergency	Numeric	Number of emergency visits of the patient in the year preceding the encounter	0	0
Number_inpatient	Numeric	Number of inpatient visits of the patient in the year preceding the encounter	0	0
Diag_1	String	The primary diagnosis (coded as first three digits of ICD9); 848 distinct values	0	0

(Continued)

TABLE 6.1 (Continued)

Listing of Features and Their Descriptions Inside the Preliminary Dataset

Feature Name	Type	Feature's Description and Values	Missing (%)	Distinct (%)
Diag_2	String	Secondary diagnosis (coded as first three digits of ICD9); 923 distinct values	0	0
Diag_3	String	Additional secondary diagnosis (coded as first three digits of ICD9); 954 distinct values	1	0
Num_diagnoses	Numeric	Number of diagnoses entered to the system	0	0
Max_glu_serum	Nominal	Indicates the range of the result or if the test was not taken. Values: '>200,' '>300,' 'normal,' and 'none' if not measured	0	0%
A1Cresult	Nominal	'Was larger than 8%, '>7' if the result was larger than 7% but less than 8%, 'normal' If the result was less than 7%, and 'none' if not measured.'	0	0
Change	Nominal	'Indicates if there was a change in diabetic medications (either dosage or generic Name). Values: 'change' and 'no' change'	0	0
Diabetes_med	Nominal	Indicates if there was any diabetic medication prescribed. Values: 'yes' and 'no'	0	0
Readmitted	Nominal	'30 days, '>30' if the patient was readmitted in more than 30 days, and 'No' for no record of readmission'	0	0
24 features for medications	Nominal	The generic names: 'metformin, repaglinide, nateglinide, chlorpropamide, glimepiride, acetohexamide, glipizide, glyburide, tolbutamide, pioglitazone, rosiglitazone, acarbose, miglitol, troglitazone, tolazamide, examide, sitagliptin, insulin, glyburide-metformin, glipizide-metformin, glimepiride-pioglitazone,metformin-rosiglitazone, and metformin-pioglitazone,' the feature indicates whether The drug was prescribed or there was a change in the dosage. Values: 'up' if the dosage was increased during the encounter, 'down' if the dosage was decreased, 'steady' if the dosage did not change, and 'no' if the drug was not prescribed	0	0

TABLE 6.2
List of Features and Their Data Types

Missing Data	Data Type	Feature
0% missing data	NUMERIC	AF3
0% missing data	NUMERIC	F7
0% missing data	NUMERIC	F3
0% missing data	NUMERIC	FC5
0% missing data	NUMERIC	T7
0% missing data	NUMERIC	P7
0% missing data	NUMERIC	O1
0% missing data	NUMERIC	O2
0% missing data	NUMERIC	P8
0% missing data	NUMERIC	T8
0% missing data	NUMERIC	FC6
0% missing data	NUMERIC	F4
0% missing data	NUMERIC	F8
0% missing data	NUMERIC	AF4
0% missing data	{0,1}	Eye Detection

as the most usable classification technique for modelling classification problems. One of the simplest decision tree algorithms is a j48 classifier, which creates a binary tree. Naïve Byes is a simple classification algorithm that builds on the probabilities concept, so it is known as a probabilistic classifier. A set of probabilities is calculated by Naïve Bayes for counting frequency and calculations of given dataset values. One of the most accurate supervised machine learning algorithms is SVM. Classification and regression problems can use SVM. Decision tree, Naïve Bayes, and SVM have mostly used classification algorithms, which ignore missing [24].

6.5.2 Results Discussion

Data represent patient, and hospital outcomes as shown in Figure 6.4 data become extracted from the database for diabetic encounters. Dataset is glad the subsequent criteria:

I. It is an inpatient come across (a clinic admission).
II. It's far a diabetic stumble upon, this is, one at some stage in which any form of diabetes become entered to the device as an analysis.
III. The duration of life turned into at the least 1 day and most 14 days.
IV. Laboratory assessments have been carried out at some point of the stumble upon.
V. Medicines had been administered in the course of the stumble upon.

As for biological data its processing is considered the imperative a part of the global computing. Thus, biological research has a first rate impact are in statistics association,

No.	1: encounter_id Numeric	2: patient_nbr Numeric	3: race Nominal	4: gender Nominal	5: age Nominal	6: weight Nominal
1	2278392.0	8222157.0	Cauca...	Female	[0-10)	
2	149190.0	5.5629189E7	Cauca...	Female	[10-20)	
3	64410.0	8.6047875E7	Africa...	Female	[20-30)	
4	500364.0	8.2442376E7	Cauca...	Male	[30-40)	
5	16680.0	4.2519267E7	Cauca...	Male	[40-50)	
6	35754.0	8.2637451E7	Cauca...	Male	[50-60)	
7	55842.0	8.4259809E7	Cauca...	Male	[60-70)	
8	63768.0	1.14882984E8	Cauca...	Male	[70-80)	
9	12522.0	4.8330783E7	Cauca...	Female	[80-90)	
10	15738.0	6.3555939E7	Cauca...	Female	[90-100)	
11	28236.0	8.9869032E7	Africa...	Female	[40-50)	
12	36900.0	7.7391171E7	Africa...	Male	[60-70)	
13	40926.0	8.5504905E7	Cauca...	Female	[40-50)	
14	42570.0	7.7586282E7	Cauca...	Male	[80-90)	
15	62256.0	4.9726791E7	Africa...	Female	[60-70)	
16	73578.0	8.6328819E7	Africa...	Male	[60-70)	
17	77076.0	9.2519352E7	Africa...	Male	[50-60)	
18	84222.0	1.08662661E8	Cauca...	Female	[50-60)	
19	89682.0	1.07389323E8	Africa...	Male	[70-80)	
20	148530.0	6.9422211E7		Male	[70-80)	
21	150006.0	2.2864131E7		Female	[50-60)	
22	150048.0	2.1239181E7		Male	[60-70)	
23	182796.0	6.3000108E7	Africa...	Female	[70-80)	

FIGURE 6.4 A part of first used dataset.

evaluation and dimension because of its sturdy mechanical and automated strategies. Mining strategies are massive for retrieving meaningful data from the organic information. Its miles a dynamic and systematic demonstration. But, extra powerful strategies, algorithms, software and incorporated gear are required for the organic processing. System studying is one of the key techniques for managing organic datasets [35].

Figure 6.5 represents data after preprocessing by converting string format into the nominal format, convert nominal data into numeric ones. Nominal data convert numeric on using one to one transformation. It also reflects removing more than 40% missing data attributes.

Figure 5.3 represents proposed GenRSAR as main filtering technique. It helps to extract the most relevant attributes for machine learning algorithm. Since genetic algorithms are powerful to fast search for large data, nonlinear also poorly seen spaces. Dissimilar to traditional selection of feature methodologies where the result will be optimized, some results could be altered at the same time. It can result in several optimal feature subsets as output.

Figure 6.6 represents proposed GenRSAR as main filtering technique. It helps to extract the most relevant attributes for machine learning algorithm. Since genetic algorithms are powerful to fast search for large data, nonlinear also poorly seen spaces. Dissimilar to traditional selection of feature methodologies where the result will be optimized, some results could be altered on the identical time. It can result in several top-quality characteristic subsets as output.

Relation: diabetic_data-weka

No.	1: race Numeric	2: gender Numeric	3: age Numeric	4: admission_type_id Numeric	5: dis1arge_disposition_id Numeric	6: admission_source_id Numeric
1	1.0	1.0	1.0	6.0	25.0	1.0
2	1.0	1.0	2.0	1.0	1.0	7.0
3	2.0	1.0	3.0	1.0	1.0	7.0
4	1.0	2.0	4.0	1.0	1.0	7.0
5	1.0	2.0	5.0	1.0	1.0	7.0
6	1.0	2.0	6.0	2.0	1.0	2.0
7	1.0	2.0	7.0	3.0	1.0	2.0
8	1.0	2.0	8.0	1.0	1.0	7.0
9	1.0	1.0	9.0	2.0	1.0	4.0
10	1.0	1.0	10.0	3.0	3.0	4.0
11	2.0	1.0	5.0	1.0	1.0	7.0
12	2.0	2.0	7.0	2.0	1.0	4.0
13	1.0	1.0	5.0	1.0	3.0	7.0
14	1.0	2.0	9.0	1.0	6.0	7.0
15	2.0	1.0	7.0	3.0	1.0	2.0
16	2.0	2.0	7.0	1.0	3.0	7.0
17	2.0	2.0	6.0	1.0	1.0	7.0
18	1.0	1.0	6.0	1.0	1.0	7.0
19	2.0	2.0	8.0	1.0	1.0	7.0
20		2.0	8.0	3.0	6.0	2.0
21		1.0	6.0	2.0	1.0	4.0
22		2.0	7.0	2.0	1.0	4.0

FIGURE 6.5 A part of first used dataset after data preprocessing.

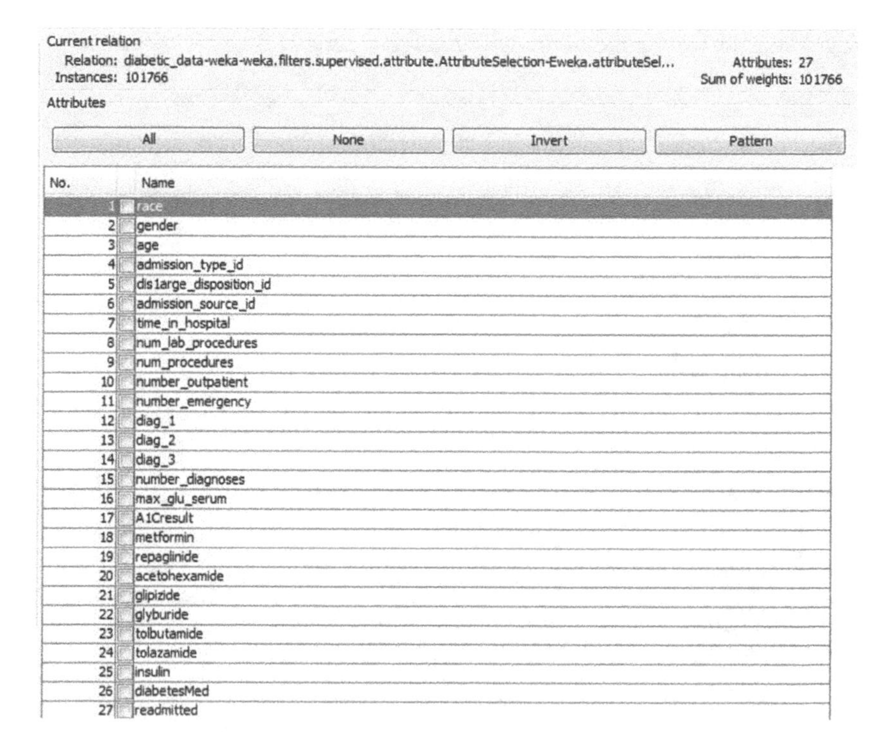

FIGURE 6.6 The effect of GenRSAR on the first dataset.

TABLE 6.3
The Effect of Attribute Selection Algorithms on Two Datasets

Dataset	Algorithm	Number of Selected Features	J48 TP (%)	J48 TN (%)	SVM TP (%)	SVM TN (%)	NAÏVE BYES TP (%)	NAÏVE BYES TN (%)
Dataset No. 1	GenRSAR	27	60	40	64.6	35.4	64.3	35.7
	Consistency	22	54.4	45.6	56.8	43.2	56.7	43.3
	Cfs subset Eval	11	65.2	34.8	65.1	34.9	64.8	35.1
Dataset No. 2	GenRSAR	15	84.5	15.5	55.1	44.8	46.8	53.2
	Consistency	15	84.5	15.5	55.1	44.8	46.8	53.2
	Cfs subset Eval	15	84.5	15.5	55.1	44.8	46.8	53.2

As illustrated in Table 6.3 and Figures 6.7 and 6.8 we perform three traditional attribute selection algorithms to reduce the attributes. Results showed that Genetic Rough Set Attribute Reduction (GenRSAR) and Correlation Feature Selection (CFS) achieved the better effect on dataset 1, but dataset 2 is preprocessed dataset. As well as GenRSAR save data simply because it selected 27 feature but CFS selected only 11 feature. So we chose GenRSAR as feature selection algorithms.

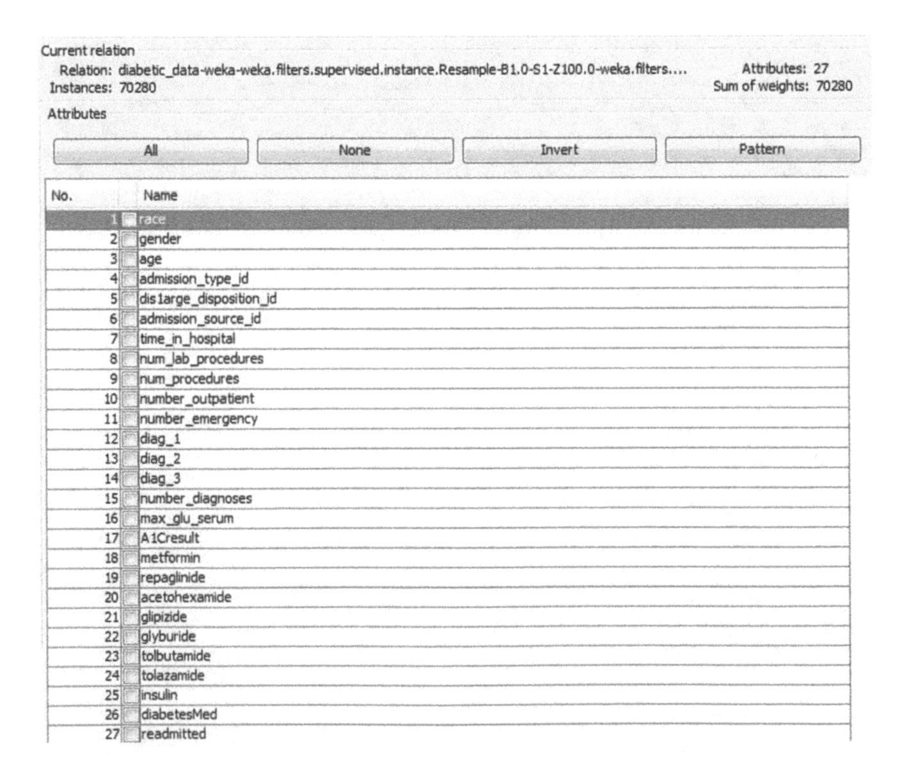

FIGURE 6.7 The effect of 5NN for removing misclassified instance on first used dataset.

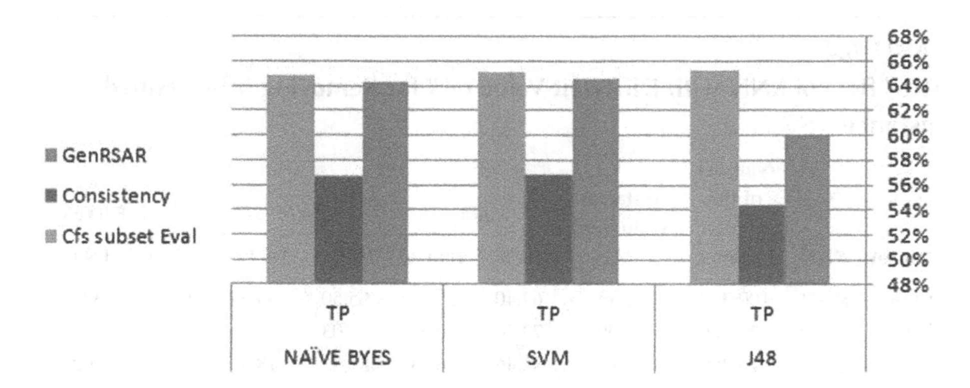

FIGURE 6.8 The effect of attribute selection algorithms on first dataset.

Figure 6.9 represents removing misclassified based on 5NN in first dataset. We realized that the instance is reduced, the misclassified instances are reduced, the missing data is reduced, and classification accuracy is increased (Table 6.4).

After performing KNN on two datasets, we concluded that when $K = 5$ is the most suitable case because it has a significant effect on three algorithms as illustrated in Figure 6.9 and Tablc 6.5.

After performing 5NN, voting feature intervals as shown in Figure 6.10 and Fuzzy-rough 5NN on two datasets, we concluded that Fuzzy-rough 5NN is the more suitable case because it has a significant effect on three algorithm performance as shown in Figure 6.11.

Figure 6.10 shows that the effect of voting feature intervals for removing misclassified instance on first used dataset. In Figure 6.11, the result proves how FuzzyRoughSet is the most efficient algorithm because it saves data. Figure 6.11 shows the relation between removing misclassification algorithms and a number of instances.

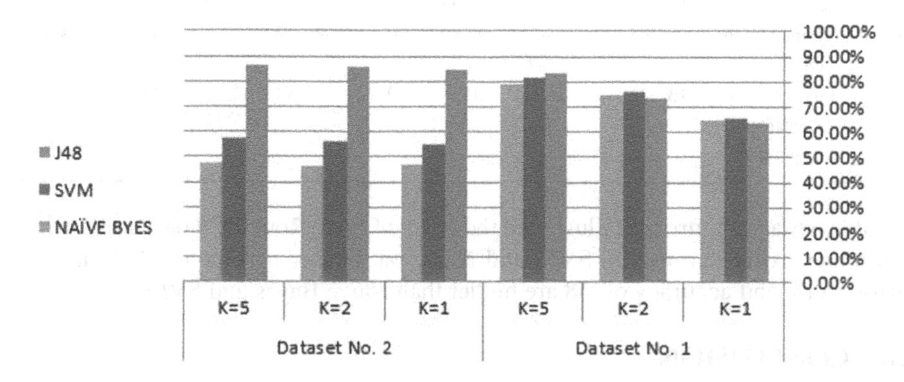

FIGURE 6.9 The effect of KNN with different value of K for removing misclassified instance on different classification algorithms.

TABLE 6.4
The Effect of KNN with Different Value of *K* for Removing Misclassified Instance

Dataset	KNN	Number of the Selected Instance	Instance Reduction (%)	J48		SVM		NAÏVE BYES	
				TP (%)	TN (%)	TP (%)	TN (%)	TP (%)	TN (%)
Dataset No. 1	$K=1$	100,320	1.42	63.40	36.60	65.50	34.50	64.60	35.40
	$K=2$	74,706	26.59	73.30	26.69	76.03	23.96	74.41	25.58
	$K=5$	70,280	30.93	83.46	16.53	81.6	18.4	79.2	29.80
Dataset No. 2	$K=1$	14,980	0	84.50	15.50	55.13	44.87	46.77	53.23
	$K=2$	13,587	1	85.77	14.24	55.94	44.06	46.04	53.96
	$K=5$	13,036	12.9	86.47	13.53	57.35	42.65	47.16	52.83

TABLE 6.5
The Comparison of Effect 5NN, Voting Feature Intervals and fuzzy-Rough 5NN on the Performance of the j48, SVM and NAÏVE BYES for Two Datasets to Remove Misclassified Instances as for DATA SET 1: Diabetes Data

Dataset	Methods	Number of the Selected Instance	Instance Reduction (%)	J48		SVM		NAÏVE BYES	
				TP (%)	TN (%)	TP (%)	TN (%)	TP (%)	TN (%)
Dataset No. 1	KNN ($K=5$)	70,280	30.9	83.46	16.53	81.6	18.4	79.2	29.8
	Voting feature intervals	59,753	41.3	89.42	10.57	84.8	15.12	83.08	16.91
	Fuzzy-rough K-nearest neighbors	73,302	28	86.4	13.6	84.8	15.1	81.8	18.2
Dataset No. 2	KNN ($K=5$)	13,036	12.9	86.47	13.53	57.35	42.65	47.16	52.83
	Voting feature intervals	6,984	53.4	99.89	0.1	96.29	3.7	96.05	3.95
	Fuzzy-rough K-nearest neighbors	13,321	11	90.5	9.5	57.34	42.66	43.76	56.24

Table 6.6 and Figure 6.12 illustrated the effect of fuzzyRoughNN on classification techniques. After acting j48, SVM and Naïve on datasets, we concluded that the performance and accuracy of j48 are higher than Naïve Bayes and SVM.

6.6 CONCLUSION

Misclassification is the biggest problems faced us when we tried to classify and handle our two biomedical big datasets. Corruptions in data cause low accuracy. In

Current relation
Relation: diabetic_data-weka-weka.filters.supervised.attribute.AttributeSelection-Eweka.attributeSele... Attributes: 27
Instances: 59753 Sum of weights: 59753
Attributes

All None Invert Pattern

No.	Name
1	race
2	gender
3	age
4	admission_type_id
5	dis1arge_disposition_id
6	admission_source_id
7	time_in_hospital
8	num_lab_procedures
9	num_procedures
10	number_outpatient
11	number_emergency
12	diag_1
13	diag_2
14	diag_3
15	number_diagnoses
16	max_glu_serum
17	A1Cresult
18	metformin
19	repaglinide
20	acetohexamide
21	glipizide
22	glyburide
23	tolbutamide
24	tolazamide
25	insulin
26	diabetesMed
27	readmitted

FIGURE 6.10 The effect of voting feature intervals for removing misclassified instance on first used dataset.

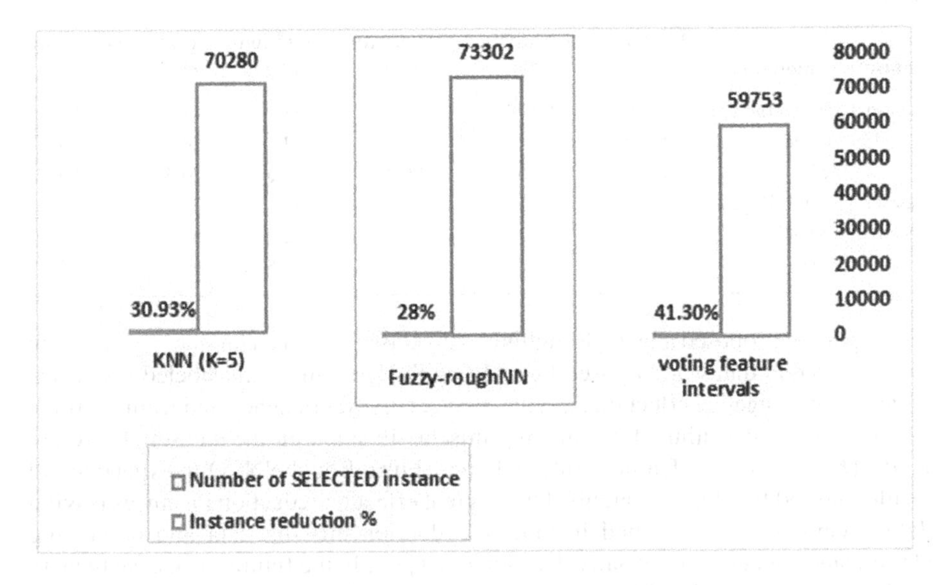

FIGURE 6.11 Relation between removing misclassified instances algorithm and number of instances.

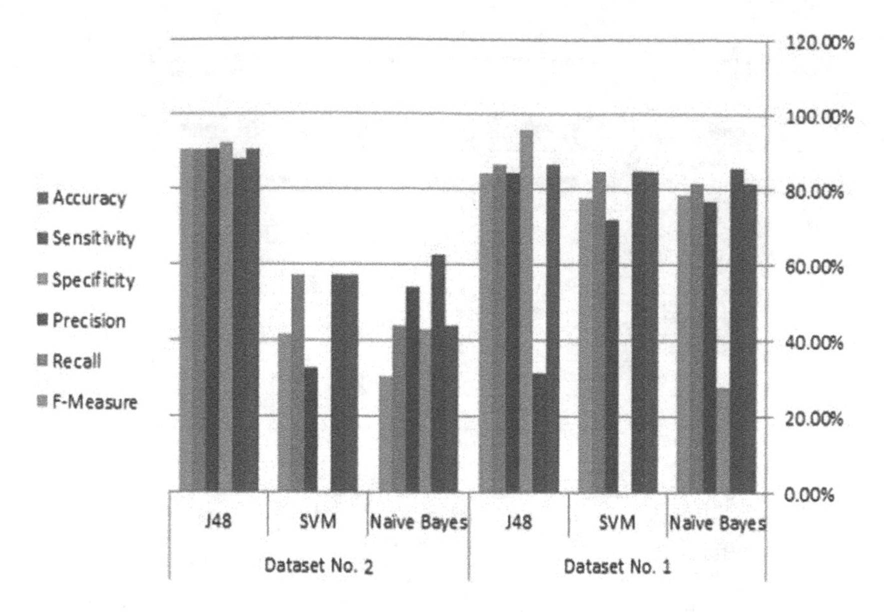

FIGURE 6.12 Effect of applying fuzzyroughNN in diabetes dataset on classification techniques performance.

TABLE 6.6
The Comparison of j48, SVM and Naïve Bayes with Other Classification Technique Performance for Two Reduced Datasets As for DATA SET 1: Diabetes Data

Dataset	Methods	Accuracy (%)	Sensitivity (%)	Specificity (%)	Precision (%)	Recall (%)	F-Measure (%)
Dataset No. 1	Naïve Bayes	81.79	85.78	27.8	77	81.8	78.8
	SVM	84.85	84.85	0	72	84.8	77.9
	J48	86.4	31.6	96.2	84.3	86.4	84.5
Dataset No. 2	Naïve Bayes	43.78	62.5	42.89	54.2	43.8	30.4
	SVM	57.34	57.3	0	32.9	57.3	41.8
	J48	90.5	87.8	92.4	90.5	90.5	90.5

this paper, we proposed a FuzzyRoughNN with classification techniques for applying to two different datasets. We used FuzzyRoughNN to remove mislabeled instances, which have a negative effect on classification accuracy, efficiency and training time. We used three algorithms for removing misclassification on the dataset. There are KNN ($K = 5$), voting features intervals and FuzzyRoughNN. Our experimental results showed that FuzzyRoughNN has a great effect on reduction instances (saving data), removing misclassified instances, reduction missing data and enhancing classification accuracy especially decision tree (j48). In the future works, we hope to apply our model based on MapReduce for processing. We want to utilize MapReduce methodology for processing data parallel not sequential. Performing parallel rough set for feature selection processing helps for saving time and data and in performing the better reduction. We also hope to improve the performance of classification techniques.

REFERENCES

1. R. Gupta, 2014. Journey From Data Mining To Web Mining To Big Data. *International Journal of Computer Trends and Technology* 10.1, 18–20. Web, 2014.
2. S. Sagiroglu and D. Sinanc, 2013. Big Data A Review. *Collaboration Technologies and Systems (CTS), 2013 International Conference On*. San Diego, CA: IEEE. Print.
3. W. Shuliang, D. Gangyi and M. Ming, 2013. Big Spatial Data Mining. *Big Data, 2013 IEEE International Conference On*. Silicon Valley, CA: IEEE. Print.
4. M. Abdrabo, M. Elmogy, G. Eltaweel and S. Barakat, 2016. Enhancing Big Data Value Using Knowledge Discovery Techniques. *International Journal of Information Technology and Computer Science* 8.8, 1–12, s2016.
5. P. Otero, W. Hersh and A. Jai Ganesh, 2014. Big Data: Are Biomedical and Health Informatics Training Programs Ready? on 2014-08-20.
6. N. Shoaip, M. Elmogy, A. Riad and F. Badria, 2015. Missing Data Treatment Using Interval-valued Fuzzy Rough Sets with SVM, *International Journal of Advancements in Computing Technology (IJACT)*, 7(5).
7. M. Sebban and R. Nock, 2000. *Contribution of Dataset Reduction Techniques to Tree-Simplification and Knowledge Discovery*, Springer, Berlin Heidelberg, 1910, 44–53.
8. S. Verbaeten and A. Van Assche, 2003. Ensemble Methods for Noise Elimination in Classification Problems, *4th International Conference on Multiple Classifier Systems, MCS'03*, Berlin, Heidelberg: Springer-Verlag, pp. 317–325.
9. X. Zeng and T. R. Martinez, A Noise Filtering Method Using Neural Networks, *The International Workshop on Soft Computing Techniques in Instrumentation, Measurement, and Related Applications*, 2003.
10. N. Segata, E. Blanzieri and P. Cunningham, A Scalable Noise Reduction Technique for Large Case-Based Systems, *8th International Conference on Case-Based Reasoning: Case-Based Reasoning Research and Development*, pp. 328–342, 2009.
11. P. Jeatrakul, K. Wong and C. Fung, 2010. Data Cleaning for Classification Using Misclassification Analysis, *Journal of Advanced Computational Intelligence and Intelligent Informatics*, 14(3).
12. R. Jensen and C. Cornelis, 2011. Fuzzy-Rough Nearest Neighbour Classification. *Transactions on Rough Sets XIII*, 56–72.
13. S. Ougiaroglou and G. Evangelidis, 2012. A Simple Noise-Tolerant Abstraction Algorithm for Fast k-NN, The Classification, *HAIS'12 Proceedings of the 7th international conference on Hybrid Artificial Intelligent Systems*, Springer-Verlag, Berlin, Heidelberg ©2012, ISBN: 978-3-642-28930-9, doi: 10.1007/978-3-642-28931-6_20.
14. M. Smith and T. Martinez, 2011. Improving Classification Accuracy by Identifying and Removing Instances that Should Be Misclassified, *Neural Networks (IJCNN), The 2011 International Joint Conference on*, pp. 2690–2697.
15. S. Meng, Data Mining for Diabetes Readmission Rate Prediction. [Online]. Available: https://github.com/siyuan1992/EE660_Project (Accessed on 20 May 2016), 2015.
16. C. Arunkumar and S. Ramakrishnan, Modified Fuzzy Rough Quick Reduct Algorithm for Feature Selection in Cancer Microarray Data, *Asian Journal of Information Technology*, doi: 10.3923/ajit.2016.199.210, 2016.
17. A. Chaudhuri, Fuzzy Rough Support Vector Machine for Data Classification, *International Journal of Fuzzy System Applications (IJFSA)*, doi: 10.4018/IJFSA.2016040103, 2016.

18. M. Amiri and R. Jensen, 2016. Missing Data Imputation Using Fuzzy-Rough Methods, *Neurocomputing* 205, 152–164. Web. 27 Sept. 2016.
19. UCI, [online] Diabetes 130-US Hospitals For Years 1999-2008 Data Set, https://archive.ics.uci.edu/ml/datasets/Diabetes+130-US+hospitals+for+years+1999-2008 (Accessed 22 May 2016), 2014.
20. UCI, [online] EEG Eye State Data Set, https://archive.ics.uci.edu/ml/datasets/EEG+Eye+State (Accessed 10 May 2016), 2013.
21. E. Dimitrova, P. Licona, J. McGee and R. Laubenbacher, Discretization of Time Series Data, *Journal of Computational Biology.* [Online]. Available: http://www.ncbi.nlm.nih.gov/pmc/articles/PMC3203514/ (Accessed 22 May 2016), 2010.
22. B. Strack, J. DeShazo, C. Gennings, J. Olmo, K. Cios and J. Clore, Impact of HbA1c Measurement on Hospital Readmission Rates: Analysis of 70,000 Clinical Database Patient Records, *Biomedical Research International.* [Online]. Available: http://www.ncbi.nlm.nih.gov/pmc/articles/PMC3996476/ (Accessed 15 Apr 2016), 2014.
23. N. Suguna and K. Thanushkodi, 2010. A Novel Rough Set Reduct Algorithm for Medical Domain Based on Bee Colony Optimization, *Journal of Computing*, 2(6).
24. T. Patil and S. Sherekar, 2013. Performance Analysis of Naive Bayes and J48 Classification Algorithm for Data Classification, *International Journal of Computer Science and Applications*, 6(2).
25. W. Zhu, N. Zeng and N. Wang, *Sensitivity, Specificity, Accuracy, Associated Confidence Interval and ROC Analysis with Practical SAS® Implementations*, NESUG 2010, 2010.
26. M. Smith and T. Martinez, An Extensive Evaluation of Filtering Misclassified Instances in Supervised Classification Tasks, 1, 2013.
27. S. Rissino and G. Lambert-Torres, February 2009. Rough Set Theory – Fundamental Concepts, Principals, Data Extraction, and Applications. *Data Mining and Knowledge Discovery in Real Life Applications*, Book edited by Julio Ponce and Adem Karahoca, (pp. 438), I-Tech, Vienna, Austria, 2009.
28. R. Jensen, 2005. Combining rough and fuzzy sets for feature selection, Doctor of Philosophy, School of Informatics, University of Edinburgh.
29. A.-R. Hedar, M. Adel Omar and Adel A. Sewisy, Rough Sets Attribute Reduction Using an Accelerated Genetic Algorithm, Software Engineering, Artificial Intelligence, Networking and Parallel/Distributed Computing (SNPD), *2015 16th IEEE/ACIS International Conference on*, Japan, 2015.
30. N. Dey, A. S. Ashour, F. Shi, S. J. Fong and R. S. Sherratt, 2017. Developing Residential Wireless Sensor Networks for ECG Healthcare Monitoring. *IEEE Transactions on Consumer Electronics*, 63(4), 442–449.
31. N. Dey, A. E. Hassanien, C. Bhatt, A. Ashour and S. C. Satapathy, (Eds.). 2018. *Internet of Things and Big Data Analytics Toward Next-Generation Intelligence.* Springer.
32. G. Elhayatmy, N. Dey and A. S. Ashour, 2018. Internet of Things Based Wireless Body Area Network in Healthcare. In: *Internet of Things and Big Data Analytics Toward Next-Generation Intelligence* (pp. 3–20). Springer, Cham.
33. L. Moraru, S. Moldovanu, L. T. Dimitrievici, F. Shi, A. S. Ashour and N. Dey, 2017. Quantitative Diffusion Tensor Magnetic Resonance Imaging Signal Characteristics in the Human Brain: A Hemispheres Analysis. *IEEE Sensors Journal*, 17(15), 4886–4893.
34. D. Wang, Z. Li, L. Cao, V. E. Balas, N. Dey, A. S. Ashour, … and F. Shi, 2017. Image Fusion Incorporating Parameter Estimation Optimized Gaussian Mixture Model and Fuzzy Weighted Evaluation System: A Case Study in Time-Series Plantar Pressure Data Set. *IEEE Sensors Journal*, 17(5), 1407–1420.
35. M. S. Kamal, N. Dey and A. S. Ashour, 2017. Large Scale Medical Data Mining for Accurate Diagnosis: A Blueprint. In *Handbook of Large-Scale Distributed Computing in Smart Healthcare* (pp. 157–176). Springer, Cham.

36. N. Dey, A. S. Ashour, F. Shi and R. S. Sherratt, 2017. Wireless Capsule Gastrointestinal Endoscopy: Direction-of-Arrival Estimation Based Localization Survey. *IEEE Reviews in Biomedical Engineering*, 10, 2–11.
37. S. Chakraborty, S. Chatterjee, A. S. Ashour, K. Mali and N. Dey, 2017. Intelligent Computing in Medical Imaging: A Study. *Advancements in Applied Metaheuristic Computing*, 143–163.

7 Fuzzy C-Mean and Density-Based Spatial Clustering for Internet of Things Data Processing

Heba El-Zeheiry, Mohammed Elmogy, Nagwa Elaraby, and Sherif Barakat

CONTENTS

7.1 INTRODUCTION

The quick data evolution has more precise necessities for storing and organization. This paper concentrates on big data storage, which denotes the storage and administration of massive size data sets whereas achieving consistency and handling of data retrieving [1]. Several big data issues are encountered through upcoming appliances when manipulators and devices will require cooperating into smart manners mutually. As a portion of the upcoming Internet, IoT targets to combine, gather information and recommend provisions for a wide variety of things utilized in diverse areas. 'Things' are ordinary items for which the IoT propose a simulated existence on the Internet, gives a precise personality, and appends abilities to interconnect with other items without human interference [2]. The IoT helps in reducing costs of implementing wireless healthcare applications [3].

Commonly, the little data is saved in the data warehouse is ordinary recognized. Alternatively, considering data occurrence in our ordinal domain including a huge, complicated, and heterogeneous quantity of data, were the core aims of the big data is to produce minor data for recognition [4].

Krik [5] observed the big data's ten v's. The core notion of the V description is related to critical issues of big data: collection, cleaning, combination, storing, handling, indexing, exploration, distribution, transmission, extracting, investigation and conception of massive sizes of fast increasing composite data. Several organizations are motivated by the chances offered by analytics of big data for marketing inventions, and construction of the bottom line.

The top ten list of the big data Vs. [5]:

Volume, variety, velocity, veracity, value, validity, venue, vocabulary, variability and vagueness.

Big data processing is a procedure for multidimensional data mining from varied high-volume data sources. A big data investigation system is implemented in a manner so that it can generate valuable data from unstructured raw data. Big data techniques are now applied in medical sciences, health informatics, computer sciences and lomany other fields [6].

7.2 BIG DATA

The big data concept is associated with the science of computers, which are necessary to develop it. The volume of data which is outside the handling capacity of traditional databases and cannot be managed by usual database procedures is identified as 'big data'. Huge volumes of data need diverse methods, procedures, techniques and tools to resolve unfamiliar challenges or previous challenges in improved manners [7].

Researchers have admitted that it is not possible to handle big data with traditional machine learning and artificial intelligence techniques or tools. Thus, these data require more accurate, efficient and effective machine-learning analyses [8].

Gartner [9] reported that by entering into the world of linked devices, IoT is predicted to increase to 26 billion interconnected devices by 2020.

Storing and investigation of data are issues common to the world linked via IoT [10]. Managing all data collected from the IoT involves three basic stages: data incorporating, data storing and data analyzing. Consequently, organizations should incorporate new software like Hadoop and MapReduce. It ought to have the capacity to offer enough disk space, web services, and computing size to remain to spread of novel data, various data processing faces are as follows:

- *Handling heterogeneous data*: In most applications of IoT, numerous data are assembled from varied sensors such as cameras, drivers, passengers, and medical sensors. The handling of diverse data offers several advantages and novel opportunities for system enhancement [11].
- *Handling noisy data*: Noisy data is inappropriate data. This data often appears as a substitute for irregular data. Its denotes any data that cannot be identified and appropriately interpreted by machines, as unstructured text. Data that is gathered, saved or modified in a way that is not familiar or utilizable to a program is noisy. Statistical methods are able to process the data gathered from various sources to eliminate noisy data and simplify data mining [11].
- *Handling massive data*: Massive data represented once its capacity can be terabytes or petabytes and denoted to a big data. Overall the data managing techniques introduced are suitable to be operated at a database. The handled data in IoT are intensive data.

7.3 IoT

On a daily basis, the generated data is almost 2.5 quintillion bytes. Therefore, 90% of the world's data has been currently generated in the last 2 years. The IoT data is collected from devices employed to collect environmental information, digital images, digital videos, business reports or GPS indicators. This large amount of data is called big data. Nowadays, we have seen a significant evolution in the volume of the data collected, through the increase of social media, multimedia, and IoT [12].

The IoT is a thought that relies on linked physical items. It originates from a group of mechanisms that may produce data. Sensor devices are surrounding us in vehicles, homes and cell phones. They gather data from our world [13]. The IoT allows us to recognize items that require exchanging, fixing or recalling. These objects are able to interact and connect with their neighbours to achieve specific aims [14]. Various surrounding items are gathered in what is now known as a smart world.

IoT produces varying amounts of data identified as a 'big data'. The Big data analysis is essential for taking benefit of its opportunities for great scale displaying and information system. The ability of the data movement and storage from objective sources is what we want to analyze through utilizing big data analytical tools.

One of the challenges of IoT big data is how to recognize communications between human and intelligent devices. The Internet originally was human-to-human communication, hence the human regulates the subject to be utilized by another human. But in the IoT, the things regulate the subject. So, the effect on our lifetimes is a public concern that requires interpreting the way of IoT creation of a significant task in an intelligent live and intelligent environment [15].

In IoT, a massive volume of crude data is gathered continuously. So, it is imperative to develop new strategies able to transform raw data into useful information.

In the medical domain [16], important human biological activities can be converted, through sensors, to raw data streams. This big data produced has features such as heterogeneity, unstructured characteristics, noise, and high redundancy.

New procedures required to handle the large data volumes apply the techniques of big data analytics. Big data can be mined and modelled through analytics tools to get a better vision and to improve smart cities characteristics [17].

7.4 IoT FOR HEALTHCARE

Adapted healthcare relies on a personal special interactive, natural and common feature. These direct to high-quality consequences by converting healthcare to be cost-valuable. A viable service based on early disease discovery and homecare can be implemented instead of subsequent treatment by special medical care. IoT can manage personal care services and maintain an ordinal identification for each person. Many devices are utilized in healthcare, joining one system with another [15].

The overall structure of the IoT involves many health supervising tools. The ordinary aspects or uses of health monitoring IoT-based devices are as follows:

- They collects data from wireless sensor networks (WSNs)
- It maintains views for users
- They permit connections to the network to get into infrastructure services
- They offer forcefulness, accuracy, consistency and stability

The IoT improves our lives anyplace at anytime. Based on rapid, protected and consistent networks, healthcare and monitoring are customized. Now, typical network facilities are the widely implemented tool for the internet. In the network side, the healthcare fixed wireless systems need characteristics which are facing the forthcoming internet. Global networks and wireless sensor networks, where the sensors are organized and interconnected by the fixed in devices where facilities offer the characteristics and provide integrated entrance to the functionality of a system. These modules produce information in many healthcare facilities such as hospitals, workplaces and households and contribute to big data.

Owing to the rapid improvements in the field of medical imagery, and the growing availability of processing power in computerized systems, medical image analysis has become one of the most exciting areas of study [18].

Thus, a labelling of the IoT relied on adapted healthcare systems involves distant supervising and medical provision:

- Distant supervising permits the organized admission to definite health supervising over manipulating principal wireless services linked over the IoT to observe the patients utilizing the protected obtained data of patient health. Many complicated algorithms and sensors are deployed for analysis of data, and later remotely distribute the data over wireless connections to assist medical health professionals in their recommendations; and

- Medical provision deploys supervising systems of IoT for clinics. This system of medical provision utilizes sensing devices for gathering biological data to be saved and examined over the cloud. It provides an endless computerized information flow, which enhances the care quality and lowers its cost.

The development of IoT in medical care is ongoing. Therefore, the tools of IoT even report the vital requirements for available and reasonable provision. In the meantime, the structure of the system of IoT for mechanical and machine to machine interaction stays to be typical. The layer of services generates the whole organization of IoT. The development is considered by connection facilities and end to end handling for medical care handled by IoT. The systems for health monitoring are able to monitor signals and provide further analysis and interpretation. The opportunities and feedback of these systems may disturb the mobility wearers through checking the critical marks [19,20].

7.5 BASIC CONCEPTS

7.5.1 MapReduce

The society of big data analytics admitted MapReduce as a template of programming for handling enormous data on separated schemas. The MapReduce model has become one of the best choices for processing massive data sets. It is a paradigm for evolving a distributed clarification for complicated difficulties over enormous data sets [5]. Users identify a map task that handles a pair of key and value to generate a set of intermediary key and value sets. Then, used shuffle and sort that is the principle controller procedure recognizes how much Reduce functions there will be. In addition, it creates a reduce task which joins every intermediary value related to its similar intermediary key.

There are four stages for deploying the MapReduce model which include reading a large data set, implementing the map function, implementing the reduce function and finally returning the resulting data from the reducer. The mapper receives massive amounts of data and produces intermediate results. The reducer reads the intermediate results and produces a final result. Figure 7.1 shows the MapReduce architecture.

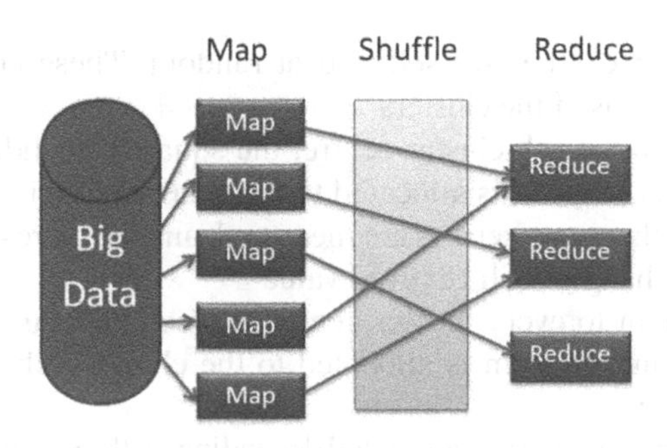

FIGURE 7.1 MapReduce architecture.

7.5.2 SPARK

Apache Spark is a very rapid cluster computing model. It depends on Hadoop MapReduce, and it expands the MapReduce framework for effectively using other computing forms, which contain collaborative requests and data flow handling. The basic characteristic of Spark is its memory cluster computing, which increases the handling rapidity of a process.

Spark is proposed for including an extensive variety of functions like group applications, iterative procedures, collaborative queries and flowing. Despite maintaining all these functions in a particular system, it decreases the administration load of supporting discrete means.

7.5.2.1 Apache Spark Characteristics

Apache Spark has the subsequent characteristics:

- *Rapidity*: Spark aids in the execution of an application in Hadoop cluster, reach to hundred stages quicker in memory, and ten stages quicker during the execution on disk. It is achieved by decreasing the total read and write processes into memory. It saves the middle processed memory data.
- *Verify various languages*: Spark offers incorporated APIs in Java, Python or Scala. So, we may create products in different languages. Spark originates with 80 extreme stage operatives for collaborative demanding.
- *Developed analytics*: Spark maintains the map and reduces functions, in addition to including SQL queries, machine learning (ML), graph algorithms, and streaming data [21].

7.5.3 K-MEANS

An easy learning procedure which explains the issue of identified clustering. The algorithm supports an easy and simple way to cluster a specified data set across a definite quantity of clusters chosen in advance. The basic concept is outlining K-centres, each cluster has one centre. These centres must be identified through a creative method owing to the fact that dissimilar positions produce dissimilar consequences. So, the improved selection is to state them as much as probably far away from each other [22]. K-means works as follows:

1. Primary cluster centres are selected at random. These characterize the 'temporary' means of the clusters
2. From each item to each cluster centre, the squared Euclidean distance is measured, and each item is allocated to the nearest cluster
3. In all clusters, the recent centres are measured, and the corresponding cluster centroid now changes each centroid value
4. From every item to every cluster centre, the squared Euclidean distance is measured, and the item is allocated to the cluster with the distance of minimum value
5. The new cluster centres are recomputed depending on the new value assignment
6. Repeat steps 4 and 5 until no items change in the clusters

7.5.4 DBSCAN

Density-based spatial clustering of applications with noise is used for determining the groups and noise in a given data set. There are suitable factors that need to be recognized, such as the Eps and MinPts of every cluster. Then, all items that are intensity-accessible from the specified item by the precise factors should be recovered.

There is an easy and operative empirical way to select the factors Eps and MinPts from the minimum thick cluster. So, DBSCAN utilizes overall estimations for Eps and MinPts, that is, the similar estimations for the overall clusters. The intensity factors of the least dense cluster are the best candidates for these overall factor estimations, indicating the smallest intensity that is not recognized to be noise [23]. DBSCAN steps are as follows:

1. Design a display of the items that we need to cluster
2. To every principle centre, C make a link from C to each item P in the ε-region of C
3. Put N to the items of the display
4. If N does not include any principle items stop
5. Choose a centre item C in N
6. Make X to be the group of items which can be moved from c by moving ahead;
   I. form a cluster including $X \cup \{c\}$
   II. $N = N/(X \cup \{c\})$
7. Endure with step 4

The advantages of DBSCAN are as follows [24]:

- It does not need a prior requirement of a number of clusters
- It is capable of recognizing noisy data during clustering
- It is capable of discovering arbitrarily sized and arbitrarily shaped clusters

7.5.5 FCM

Fuzzy c-means (FCM) is a technique used for clustering that permits single data elements to be related to two or more clusters [25].

The data set is labelled to n clusters. Each data element in the data set is associated to a cluster, which has a great level of association with that cluster. Another data element that remotely stays from the midpoint of a cluster has a small level of association with that cluster. This algorithm is regularly utilized in the identification of patterns. It relies on reducing the actual utility. The algorithmic stages for FCM clustering are as follows:

First, calculate the centre of the clusters using the following equation [26]:

$$C_j = \sum_{i=1}^{N} (M_{ij}^m * x_i)/M_{ij}^m \tag{7.1}$$

Then, the objective function is measured on the membership matrix by the following calculation:

$$J_m = \sum_{i=1}^{N} \sum_{j=1}^{C} M_{ij}^{m} \|x_i - c_j\|^2 \tag{7.2}$$

Finally, the membership value is altered by:

$$M_{ij} = 1 / \left(\sum (\|x_i - c_j\| / \|x_i - c_k\|) \right)^{2/(m-1)} \tag{7.3}$$

where m is an actual digit larger than 1, M_{ij} is the level of membership of x_i in the cluster j, x_i is the ith of d-dimensional precise data, c_j is the d-dimension centre of the cluster, and ||*|| is the similarity measure among any precise data and the centre. FCM sequentially transfers the cluster centres to the right area inside a data set. FCM clustering strategies rely on fuzzy behavior [27], and they give a method that is normal to produce a clustering where membership values have a characteristic translation but are not probabilistic at all. The advantages of FCM are as follows [28]:

- Provides the finest outcome for corresponded data and fairly improved than a K-means algorithm.
- Dissimilar to K-means, hence a data item should completely fit into one centre of clusters, the data item is given a membership to each cluster centre it results in the data item may fit into more than one centre of clusters.

7.6 RELATED WORK

Big data clustering based on Spark is an important topic. Many articles discuss it in the litterature. For example, Liu et al. [29] presented a matrix reverse procedure for massive amount matrices and its employment on Spark. They divided the massive calculation into a group of smaller sets for obtaining the LU breakdown. Regarding the upcoming work, they suggested two routes: (1) optimizing the suggested approach by reducing the communications expense with new technologies like Tachyon; and (2) proposing and applying new linear algebra software libraries depending on the procedure execution for huge level matrix computation involving singular value decomposition.

Aurelio et al. [30] presented a method concerned with the explanation of the juniper model for serious systems involving big data handling and reply time assurances. The proposed method is needed to extract from the data administering movement, put the scheduling conditions and aid in executing the interaction tools among data administering steps. These methods enable the juniper model to offer an effective structure, merging the power of Hadoop and Fpga-enabled items to tackle big data issues of the current serious organizations. This effort to enhance the juniper platform is still continuing. The authors executed the petaFuel case study to evaluate their methodology.

Li et al. [31] proposed an outline, Big Provision, which may simplify the specification of a system that contains varied features from the computing structure to the analytic model to the procedures. The basic model outline is the estimation and the formation of the operation of big data analytics models diverse, offered a group of the data model and many analytics prerequisites, as the predictable outcomes, retrieving time. Relying on the assessment and modeling outputs, big provision can produce viewing structure which can be utilized to design a full big data analytics system.

Zaharia et al. [32] found a solution for the challenges of current cluster computing models, involving iterative and collaborative computations. They thought that the main notion after data set RDDs, manage sufficient information for the formation of the data set from data existing in consistent storage, and may verify valuable in evolving other concepts to manage clusters. In upcoming work, they intend to concentrate on many areas like the illustration of the RDDs and spark properties, as well as Spark's other abstractions such as improving the RDD abstraction to permit programmers to exchange among storage cost. They described new actions to convert RDDs involving the 'shuffle' operation, which recomposed an RDD using a specified key. This process would permit them to deploy collection by an intersection, and offer better-stage of shared boundaries on the Spark, like SQL and R shells.

Gallego et al. [33] suggested the spread employment of the entropy minimization discretizer using the platform of Apache Spark. Their solution goes far from an easy parallelization, converting the iterative produced by the special offer in a one-stage calculation. The experimental outcomes of two data sets illustrated that their clarification could enhance the accuracy of the classification in addition to enhancing the operation of learning, which presented a high volume data handling challenge, concentrating on the process of high volume data sets decomposition. Through this employment, they like to participate in MLlib library by the addition of a complicated discretization procedure where just easy discretizers are offered. The experimental outcomes have confirmed the enhancement in both classification accuracy and time for both data sets used. Furthermore, their procedure showed the formation 300 epochs quicker than the consecutive form for the high volume data set.

Koliopoulos et al. [34] discussed Distributed WekaSpark; a spread Weka model that provides the current user interface. It has been executed on Spark, a Hadoop-related distributed model with rapid in-memory handling abilities and a provision for sequential calculations. By merging Weka's usage and management control of Spark, Distributed WekaSpark offers an operational model workbench for spread big data mining that realizes a near-linear level in implementing different practical level assignments. The Weka scaling efficiency is 91.4% and fits for 4x quicker than Hadoop.

Tayeb et al. [35] employed artificial intelligence to identify diseases. This was applied based on K-nearest neighbour (KNN) to identify present or probable diseases. They verified the accuracy of this algorithm on UCI data sets, specifically heart disease and chronic kidney failure. The results demonstrated a diagnostic accuracy of 90%.

Lakshmi et al. [36,37] investigated large amounts of structured and unstructured data. In this analysis, the SVM technique was shown to be more efficient and gave a normalized result.

Jin et al. [38] obtained knowledge from a large volume of information by using the MapReduce framework. The approach was based on fuzzy systems. They combined fuzzy rule interpolation and fuzzy rule inference efficiently.

Mane [39] classified heart disease through a big data method, and as big data is thought so worked on Hadoop MapReduce model. They applied enhanced K-means for clustering and a decision tree for classification; they used them together in the hybrid approach.

Tao and Ji [40] utilized the MapReduce technique to investigate various little data sets. They proposed a procedure for monstrous little data in light of the K-means clustering calculation. Their outcomes established that their method might enhance the proficiency of preparing information . They used K-means calculation for data examination using MapReduce. Then, they utilized a record for converging the data inside the cluster. The data in the same square have a high likeness when the merger is finished. The exploration will help them to plan a merger technique of little information in IoT. The DBSCAN algorithm can be suitable for big data because the number of clusters does not need to be known in the beginning.

Ang et al. [41] reviewed many assistive tools expansion in collaboration of mobile. They used smart guides tools for visually impaired people to offer a wireless assistive management approach. The 'smart guide' is a device designed as a separate handheld mechanism. It involved a camera network to increase security and reliability.

Kamal et al. [42] processed data gathered legally from Facebook users for observing the communications between thousands of users. The gathering and observing was used for managing big data with directions relied on machine learning procedures such as a maximum likelihood, dynamic source monitoring and canonical correlations. Pre-processed data are directed to big table stage for useful mapping. Mapping is utilized as big table direction where gathered, and handled data are referred to cluster consequently in the specific database. Big table allowed the regular change of the database tables to frequently renew within the definite time interval.

Morarou et al. [43] explained the many stages of principal component analysis (PCA) in addition to the use of PCA in the medical image processing area. Hence, medical images may be noisy, which is the basic challenge of medical images. Therefore, it is important to de-noise, classify and obtain patterns from medical images for efficient diagnosis. The basic aim of this paper was to explore various uses of PCA in the area of medical image processing. This work showed that PCA may be applied for image registration, fusion, segmentation, compression, redundant information removal, noise removal and feature extraction. PCA is applied ordinarly for feature extraction and image compression. The PCA technique proved its efficiency in various medical image applications.

Kamal et al. [44] suggested a classified system to isolate proteins into two varied sets. Training data sets were confirmed with a collection of mining procedures. Self-organizing maps (SOMs) handled all the data with a specific outline. It operated as a mapper in addition to an organizer. First, it outlined the related data into a specific space by considering comparative spaces between proteins. Then, it outlined the proteins into a corner regarding their angles. The investigation outcomes indicated that swarm bases SOM is better than the swarm less SOMs. The investigation confirmed that the proposed PSO-centric with SOM approach was faster than the other algorithm processes.

Maitrey et al. [45] implemented a procedure used for machine sources, and proper for usage on lots of the great computational troubles occurred at Google. Controlling the programming framework causes it simply to parallelize and spread computations. A quantity of optimizations in this system is aimed so that the quantity of data directed by the network is decreased. Thus, lead to locality optimization which permits delivering data from local disks, and making a specific replica of the intermediate data to resident disk protects network size.

Li et al. [46] proposed the Mux K-means procedure. Mux K-means receives many centroid sets to do clustering tasks. Unlike K-means, it conducts all centroid sets in parallel. Mux-K-means increase assesses, Prune, Permute and develop stages to enhance clustering accuracy and decrease runtime. They implemented Mux K-means on a MapReduce framework. They operated Mux K-means on Amazon and assets its clustering operation. The results showed that, for many cases, Mux K-means has the finest clustering accuracy. Mux K-means needs minimum operating time when handling many centre sets. Mux K-means has low accuracy values for big data.

Xu and Xun [47] outlined the MapReduce model of distributed computing. In the MapReduce instrument, they consolidated the structural planning attributes and key innovations of IoT. They led conveyed mining on information and data in the IoT world. Also, they represent stream information distribution. Customarily for valuable mining information from raw data created by IoT, analyze deficiencies of the conventional Apriori calculation. Apriori has a minor extracting proficiency and consumes a high volume of memory. The extraction technique for monstrous information in the IoT involves stream information investigation, grouping and so on. They planned to suggest a system for handling big data with a little cost and operate security of information. The suggested the system has a minor effectiveness, so it should be moved forward.

Wang et al. [48] investigated structural planning of the IoT in agribusiness and describes distributed processing and its usage there. The execution plan was on a two-tier construction using HBase. The structural planning gives constant read or access to their enormous sensor information. In addition to backing the sensor data executed by MapReduce model. XML documents put standards for the framework to bind the organizations of heterogeneous sensor data. Utilizing this framework lead the framework to lack a variety of sensor data.

Gole and Tidk [49] proposed a clustbigFIM method, which is based on the MapReduce structure for massive data mining. It is an improvement of the bigFIM procedure that is offering velocity to obtain information from massive data sets. They are relying on sequential patterns, associations, correlations, and other data mining missions. It needs to employ regular elements to make extraction calculation and MapReduce system on a flow of information.

Ghit et al. [50] outlined the problem of designing huge processing schemes as grids and clouds. MapReduce was a basis for a big data distributing. Reachable performance procedures for MapReduce operate only by particular jobs that handle a little part of the data set. So, it fails to examine the abilities of the MapReduce model under weighty loads that handle exponentially rising data sizes.

Martha et al. [51] suggested h-MapReduce based on MapReduce to report the scarcity of load consideration in the MapReduce model. Assignment consideration in h-MapReduce is reached by dividing weight jobs. The results present many problems, such as blocks and inheritance faults. H-MapReduce tackles these problems. Their tests were based on many networks containing document-term and social networks. The experiments also discovered the poor results of h-MapReduce with weighty jobs. Their future work will further explore enhancements to h-MapReduce's performance.

Matallah et al. [52] suggested a model to enhance metadata of the service for Hadoop to provide stability with no much metadata scalability and efficiency by assuming a joined explanation among metadata concentration and sharing to improve the efficiency of the approach. The enormous raises have caused in the deployment of recent tools and reached to data. The last stage in this development has advanced recent techniques: cloud and big data. It relies on the isolation of the centralization metadata and the storage server's metadata.

7.7 THE PROPOSED SYSTEM

7.7.1 Data Preparation and Storing

Proper clustering accuracy depends on the data quality; however, IoT data is noisy, incomplete, heterogeneous and massive. So, it should be converted to a more appropriate formula. There are various procedures utilized in data preparation, such as data cleaning, integration and discretization.

At this stage, we gather IoT raw data, construct it in a table shape and remove duplicated data. Also, handling missing data it is a very serious stage that should be handled wisely to prevent a negative effects on the classification's accuracy.

7.7.2 Data Cleaning

The procedure for cleaning data is not easy. The confusion may reach more than 30% of real data, which could be questionnable. In addition, it has exceptional cost. Data can be cleaned based on procedures, such as filling in missing values, smoothing the noisy data or solving the inconsistencies in the data. Numerous methods have been utilized to deal with omitted data, such as:

- *Removing*: It eliminates the omitted data and consumes the rest of the data in the analysis. This deletion can be inefficient as it decreases the data set size and may delete valuable data
- *Replacement*: It fills in the omitted values with the support of some methods, such as:
    - *Mean or Mode*: Fills the missing data by replacing it with the mean of a numeric attribute, or mode for a nominal attribute of all data
    - *KNN Imputation*: It deploys KNN algorithms to fill in the missing data. It can treat it with discrete and continuous attributes. KNN searches all the data to discover the highest related instances. It can select a possible value from the data set (Figure 7.2)

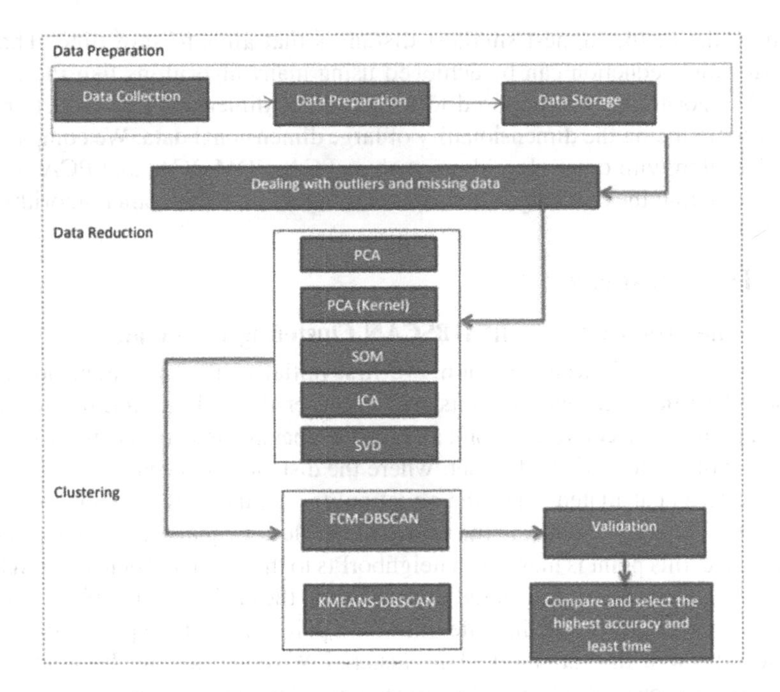

FIGURE 7.2 Proposed system for IoT big data storage.

7.7.3 Data Reduction

A massive amount of information is becoming available from different sources, for example, logistics insights, cameras, receivers and RFID.

High-dimensional information gets extraordinary difficulties terms of many-sided computational quality and characterization execution. Along these lines, it is important to obtain a low-dimensional component space from high dimensional component space to outline a learner with great execution.

Figure 7.3 shows that the cleaned data is the input for the data reduction stage. Data reduction separated to numericity reduction and dimensionality reduction. The data numericity reduction can be applied using regression or sampling. The used sampling algorithm is Kennard sample. Kennard sample reduces the number of iterations by

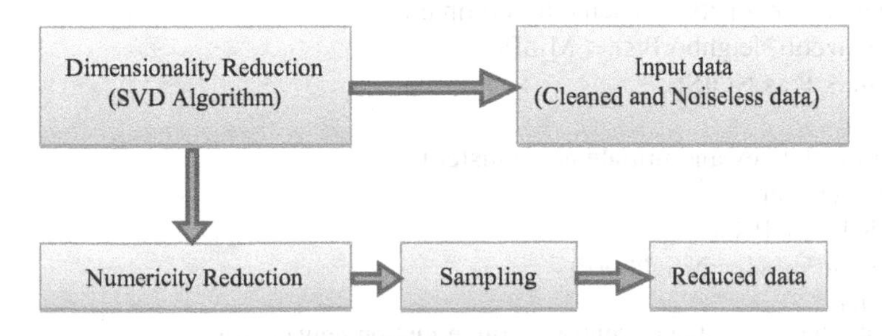

FIGURE 7.3 Block diagram of the data reduction steps.

viewing a list of the highest smallest distances that aims to save time. The data dimensionality reduction can be achieved using many algorithms like PCA, SOM and SVD algorithms. We proposed to use SVD for dimensionality reduction. It is suitable for reducing the dimensionality of large dimensional data. We compared the SVD algorithm with other algorithms such as PCA, SOM, ICA, and PCA (kernel). We conclude that the SVD algorithm operates in less time than other algorithms.

7.7.4 Data Clustering

7.7.4.1 The Proposed Kmeans-DBSCAN Clustering Technique

In Kmeans-DBSCAN Map function, we first initialize the minimum points that represent the minimum value of points in each cluster, the epsilon value that represents the distance between centre and point and the membership matrix. Then, we calculate the centres of clusters of the data set, where the distance between points and centre of the cluster is calculated using the equation $d = \sum\sqrt{((x_1 - y_1)^2 + \cdots (x_n - y_n)^2)}$. If the distance between a point and the centre of the cluster equals or is greater than the epsilon value, this point is marked as neighborPts to this cluster. Then, the neighbour points for each centre are calculated depending on the epsilon value. If the neighbor points to any cluster are less than the minimum points, then these points are marked as noise, otherwise they are marked as clustered. We determine the key and create a new cluster. It repeats until reach to convergence state. Finally, emit each point and each belonging cluster.

KM-DBSCAN Map function

Kmeans-DBSCAN(D, eps, MinPts, K)
  Initialize a number of clusters.
  For each point P in data set D
    if P is visited
    indicate P as visited
    continue next point
calculate the centre of clusters
calculate distance using equation $d = \sum\sqrt{(x_1 - y_1)^2 + \cdots (x_n - y_n)^2}$

If d<= eps mark p as neighborPt
Calculate neighborPts for each c based on eps
    if sizeof(NeighborPts) < MinPts
    mark P as NOISE
    else
Determine the key and initiate new cluster C.
C = next cluster
expandCluster(P, C)
C.neighborPoints = NeighborPts
For each c
Calculate the new value of centres using equation new centres $= \frac{\sum_{i=1}^{N} x_i}{N}$

```
if new centre=old centre then Break
End for
Emit(key, c)
End for
End function
```

In the Kmeans-DBSCAN Reduce function, the inputs are minimum points, epsilon value, clusters and keys. For each C cluster, the final cluster points equal the previous cluster points in addition to the current cluster points. For all points in the cluster, if a point is marked as unvisited, then this point is marked as visited. The neighbour points are calculated and compared with minimum points. If neighbour points are greater or equal to a minimum point, then the neighbor points are equal to neighbour points and cluster points. Finally, the output is a set of a cluster of data.

KM-DBSCAN Reduce function

```
Kmeans-DBSCAN Reduce function (key, clusters, eps, MinPts)
For every C cluster execute
Set final C.Points equal finalC.points ∪ C.points
For every Point P in C.neighborPoints do
  if P′ is not visited
  Indicate P′ as visited
Calculate NeighborPts' for each P′ based on eps
If size of NeighborPts' >= MinPts
 Set NeighborPts equal NeighborPts ∪ NeighborPts'
End if
If P′ is not yet a member of any cluster
add P′ to cluster C.
End if
End for
End for
Output: Set of clusters of data.
```

7.7.4.2 The Proposed FCM-DBSCAN Clustering Technique

In FCM-DBSCAN Map function, we first initialize the minimum points that represent the minimum value of points in each cluster, epsilon value that represent the distance between centre and point, and membership matrix. Then, we calculate the centres of clusters using the following equation $\frac{\sum_{i=1}^{D} M_{ij} \times x_i}{\sum_{i=1}^{D} M_{ij}}$. For each point in the data set, the distance between points and centre of the cluster is calculated using the equation $d = \sum_{i=1}^{N} \sum_{j=1}^{C} M_{ij}^{m} \|x_i - c_j\|^2$. If the distance between point and centre of cluster equal or is greater than the epsilon value, this point is marked as neighborPts to this cluster.

Then, the neighbor's points for each centre are calculated depending on the epsilon value. If neighbor points to any cluster are less than the minimum points, then the point is marked as a noise otherwise the point is marked as clustered. We determine the key and create a new cluster. It repeats until reach to convergence state. Finally, emit each point and each belonging cluster.

In FCM-DBSCAN Reduce function, the inputs are minimum points, epsilon value, clusters, and keys. For each C cluster, the final cluster points equal to previous cluster points in addition to the current cluster points. For all points in the cluster if a point marked as unvisited, then this point is marked as visited. The neighbor points are calculated and compared with minimum points. If neighbor points are greater or equal to a minimum point, then neighbor points are equal to neighbor points and cluster points. Finally, the output is a set of a cluster of data.

FCM-DBSCAN Map function

FCM-DBSCAN(D, eps, MinPts, M)
 Initialize a number of clusters.
 For every point P in data set D
  If P is visited
  Continue next point
  Indicate P as visited

Calculate the centre of clusters by equation $\frac{\sum_{i=1}^{D} M_{ij} \times x_i}{\sum_{i=1}^{D} M_{ij}}$

Calculate distance using equation $d = \sum_{i=1}^{N} \sum_{j=1}^{C} M_{ij}^{m} \|x_i - c_j\|^2$.

For each p in D
 Calculate neighborPts for each c based on eps
 If d<= eps mark p as neighborPt
  If sizeof (NeighborPts) < MinPts
  Indicate P as NOISE
  else Determine the key and initiate new cluster C.
 C = next cluster

expandCluster(P, C)

C.neighborPoints = NeighborPts

For each c

Calculate the new value of membership by equation $M_{ij} = 1/(\sum(\|x_i - c_j\| / \|x_i - c_k\|))^2/(m-1)$

Calculate the centre of clusters by equation $\frac{\sum_{i=1}^{D} M_{ij} \times x_i}{\sum_{i=1}^{D} M_{ij}}$

if new centre=old centre then Break

End for

Emit(key, c)

End for

End for

End function

FCM-DBSCAN Reduce function

FCM-DBSCAN Reduce function (key, c, eps, MinPts)
For all C clusters do
Set finalC.Points equal finalC.points ∪ C.points
For all P in C.neighborPoints do
  If P′ is not visited
Indicate P′ as visited
Calculate NeighborPts' for each P′ based on eps
If size of NeighborPts' >= MinPts
  set NeighborPts equal NeighborPts ∪ NeighborPts'
End if
If P′ is not yet a member of any cluster
add P′ to cluster C
End if
End for
End for
Output: Set of clusters of data.

7.8 RESULTS

The data set includes ordinary IADL housekeeping activities [53]. These activities are ironing, vacuuming, brooming, dusting, watering plants, cleaning windows, washing dishes, making the bed, setting the table, and mopping (Figure 7.4).

Row No	Act	Acc1	Acc2	Acc3	Tmp	Lgt1	Lgt2	Tlt	Rtn	Rtc	Time
55208	4	149	128	125	88	23	176	10010111	0	?	0
55209	4	148	128	125	88	24	175	10010111	0	?	0
55210	4	148	129	122	88	23	175	10010111	0	?	0
55211	4	149	128	121	88	23	176	10010111	0	?	0
55212	4	150	128	123	88	23	174	10111	0	?	0
55213	4	153	128	123	88	22	175	10010111	0	?	0
55214	4	152	129	123	88	22	175	10010111	0	?	0
55215	4	150	128	125	88	21	176	10010111	0	?	0
55216	4	149	127	125	88	21	176	10010111	0	?	0
55217	4	148	127	125	88	20	176	10010111	0	?	0
55218	4	148	128	126	88	20	176	10010111	0	?	0
55219	4	148	127	124	88	20	176	10010111	0	?	0
55220	4	148	126	126	88	19	175	10010111	0	?	0
55221	4	149	125	127	88	18	175	10010111	0	?	0
55222	4	150	125	129	88	18	174	10010111	0	?	0
55223	4	152	125	127	88	17	175	10010111	0	?	0
55224	4	153	125	130	88	16	175	10010111	0	?	0
55225	4	152	124	132	88	15	175	10010111	0	?	0
55226	4	150	123	134	88	14	176	10110111	0	?	0
55227	4	150	122	137	88	13	174	110110111	0	?	0
55228	4	154	125	137	88	12	174	110110111	0	?	0

FIGURE 7.4 Part of the data set.

Row No.	outlier	Act	Acc1	Acc2	Acc3
1	false	4	121	142	107
2	false	4	122	141	107
3	false	4	117	139	112
4	false	4	116	136	112
5	false	4	122	153	121
6	false	4	122	155	121
7	false	4	124	158	123
8	false	4	116	155	107
9	false	4	127	146	87
10	true	4	131	140	98
11	true	4	150	143	110
12	false	4	141	106	140
13	false	4	124	177	147
14	false	4	171	131	145
15	false	4	144	74	172
16	false	4	113	132	102
17	false	4	127	119	105
18	false	4	125	126	108
19	false	4	125	172	107
20	true	4	114	131	110
21	true	4	127	126	133

FIGURE 7.5 Outlier detection.

The general interval of the data set is 240 minutes. The intervals differ amongst various activities, indicating the usual distribution of behaviors in daily life. They utilized the Porcupine sensor along with the iBracelet to register the acceleration and RFID tag recognition. The data set consists of 1,048,576 records. We implemented the proposed technique on the data set using Radoop, KNIME and MATLAB® 2015b on a Core (TM) 2 Due, 2 GH processor, and 3 GB RAM.

Figure 7.5 shows the outlier detection. A new field called outlier appears. In the state of finding an outlier, the value of this field is true otherwise the value is false. The outlier is true when an observation is well outside of the expected scope of values in an experiment. An outlier arises from variability in the measurement or from an experimental error indication. The outliers are excluded from the data set.

Figure 7.6 shows that the outlier property has the values false for all the tuples and the missing values are replaced by the most probable value depending on KNN regression.

Figure 7.7 shows that applying of the SVD algorithm results in a reduction of the data set. The data are represented using a smaller number of properties. The attribute with high singular value has the priority to be presented. SVD1 has the highest probability to present the data.

Row No.	outlier	Act	Acc1	Acc2
1	false	0.444	0.561	0.660
2	false	0.444	0.568	0.650
3	false	0.444	0.535	0.631
4	false	0.444	0.529	0.602
5	false	0.444	0.568	0.767
6	false	0.444	0.568	0.786
7	false	0.444	0.581	0.816
8	false	0.444	0.529	0.786
9	false	0.444	0.600	0.699
10	false	0.444	0.690	0.311
11	false	0.444	0.581	1
12	false	0.444	0.884	0.553
13	false	0.444	0.710	0
14	false	0.444	0.510	0.563
15	false	0.444	0.600	0.437
16	false	0.444	0.587	0.505
17	false	0.444	0.587	0.951
18	false	0.333	0.503	0.709
19	false	0.889	0.258	0.816
20	false	0.889	0.574	0.718
21	false	0.889	0.594	0.447

FIGURE 7.6 Outlier removal and missing data replacement.

Component	Singular Val...	Proportion ...	Cumulative ...	Cumulative ...
SVD 1	4097998668	1.000	4097998668	1.000
SVD 2	1325668277	0.000	4099324336	1.000
SVD 3	5070.077	0.000	4099324341	1.000
SVD 4	3094.018	0.000	4099324344	1.000
SVD 5	707.756	0.000	4099324345	1.000
SVD 6	534.296	0.000	4099324345	1.000
SVD 7	381.067	0.000	4099324346	1.000
SVD 8	325.929	0.000	4099324346	1.000
SVD 9	108.318	0.000	4099324346	1.000
SVD 10	56.125	0.000	4099324346	1.000
SVD 11	0.000	0.000	4099324346	1.000

FIGURE 7.7 SVD deployment.

7.9 EVALUATION

The evaluation measures the time and accuracy of preprocessing the data set.

Table 7.1 shows that the precision value is 99.3%, the sensitivity value is 99.53%, which represents the true positive rate, and the value of specificity is 85.52%, which represents the true negative rate. From the previous results and evaluation, we conclude that the reduction step and FCM-DBSCAN enhanced the accuracy of the big data set to 98.9%.

TABLE 7.1
The Performance Measure of Our Proposed System

Sensitivity	Precision	Specifity	Accuracy	F-measure
99.53%	99.3%	85.52%	98.9%	99.39%

$$\text{Sensitivity, Recall (TP rate)} = \frac{TP}{TP + FN}$$

$$\text{Precision} = \frac{TP}{TP + FP}$$

$$\text{Specificity (TN rate)} = \frac{TN}{TN + FP}$$

$$\text{Accuracy} = \frac{TP + TN}{TP + TN + FP + FN}$$

$$\text{F-measure} = \frac{2TP}{2TP + FP + FN}$$

Note: The TP refers to the true positive value, TN is the true negative value, FP is the false positive value, and FN is the false negative value.

7.10 DISCUSSION

From our studies in Tables 7.2 and 7.3, we conclude that K-means and optics have the nearest accuracy value, but optics have longer time. The EM algorithm takes a longer time than other techniques. The DBSCAN has high accuracy but takes a longer time.

Kmeans-DBSCAN requires less time to retrieve the data but has less accuracy than FCM-DBSCAN. In FCM-DBSCAN, the accuracy increased and the expected time decreased.

We compared among various techniques which are KM, Optics, EM, DBSCAN, Kmeans-DBSCAN and our proposed technique FCM-DBSCAN. We found that FCM-DBSCAN, with its varied approaches for data reduction, had the highest accuracy value. FCM-DBSCAN with SVD have the highest value of accuracy and retrieve data in little time.

The proposed algorithm FCM-DBSCAN based on the Spark model was implemented in less time than the MapReduce model. FCM-DBSCAN with SVD has the highest value of accuracy and retrieves data in a small time.

The advantage of our proposed model is that it is a hybrid of FCM-DBSCAN implemented on the MapReduce model. It is intensity established clustering algorithm set an arrangement of entities in some area. It can determine the clusters of varied forms and sizes from a massive amount of data, without detecting some clusters at the start.

The Spark model has various features than MapReduce as speed, advanced analytics, and support of different languages. The challenge of Spark is that it captures a lot of memory and subject to memory spending.

From Tables 7.1–7.3 and Figures 7.8 and 7.9, one can see that we compared between many different techniques such as KM, Optics, EM, DBSCAN, and the proposed techniques FCM-DBSCAN and KM-DBSCAN.

TABLE 7.2
The Comparison Between Different Clustering Algorithms Implemented on MapReduce Model

Reduction Algorithm / Clustering Algorithm	PAC		PCA (Kernel)		ICA		SOM		SVD	
Performance Measure	**Accuracy (%)**	**Time (Sec.)**	**Accuracy (%)**	**Time (Sec.)**	**Accuracy (%)**	**Time (Sec.)**	**Accuracy (%)**	**Time (Sec.)**	**Accuracy (%)**	**Time (Sec.)**
K-means	92	0.2	89	2	87	0.2	92.15	0.1	94.73	0.2
Optics	90.2	1.9	91	9.68	65.48	0.6	91.05	1.9	90	0.79
EM	65.82	13	75.28	2.17	94.4	2.54	66.64	8	95.21	2
DBSCAN	93.4	0.3	89.3	7.3	90.12	0.4	88.46	1	98	3.11
Kmeans-DBSCAN	92.25	0.2	90.1	1.65	90	0.32	93.4	0.982	96.43	1.34
FCM-DBSCAN	**94.5**	**0.25**	**91.6**	**5.2**	**97.48**	**0.5**	**93.5**	**2.3**	**98.9**	**1.5**

TABLE 7.3
The Comparison Between Different Clustering Algorithms Implemented on Spark Model

Reduction Algorithm / Clustering Algorithm	PAC		PCA (Kernel)		ICA		SOM		SVD	
Performance Measure	**Accuracy (%)**	**Time (Sec.)**	**Accuracy (%)**	**Time (Sec.)**	**Accuracy (%)**	**Time (Sec.)**	**Accuracy (%)**	**Time (Sec.)**	**Accuracy (%)**	**Time (Sec.)**
K-means	92	0.1	0.89	1.2	87	0.2	92.15	0.1	94.73	0.2
Optics	90.2	1.1	91	7.6	65.48	0.6	91.05	0.8	90	0.62
EM	65.82	11.32	75.28	1.34	94.4	1.67	66.64	7.2	95.21	1.4
DBSCAN	93.4	0.23	89.3	6.2	90.12	0.3	88.46	3	98	2.4
Kmeans-DBSCAN	92.25	0.1	90.1	1.3	90	0.23	93.4	0.12	96.43	0.124
FCM-DBSCAN	**94.5**	**0.2**	**91.6**	**3.2**	**97.48**	**0.25**	**93.5**	**1.3**	**98.9**	**0.7**

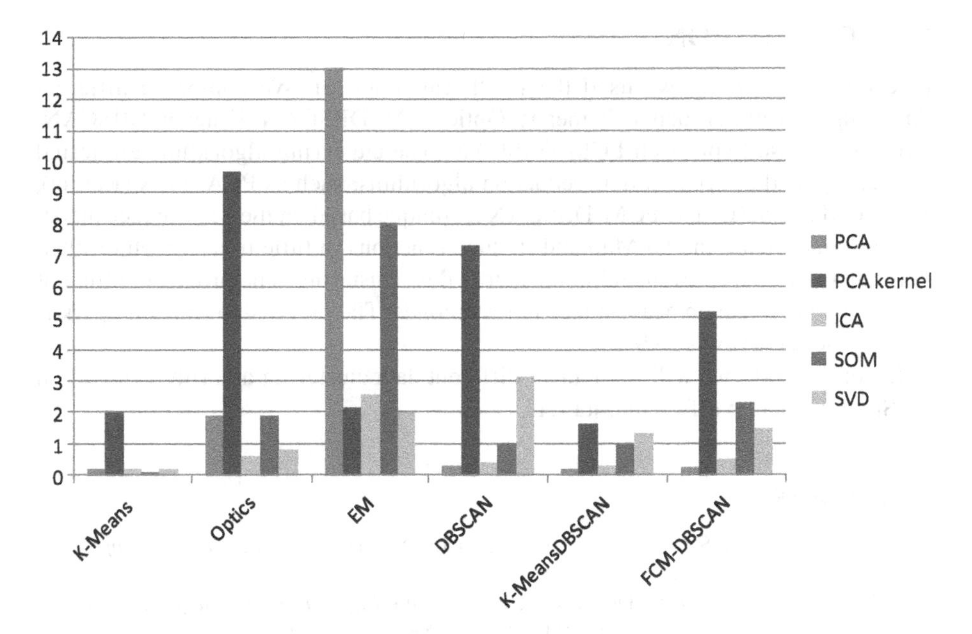

FIGURE 7.8 Time comparison between different clustering techniques based on the MapReduce model.

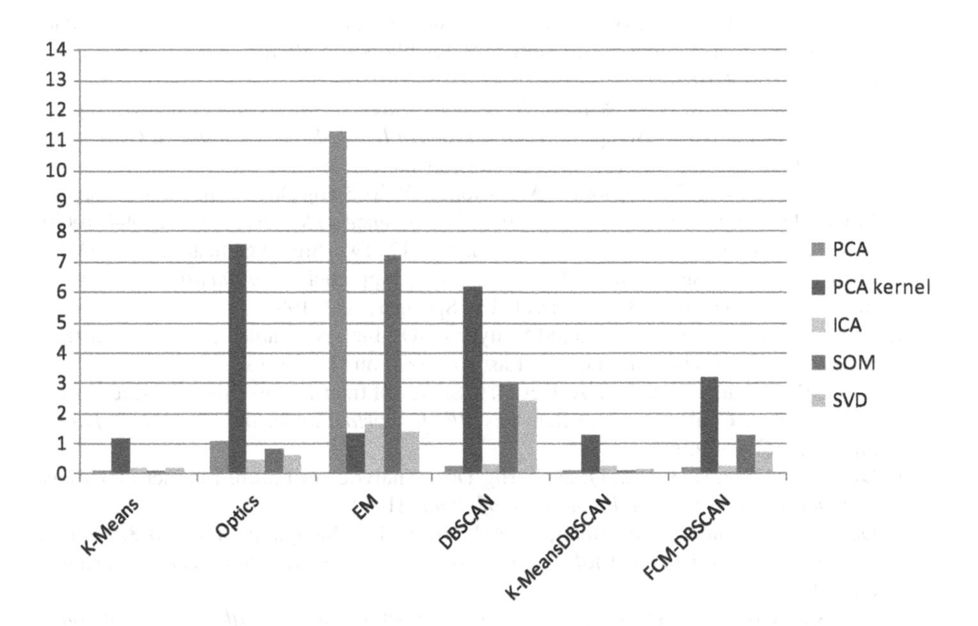

FIGURE 7.9 Time comparison between different clustering techniques based on the Spark model.

7.11 CONCLUSION

To evaluate our work, we used the IADL raw data set. We compared different clustering algorithms such as K-means, Optics, EM, DBSCAN, Kmeans-DBSCAN, and our proposed approach FCM-DBSCAN. The clustering algorithms are tested after applying the different data reduction algorithms, such as PCA, PCA (Kernel), ICA, SOM and SVD. The FCM-DBSCAN technique based on the MapReduce model achieves data clustering by Map and Reduce functions in little time, which resulted from the use of a reduction technique before data clustering. The processing time of the proposed system on MapReduce is 1.5 seconds. The processing time on Spark is 0.7, and the accuracy is 98.9%.

In future work, we will implement different data clustering algorithms based on the Spark model on different data sets.

REFERENCES

1. Chen, M., Mao, S. & Liu, Y. 2014. Big Data: A Survey. *Mobile Network Appl*, (19), (171–209).
2. Ciobanu, R., Cristea, V., Dobre, C. & Pop, F. 2014. *Big Data Platforms for the Internet of Things*. Springer International Publishing, Switzerland, (3–34).
3. Elhayatmy, G., Dey, N. & Ashour, A. S. 2018. Internet of Things Based Wireless Body Area Network in Healthcare. In *Internet of Things and Big Data Analytics Toward Next-Generation Intelligence* (pp. 3–20). Springer, Cham.
4. Oweis, N. E., Owais, S. S., George, W., Suliman, M. G. & Snasel, V. 2015. *A Survey on Big Data, Mining: Tools, Techniques, Applications and Notable Uses*. Springer International Publishing, Switzerland, (109–119).
5. 'Top 10 List – The V's of Big Data', Datasciencecentral.com, 2016. [Online]. Available: http://www.datasciencecentral.com/profiles/blogs/top-10-list-the-v-s-of-big-data. Accessed [21/5/2016].
6. Kamal, M. S., Dey, N. & Ashour, A. S. 2017. Large Scale Medical Data Mining for Accurate Diagnosis: A Blueprint. In *Handbook of Large-Scale Distributed Computing in Smart Healthcare* (pp. 157–176). Springer, Cham.
7. Kaisler, S., Armour, F., Espinosa, J. A. & Money, W. 2013. Big Data: Issues and Challenges Moving Forward. *46th Hawaii International Conference on System Sciences*, (994–1003).
8. Dey, N., Ashour, A. S. & Borra, S. (Eds.). 2017. Two-Step Verifications for Multi-instance Features Selection: A Machine Learning Approach. *Classification in BioApps: Automation of Decision Making* (Vol. 26). Springer, (173–198).
9. Gartner, 'BigDataManagementandAnalytics,' [Online]. Available: http://www.gartner.com/technology/topics/big-data.jsp. Last Accessed on [2/11/2015].
10. Singh, D., Tripathi, G. & Jara, A. J. 2014. A Survey of Internet-Of-Things: Future Vision, Architecture, Challenges and Services. *IEEE World Forum on Internet of Things (WF-IoT)*, (287–292).
11. Ding, G., Wang, L. & Wu, Q. 2013. Big Data Analytics in Future Internet of Things. *National Natural Science Foundation of China*, (1–6).
12. Dey, N., Hassanien, A. E., Bhatt, C., Ashour, A. S. & Satapathy, S. C. 2018. *Internet of Things and Big Data Analytics Toward Next-Generation Intelligence*. Springer, (381–405).
13. Agrawal, S. & Vieira, D. 2013. A Survey on Internet of Things. *Abakós, Belo Horizonte*, 1(2), (78–95).

14. Atzori, L., Iera, A. & Morabito, G. 2010. The Internet of Things: A Survey. *Computer Networks* 54, (2787–2805).
15. Bhatt, C., Dey, N. & Ashour, A. S. 2017. Internet of Things and Big Data Technologies in Next Generation Healthcare. *Studies in big data*, Springer, (23), (3–10).
16. Acharjya, D. & Anitha, A. 2017. A Comparative Study of Statistical and Rough Computing Models in Predictive Data Analysis. *International Journal of Ambient Computing and Intelligence Archive*, 8(2), (32–51).
17. Dey, N., Hassanien, A. E., Bhatt, C., Ashour, A. & Satapathy, S. C. (Eds.). 2018. *Internet of Things and Big Data Analytics Toward Next-Generation Intelligence*. Springer, pp. 359–379.
18. Hore, S., Chakroborty, S., Ashour, A. S., Dey, N. & Ashour, A. S. (Eds.). 2015. Finding Contours of Hippocampus Brain Cell Using Microscopic Image Analysis. *Journal of Advanced Microscopy Research*, 10(2), (93–103).
19. Dey, N., Ashour, A. S., Shi, F., Fong, S. J. & Sherratt, R. S. 2017. Developing residential wireless sensor networks for ECG healthcare monitoring. *IEEE Transactions on Consumer Electronics*, 63(4), (442–449).
20. Dey, N., Ashour, A. S., Shi, F. & Sherratt, R. S. 2017. Wireless Capsule Gastrointestinal Endoscopy: Direction-of-Arrival Estimation Based Localization Survey. *IEEE Reviews in Biomedical Engineering*, 10, 2–11.
21. www.tutorialspoint.com. 2016. Apache Spark Quick Guide. [online] Available at: http://www.tutorialspoint.com/apache_spark/apache_spark_quick_guide.htm Accessed [21/5/2016].
22. KMeans Clustering, Clustering and Classification Lecture8, jonathantemplin.com/files/clustering/psyc993_09.pdf.
23. Ester, M., Kriegel, H. P., Sander, J. & Xu, X. 1996. A Density-Based Algorithm for Discovering Clusters in Large Spatial Databases with Noise. *KDD-96 Proceedings*, (226–231).
24. Bijuraj, L.V. 2013. Clustering and its Applications. *Proceedings of National Conference on New Horizons in IT*, (169–172).
25. http://home.deib.polimi.it/matteucc/Clustering/tutorial_html/cmeans.html Accessed on [21/5/2016].
26. Soumi, G. & Sanjay, K.D. 2013. Comparative Analysis of K-Means and Fuzzy CMeans Algorithms. *International Journal of Advanced Computer Science and Applications*, 4(4), (35–39).
27. Dey, N. & Ashour, A. S. 2018. Computing in Medical Image Analysis. In *Soft Computing Based Medical Image Analysis*, (pp. 3–11).
28. Suganya, R. & Shanthi, R. 2012. Fuzzy C-Means Algorithm-A Review. *International Journal of Scientific and Research Publications*, 2(11), (1–3).
29. Liu, J., Liang, Y. & Ansari, N. 2016. Spark-based Large-scale Matrix Inversion for Big Data Processing. *IEEE Access*, (99), (1–10).
30. Silva, M. A. A., Cheptsov, A. & Adam, L. 2014. JUNIPER: Towards Modeling Approach Enabling Efficient Platform for Heterogeneous Big Data Analysis. *10th Central and Eastern European Software Engineering Conference in Russia*, p. 12.
31. Li, H., Lu, K. & Meng, S. 2015. BigProvision: A Provisioning Framework for Big Data Analytics. *IEEE Network*, (50–56).
32. Zaharia, M., Chowdhury, M., Franklin, M. J., Shenker, S. & Stoica, I. 2010. Spark: Cluster Computing with Working Sets. *2nd USENIX Conference on Hot Topics in Cloud Computing*, (1–7).
33. Gallego, S. R., Garca, S., noTalın, H. M. & Rego, D. M. 2015. Distributed Entropy Minimization Discretizer for Big Data Analysis under Apache Spark. *IEEE Trustcom BigDataSE -ISPA*, (33–40).

34. Koliopoulos, A. K., Yiapanis, P., Tekiner, F., Nenadic, G. & Keane, J. 2015. A Parallel Distributed Weka Framework for Big Data Mining using Spark. *IEEE International Congress on Big Data*, (9–16).
35. Tayeb, S., Pirouz, M., Sun, J., Hall, K. & Chang, A. (Eds.). 2017. Toward Predicting Medical Conditions Using k-Nearest Neighbors. *IEEE International Conference on Big Data (BIGDATA)*, (3897–3903).
36. Lakshmi P.R.V., Shwetha, G. & Raja, N. S. 2017. Preliminary Big Data Analytics of Hepatitis Disease by Random Forest and SVM Using R- Tool. *3rd International Conference on Biosignals, images and instrumentation (ICBSII)*, (1–24).
37. Chakraborty, S., Chatterjee, S., Ashour, A. S., Mali, K. & Dey, N. 2017. Intelligent Computing in Medical Imaging: A Study. *Advancements in Applied Metaheuristic Computing*, 143.
38. Jin, S., Peng, J. & Xie, D. 2016. Towards MapReduce Approach with Dynamic Fuzzy Inference/Interpolation for Big Data Classification Problems. *16th International Conference on Cognitive Informatics and Cognitive Computing*, (407–413).
39. Mane, T. U. 2017. Smart heart disease prediction system using Improved K-Means and ID3 on Big Data. *International Conference on Data Management, Analytics and Innovation (ICDMAI)*, (239–245).
40. Tao, X. & Ji, C. 2014. Clustering massive small data for IoT. *2nd International Conference on Systems and Informatics(ICSAI)*, Shanghai, (974–978).
41. Ang, L.M., Seng, K.P. & Heng, T.Z. 2016. Information Communication Assistive Technologies for Visually Impaired People. *International Journal of Ambient Computing and Intelligence (IJACI)*, 7(1), (45–68).
42. Kamal, S., Dey, N., Ashour, A.S., Ripo, S., Balas, V.E. & Kaysar, M.S. 2017. FbMapping: an automated system for monitoring facebook data. *Neural Network World* 1, (27–57).
43. Nandi, D., Ashour, A. S., Samanta, S., Chakraborty, S., Salem, M.A.M. & Dey, N. 2015. Principal Component Analysis in Medical Image Processing: A Study. *Inernational Journal of Image Mining*, 1(1), (65–86).
44. Kamal, S., Sarowar, G., Dey, N., Ashour, A.S. & Ripon, S.H. (Eds.). 2017. *Self-Organizing Mapping Based Swarm Intelligence for Secondary and Tertiary Proteins Classification*. Springer-Verlag GmbH, Germany, (1–24).
45. Maitrey, S. & Jha, C.K. 2015. Handling Big Data Efficiently by using MapReduce Technique. *IEEE International Conference on Computational Intelligence and Communication Technology*, (703–708).
46. Li, C., Zhang, Y., Jiao, M. & Yu, G. 2014. Mux-Kmeans: Multiplex Kmeans for Clustering Large-scale Data Set. *ACM*, (25–32).
47. Liancheng, X. & Jiao, X. 2014. Research on distributed data stream mining in Internet of Things. *International Conference on Logistics Engineering, Management and Computer Science (LEMCS 2014)*, Atlantis Press, (149–154).
48. Wang, H.Z., Lin, G. W. & Wang, J. Q. (Eds.). 2014. Management of Big Data in the Internet of Things in Agriculture Based on Cloud Computing. *Applied Mechanics and Materials (AMM)*, 548(549), (1438–1444).
49. Gole, S. & Tidke, B. 2015. Frequent itemset mining for big data in social media using ClustBigFIM algorithm. *IEEE International Conference on Pervasive Computing (ICPC)*, Pune, (1–6).
50. Ghit, B., Iosup, A. & Epema, D. 2013. Towards an Optimized Big Data Processing System. *13th IEEE/ACM International Symposium on Cluster, Cloud, and Grid Computing*, (83–86).
51. Martha, V. S., Zhao, W. & Xu, X. 2013. h-MapReduce: a framework for workload balancing in MapReduce. *27th IEEE International Conference on Advanced Information Networking and Applications*, (637–644).

52. Matallah, H., Belalem, G. & Bouamrane, K. 2017. Towards a New Model of Storage and Access to Data in Big Data and Cloud Computing. *International Journal of Ambient Computing and Intelligence*, 8(4), (1–14).
53. Ess.tu-darmstadt.de: ADL Recognition Based on the Combination of RFID and Accelerometer Sensing | Embedded Sensing Systems–www.ess.tu-darmstadt.de. http://www.ess.tudarmstadt.de/data sets/PHealth08-ADL. Last Accessed on [17/8/2016].

8 Parallel Data Mining Techniques for Breast Cancer Prediction

R. Bhavani and G. Sudha Sadasivam

CONTENTS

8.1 INTRODUCTION

The past two decades have witnessed rapid advances in the area of genomics and proteomics. This has resulted in an explosive growth of biological data. It is not easy to manually organize and extract useful information from such large amounts of data (Needleman and Wunsch 1970, Waterman et al. 1976). The necessity of automated biological data analysis methods has led to the emergence of a new field called bioinformatics. Bioinformatics is an interdisciplinary field applying computational and statistical techniques to extract useful information from biological data. The objectives of bioinformatics are threefold. The first objective is to organize data in a way that is easy to access and update. The second objective is to develop algorithms and tools that aid in analysing the biological data. The third objective is to use the developed algorithms and tools for analysing and interpreting the data and the results in a biologically meaningful manner (Baxevanis and Ouellette 2004). The application and development of data mining techniques is to extract meaningful information from biological data, which helps in drug discovery, personal medicine, crop improvement, insect resistance and so on (Acharjya and Anitha 2017, Ang et al. 2017, Benson et al. 2013, Nandi et al. 2015, Karaa 2015, Hore et al. 2015, Dey and Ashour 2018).

As the size of the biological data varies from terabytes to petabytes, general data mining algorithms will not generate results in a stipulated amount of time. Sequential algorithms on a single core further slow down the processing. Hence, there is a need for processing data in parallel. Parallel data mining reduces the sequential bottleneck in the current mining algorithms, as it scales massive data sets and improves the response time (Belarbi et al. 2017, Marucci et al. 2014, Matallah et al. 2017, Moraru et al. 2017a,b).

Parallelism can be achieved in many ways like multi-core computing, graphics processing unit computing (GPU), OpenMP, message passing interface (MPI), MapReduce programming and so on. Kraus and Kestler (2010), Marcais and Kingsford (2011) and Deb and Srirama (2013) used multi-core processing to perform instruction level parallelism. They have good processing speed due to the multiple cores operating simultaneously on instructions. The performance of a multi-core processor is decided based on the level of parallelism. Mrozek et al. (2014) and Brito et al. (2017) proposed a GPU-based implementation of similarity search in 3D protein structure and human activity detection, respectively. They found that the execution time is reduced due to parallelization of the search processes on many-core GPU devices. However, it cannot process exhaustive data sets because of its small memory size.

Khan et al. (2015) used OpenMP based parallel algorithms. The main advantage of OpenMP is that it uses a multi-threading concept. But it can run only in shared memory computers. Concurrent access to memory may degrade performance. So, it fails when data sets are exhaustive. Kalegari and Lopes (2013), Wu et al. (2013) and Hulianytskyi and Rudyk (2014) used MPI to achieve parallelism for their tasks. Each processor can rapidly access its own memory without interference. But, a programmer is responsible to explicitly define how the data is communicated to synchronize between tasks. Splitting the exhaustive data manually by the users is a complex task.

Karim et al. (2012), Wei-Chun et al. (2013), Kumar et al. (2016) and Dey et al. (2017a,b) used MapReduce programming which uses data parallelism to perform their task. MapReduce programming exhibits good scalability when the input size is exhaustive. So, it is best suited for processing biological data sets. Sudha and Baktavatchalam (2009) performed multiple sequence alignment of DNA sequences using a MapReduce programming model of Hadoop framework. Permutations of DNA sequences are first generated and are stored in HDFS, which are then processed using the Hadoop data grid. It was observed that there is reduction in the execution time when the block size of the data increases. Zou et al. (2013) performed a survey on MapReduce frame operations in bioinformatics. They explored how MapReduce is used for next-generation sequencing, sequence alignment, gene expression analysis and SNP analysis, and found that it is scalable, efficient and fault-tolerant. Hung and Lin (2013) proposed a parallel protein structure alignment service over Hadoop. The protein structure alignment algorithm is performed using the MapReduce programming model. They found that the performance of the proposed alignment model is proportional to the number of processors. Kamal et al. (2017a,b,c,d) proposed MapReduce-based classification of metagenomic data. In Map Phase, features were extracted and in the reduce phase, classification of DNA sequences were carried out. Improvements in time and space were observed. MapReduce programming uses data parallelism to perform their task. MapReduce programming exhibits good scalability when the input size is exhaustive. So, it is best suited for processing biological data sets.

In order to improve the performance of traditional clustering and classification approaches, they are hybridized with evolutionary algorithms. Chakraborty et al. (2017), Dey et al. (2018a,b), Xiao et al. (2003) proposed a hybrid clustering approach based on Self Organizing Maps (SOM) and PSO to cluster genes with related expression levels in gene expression data. PSO is used to evolve the weights for SOM and hence performed better than traditional SOM. Deng et al. (2005) and Kamal et al. (2017d) proposed a hybrid clustering approach based on a K-means clustering algorithm and PSO. Initially nuclei clusters are formed by the combination of two particles. Each data vector is taken as the particle for PSO. Next level clusters are formed when one cluster or particle combines with another cluster. Lastly, a one step K-means operation is carried out to reduce the number of clusters which fail to merge with other clusters to form mega clusters. PSO-K-means clustering performed better than traditional K-means clustering.

Di Gesu et al. (2005) used a genetic algorithm for clustering gene expression data. Each centroid has its own variance. The task is to form clusters such that the total internal variance is minimized. He and Hui (2009) explored gene expression data clustering and associative classification using ant-based algorithms. Minimum spanning tree is formed using ant colony optimization (ACO). Next, the edges which have values greater than the threshold are removed to form clusters. Yin et al. (2004) employed a genetic algorithm to find out the best regulatory network in a very large solution space of Bayesian networks. Du and Lin (2005), Tan and Pan (2005) identified the group of genes that has similar expression patterns in gene expression data using hierarchical clustering. Ooi and Tan (2003) applied a genetic algorithm and maximum likelihood classification methods to predict the classes in gene expression data. A genetic algorithm is used to select the genes used to classify the samples, and the classification is performed using a maximum likelihood classifier. Kamal et al. (2017a,b,c,d) proposed PSO-centric with SOM approach for protein classification. Wang et al. (2017) used fuzzy systems for evaluating time-series data. Fowdur et al.

(2018) presented an overview of the various open-source tools used for analysing and learning from big data. They include decision trees, random forest, and multinomial logistic regression classification algorithms. Elhayatmy et al. (2018) developed IoT based on wireless techniques for healthcare applications.

This book addresses recent advances in mining, learning, and analysis of big volumes of medical data with a variety of data formats, data types, complex and large set of features having large and complex dimensions. Biological data are available in various forms such as sequence data, gene expression data, microarray images and gene annotations. This chapter considers processing of the gene expression data in a parallel manner for breast cancer prediction. The parallel implementation of clustering and classification algorithms, using the MapReduce programming model, is explained in the remainder of this chapter.

8.2 GENE EXPRESSION DATA

A gene expression data set is represented by a real-valued matrix M where the rows form the expression patterns of genes, the columns form the expression profiles of samples, and each cell w_{ij} is the measured expression level of gene i in experiment sample j. Therefore, M is defined as shown in Figure 8.1. Analysing the gene expression data for useful information is a challenge in the field of bioinformatics.

Gene expression data is used to

- Identify differentially expressed genes, which aids in detailed molecular studies or drug treatments;
- Cluster genes that finds groups of genes having similar expression patterns, which helps in discovering new cancer subtypes; and
- Build classifiers, with machine learning and deep learning techniques, which could be used in the diagnosis of cancer patients.

This chapter discusses the clustering and classification of gene expression data to predict breast cancer using parallel K-means clustering, parallel K-means clustering hybridized with DE and ACO and parallel K-NN classification hybridized with PSO.

8.3 DATA SET DESCRIPTION

The data set used in the experiments are the breast cancer data set (GSE26304) with four clusters. The data set was downloaded from the NCBI repository. Breast cancer samples consist of 31 pure ductal carcinoma *in situ* patients, 36 invasive ductal carcinomas, 42 mixed and 6 normal (Muggerud et al. 2010).

$$M = \begin{bmatrix} w_{11} & w_{12} & \cdots & w_{1m} \\ w_{21} & w_{22} & \cdots & w_{2m} \\ \vdots & \vdots & \ddots & \vdots \\ w_{n1} & w_{n2} & \cdots & w_{nm} \end{bmatrix}$$

FIGURE 8.1 Gene expression data representation.

8.4 EXPERIMENTAL SETUP

The parallel clustering and classification algorithms were experimented in a Hadoop cluster of 8 machines with each machine having the configuration of 1 GB RAM, Intel core 2 duo processor and 75 GB hard disk drive. Block size in HDFS is set to 64MB.

8.5 PARALLEL DATA MINING: AN INTRODUCTION

As the data sizes increase, from terabytes to petabytes, sequential data mining algorithms may not produce results in a stipulated amount of time (Gebali 2011). The main memory of a single processor alone cannot hold all the data. Sequential algorithms on a single core further slow down the processing. Hence, there is a need for processing data in parallel. Parallel data mining helps current mining methods by overcoming the sequential bottleneck and provides the ability to scale large data sets by improving the response time. The three main parallel programming models are shared memory, message passing and data parallelism.

8.5.1 Shared Memory

Shared memory multi-processing programming is a programming model where the processors share the common address space. All the processors access the shared memory space through an interconnection network. OpenMP is an application programming interface (API) to produce multi-threaded code for shared memory machines.

8.5.2 Message Passing

In this programming model, each processor has its own memory area. Using a message passing mechanism a processor accesses the memory area associated with other processors. Message passing interface (MPI) is a communication protocol for programming in parallel computers. It supports both point-to-point and collective communication.

8.5.3 Data Parallelism

In this programming model, data is distributed across many parallel-computing nodes. Each processor performs the same task on different pieces of distributed data. Hadoop MapReduce is a software framework for processing large amounts of data in parallel. The following section briefly explains the Hadoop framework since this thesis aims at data parallelism.

8.5.3.1 Hadoop

Hadoop is an open-source software framework for distributed storage and processing of very large data sets on clusters of computers (White 2012). The storage part of the Hadoop framework is known as Hadoop distributed file system (HDFS), and the processing part is called MapReduce. Hadoop chunks files into large blocks and

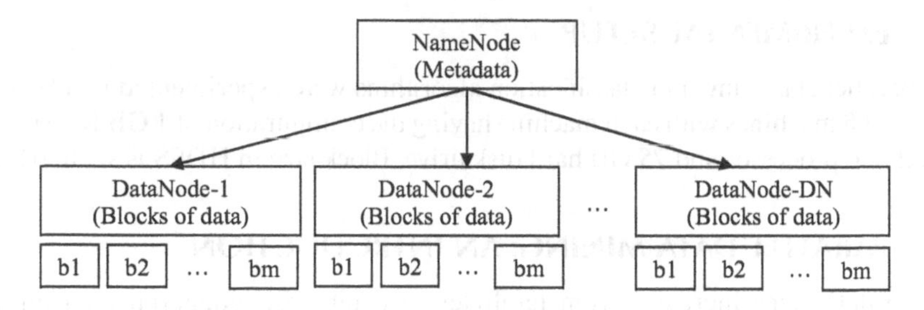

FIGURE 8.2 HDFS architecture.

distributes the blocks across nodes in a cluster. The main advantage of Hadoop is data locality, where data is stored in the nodes that manipulate the data they have access to.

8.5.3.2 HDFS

The HDFS is based on the Google file system (GFS) and provides a distributed file system designed to run on a large cluster of computers in a reliable, fault-tolerant manner. HDFS uses a master–slave architecture where the master is a single name node that manages the metadata of the file system and slaves are data nodes storing the actual data, as depicted in Figure 8.2.

A file in an HDFS is chunked into many blocks which are stored in a set of data nodes. The name node keeps a mapping of the blocks to the data nodes. The data nodes operate on the block with read and write operation in the HDFS. They are also responsible for creating blocks, deleting blocks and keeping replicas. They work based on the instructions given by the name node.

8.5.3.3 MapReduce

MapReduce is a programming model for distributed computing based on java (White 2012).

With MapReduce it is easy to process massive data in parallel on a large cluster of computers in a reliable, fault-tolerant manner. Typically the input and the output of the MapReduce job are stored in the HDFS. The mapper processes the independent chunks of input file in a completely parallel manner. The output of the mapper is first sorted and is given as input to the reducer. The framework is responsible for scheduling tasks, monitoring them and re-executing the failed tasks.

The name node consists of the job tracker and each data nodes consist of the task tracker. The job tracker takes care of scheduling the tasks on the slaves, monitoring them and re-executing the failed tasks. The task tracker executes the tasks as directed by the job tracker. The MapReduce programming model operates on key-value pairs. It views the input to the job as a set of key-value pairs and produces another set of key-value pairs as the output of the job, as depicted in Figure 8.3.

8.6 CLUSTERING OF BREAST CANCER GENE EXPRESSION DATA

Clustering is an important task in data mining, where knowledge is obtained from unlabelled data. It is an unsupervised learning technique which groups a set of data objects into classes of similar data objects. The objects are clustered or grouped

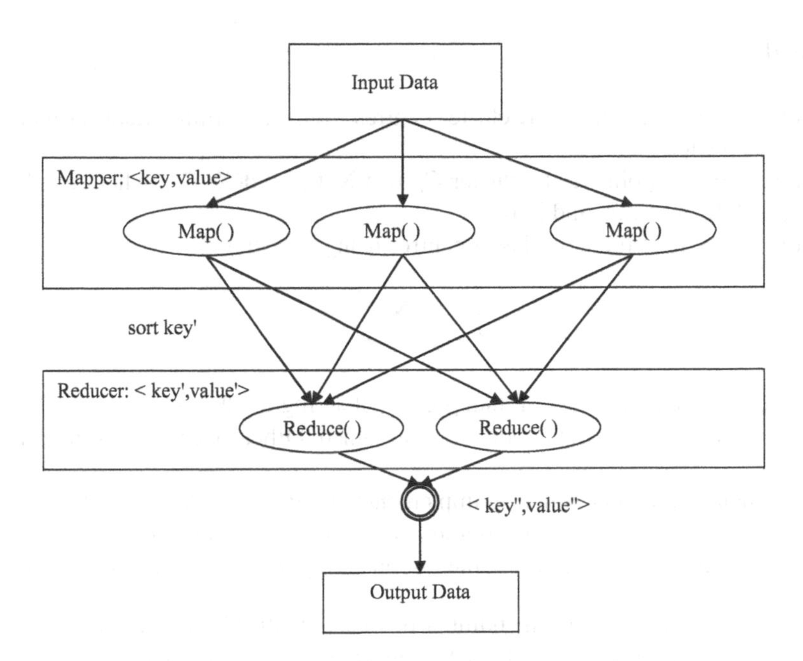

FIGURE 8.3 Structure of MapReduce programming model.

such that the similarity with other classes is maximized and the similarity with-in the class is minimized.

Clustering algorithms are broadly classified as partition-based clustering and hierarchical clustering approaches (Han et al. 2011). Partition-based clustering methods partition the given data set into a number of partitions such that each partition contains at least one object, and each object must belong to exactly one partition. The hierarchical method works by constructing a tree for grouping data objects. It is further classified as either agglomerative or divisive. Agglomerative hierarchical clustering is a bottom-up approach which first places each object in its own cluster, and at each stage it merges these small clusters into big clusters. Divisive hierarchical clustering is a top-down approach that starts with one cluster having all objects in it, and at each stage it is subdivided into smaller clusters.

The most popular partition-based clustering method is the K-means clustering algorithm (Hartigan and Wong 1979, Dembele and Kastner 2003, Othman et al. 2004). Given K an integer, K-means clustering algorithm clusters the data set into K non-overlapping clusters. The first K data points are randomly chosen as initial centroids of each cluster. After each iteration, the data objects are reassigned to the closest centroid. Finally, a cluster contains all data points that are closest to a given centroid. They are best suited for real-valued features. So, in this chapter, K-means clustering is used to cluster the gene expression data which have real-valued features.

The steps followed in the K-means clustering algorithm are as follows:

Input: Training data set (X_i) // $X_i \in R^n$
K //number of centroids
Output: (X_i, y_j) // $X_i \in R^n$; $y_j \in \{c_1, c_2, \ldots c_K\}$

Algorithm

Step 1: Randomly choose K cluster centres from the training data set with N data points.
Step 2: Assign point X_i, to cluster C_j, if $d(X_i\text{-}C_j) < d(X_i\text{-}C_p)$, where $i = 1..N$, $j = 1..K$, $p = 1..K$ and $j \neq p$.
Step 3: Calculate the new cluster centres using Equation 8.1

$$c_j^k = \frac{1}{n_j}\sum_{i=1}^{nj} X_i \tag{8.1}$$

where n_j is the number of data points belonging to cluster c_j.
Step 4: Stop when no data point was reassigned. Otherwise repeat from step 2.

Even though, it is very simple to implement with time complexity is $O(N \cdot K \cdot I_{max})$ where N is the number of data points, K is the number of centroids. I_{max} is the maximum number of iterations, K-means clustering has two main disadvantages:

- The user must specify the number of cluster centroids (K) in advance.
- The clustering result depends on the initial random assignment of cluster centres.

Even though the time complexity of K-means clustering is linear, if the size of the file is gigabytes or terabytes, the parallel implementation reduces the running time of the algorithm.

8.6.1 Parallel K-Means Clustering

Parallel K-means clustering using the MapReduce programming model computes the distance between the centroids and the data set in the map phase followed by the assignment of the data set to the clusters. The reduce phase evaluates the new cluster centres. The flowchart of the proposed parallel K-means clustering algorithm using MapReduce is shown in Figure 8.4.

The steps followed in Map phase are as follows:

Map<key,value>
key: record number
value: gene vector from gene expression data

- Find the Euclidean distance between cluster centroids and the gene vector.
- For each gene vector find the closest cluster centroid.

Output<key',value'>.
key': cluster centre number
value': gene vector

The input gene expression matrix is stored in HDFS as key–value pairs. The key is the record number of the feature vector and the value is the feature vector in the

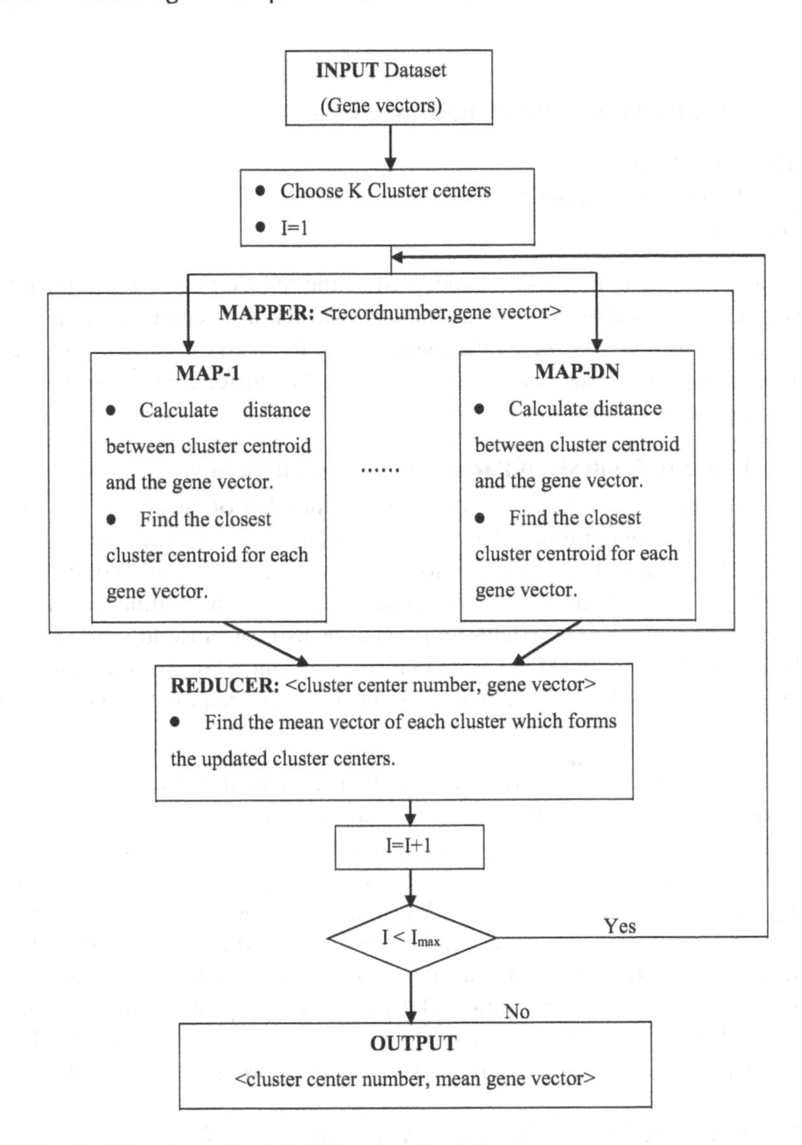

FIGURE 8.4 Flowchart of parallel K-means clustering.

data set. The input is chunked and passed to the mappers. Each mapper processes the value received by calculating the distance between the cluster centres and the feature vector. Next, the feature vector is assigned to the cluster centre with which its distance is minimal. Finally, the output key' of the mapper is the cluster centre number and the value' is feature vector.

The steps followed in the reduce phase are as follows:

Reduce<key',value'>
key': cluster centre number
value': gene vector

For each key'

- Calculate the mean value of the feature vector.

Output<key",value">
key": cluster centre number
value": mean gene vector

The output of the map phase is processed by the reduce phase. The reduce phase receives the cluster centre number as the key' and feature vector as its value' from the map phase. It then calculates the mean vector for each cluster from the feature vectors. Finally, the output key" of the reducer is the cluster centre number and the value" is the mean feature vector.

8.6.1.1 Runtime Analysis of Parallel K-Means Clustering

Let N be the number of feature vectors, K be the number of cluster centres and I_{max} denotes the number of iterations. The time complexity of a simple K-means algorithm is $O(N \cdot K \cdot I_{max})$, where the major operation involved is the calculation of the distance between the K cluster centres and N data points. Let D be the number of data points in a block of data chunked by HDFS for processing using MapReduce programming. The default block size of 64MB is used in all experiments. The minimum number of map functions required is N/D. If m is the number of maps required, the total running time of the input file is $\sum_{i=1}^{m} \frac{K \cdot N}{D}$.

The reducer takes constant time C to output the calculated mean vector. If CT is the communication time between the mapper and the reducer, then the total time taken by the proposed MapReduce-based K-means clustering algorithm is $I_{max} \cdot ((K \cdot N/D)+CT+C)$. Hence, the time complexity of the proposed algorithm is $O(I_{max} \cdot K \cdot N/D)$ since $CT << K \cdot N/D$ and $C << K \cdot N/D$. To overcome the disadvantages of K-means clustering of specifying the K-value, evolutionary optimization algorithm-based clustering is used in the literature. This chapter discusses how differential evolution (DE) is used to optimize the K-value and ACO is used to improve the rate of convergence of the optimum solution. Parallel implementation was carried out using the MapReduce programming model.

8.6.2 Parallel K-Means Clustering with DE and ACO

K-means clustering algorithm has linear time complexity. But, K-means clustering algorithm converges to the nearest local optimum due to the random initialization of cluster centroid in its search process. To overcome the local optimum problem, evolutionary algorithms are used. The proposed approach uses evolutionary algorithm such as DE and ACO with K-means clustering with MapReduce programming for parallelization. DE is a population-based global optimization algorithm (Storn and Price 1997). DE is used to optimize the K-value while K-means is used for grouping. ACO was developed based on the ants finding the shortest path from nest to food with the aid of pheromones (Dorigo et al. 2006). Here, ACO is used to improve the rate of convergence of the clusters. The flowchart of the proposed parallel K-means clustering with DE and ACO using MapReduce is shown in Figure 8.5.

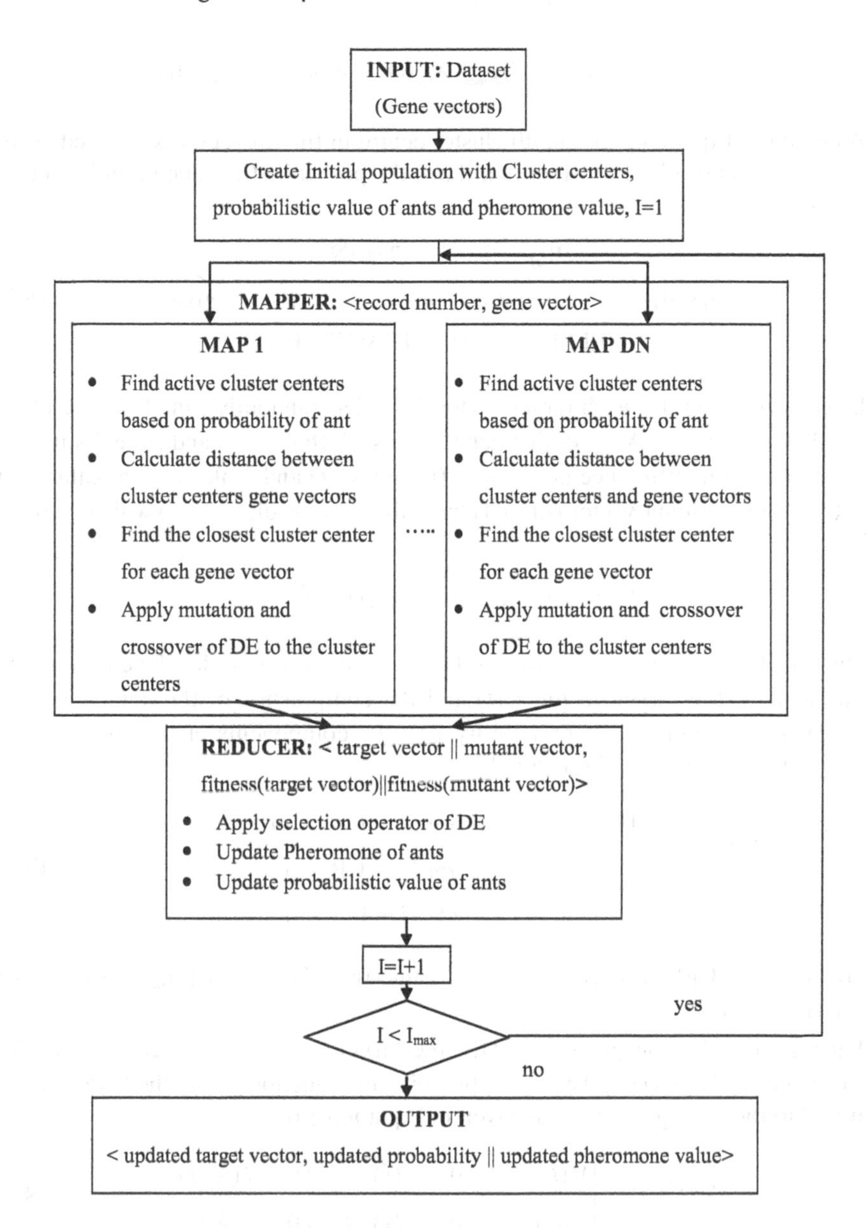

FIGURE 8.5 Flowchart of parallel K-Means clustering with DE and ACO.

The vectors for DE are represented as real numbers of dimension $K_{max} + K_{max} \times d$, where K_{max} is the maximum is number of cluster centres and d is the dimension of the data point (Sen et al. 2007, Das et al. 2008). Activation thresholds $T_{i,j}$ in range [0,1] controls the active cluster. ACO is applied, such that the activation threshold of the cluster centre is considered as an ant. Each ant has its own pheromone value $\tau_{i,j}$ and a probabilistic value $p_{i,j}$ based on which cluster centres are activated. The target vector is represented as in Equation 8.2.

$$\vec{X}_i(t) = [T_{i,1}, T_{i,2}, \ldots, T_{i,K_{\max}}, \vec{X}_{i,1}(t), \vec{X}_{i,2}(t), \ldots, \vec{X}_{i,K_{\max}}(t)] \tag{8.2}$$

As given in Equation 8.3, the jth cluster centre in the ith vector is selected as an active cluster centroid for partitioning, if $p_{i,j}$, the probabilistic value of an is greater than 0.5.

$$\begin{array}{c} \text{IF } p_{i,j} > 0.5 \quad \text{THEN} \\ \text{the } j\text{th} \quad \text{cluster centre} \quad \text{in} \quad \vec{X}_{i,j} \quad \text{is} \quad \text{Active} \\ \text{ELSE} \quad \text{it is} \quad \text{INACTIVE} \end{array} \tag{8.3}$$

During mutation, for each target vector $\vec{X}_i(t)$, DE randomly samples three other vector $\vec{X}_x(t)$, $\vec{X}_y(t)$ and $\vec{X}_z(t)$ in that generation such that i, x, y and z are distinct. It then calculates the difference between $\vec{X}_x(t)$ and $\vec{X}_y(t)$ and scales it by a scalar F in [0,2] to create a mutant vector $\vec{U}_i(t+1)$ by adding the result to $\vec{X}_z(t)$ which is given in Equation 8.4.

$$\vec{U}_i(t+1) = \vec{X}_z(t) + F(\vec{X}_x(t) + \vec{X}_y(t)) \tag{8.4}$$

During the crossover operation of DE, a trial vector is developed from the components of the target vector $\vec{X}_i(t)$ and the components of the mutant vector, $\vec{U}_i(t+1)$. With the crossover probability CR, the components of the donor vector enter the trial vector using Equation 8.5.

$$\vec{V}_{i,j}(t+1) = \begin{cases} \vec{U}_{i,j}(t+1) & \text{if } rand_{i,j} \leq \text{CR} \quad \text{or } j = I_{rand} \\ \vec{X}_{i,j}(t) & \text{if } rand_{i,j} > \text{CR} \quad \text{or } j \neq I_{rand} \end{cases}$$
$$i = 1,2,\ldots,P \quad \text{and} \quad j = 1,2,\ldots,d \tag{8.5}$$

where $rand_{i,j} \sim$ U[0, 1], I_{rand} is a random value from [1, 2, …,d]. I_{rand} ensures that $\vec{U}_i(t+1) \neq \vec{X}_i(t)$.

During the selection process of DE, the target vector $\vec{X}_i(t)$ is compared with the trial vector $\vec{V}_i(t+1)$ and based on the objective function value, the better one is admitted to the next generation, as given in Equation 8.6.

$$\vec{X}_i(t+1) = \begin{cases} \vec{V}_i(t+1) & \text{if} \quad f(\vec{V}_i(t+1)) > f(\vec{X}_i(t)) \\ \vec{X}_i(t) & \text{if} \quad f(\vec{V}_i(t+1)) \leq f(\vec{X}_i(t)) \end{cases} \tag{8.6}$$

where $f(.)$ is the objective function to be maximized.

The objective function used during the selection process of DE is based on the overall cluster quality measure of the partitioning. The objective function is defined in Equation 8.7

$$f(\vec{X}_i(t)) = \frac{1}{\min_{j=1,2,\ldots,K\max} d(\vec{X}_{i,j}(t), \vec{X}_p(t)) + sc} \tag{8.7}$$

where $d(\vec{X}_{i,j}(t), \vec{X}_p(t))$ is the distance between the cluster centre $\vec{X}_{i,j}(t)$ and data point $\vec{X}_p$ and sc is a very small valued constant (0.0002). Therefore, maximizing this function actually results in the minimization of the distance between the data point and the cluster centre.

The choice of a solution component is done probabilistically with respect to the pheromone model given in Equation 8.8.

$$p_{i,j} = \frac{[\tau_{ij}] \cdot [\eta(c_{ij})]^{\beta}}{\sum_{c_{lk} \in N(S^P)} [\tau_{lk}]^{\alpha} \cdot [\eta(c_{lk})]^{\beta}} \tag{8.8}$$

where η is an optional weighting function that assigns at each construction step a heuristic value $\eta(c_{i,j})$ to each feasible solution component $c_{i,j} \in N(S^P)$ depending on the current construction step. The value of parameters α and β, $\alpha > 0$ and $\beta > 0$, determines the relative importance of the pheromone value and heuristic information.

The pheromone value is updated using Equation 8.9.

$$\tau_{i+1,j} = \begin{cases} \rho\tau_{i,j} + \dfrac{Q}{d_{\max}(\vec{X}_{i,j})} & \text{if } \vec{X}_{i,j} \quad \text{is mutated} \\ \rho\tau_{i,j} & \text{otherwise} \end{cases} \tag{8.9}$$

where $\tau_{i,j}(t)$ is the pheromone value for each cluster centre j in the tth generation, $\rho \in [0,1]$ is the evaporation rate and Q is a constant in [0,1].

During each iteration, the activation threshold is updated using Equation 8.10.

$$T_{i,j} = \frac{\tau_{i,j}}{\sum_{m=1}^{K_{\max}} \tau_{i,m}} \tag{8.10}$$

where $\tau_{i,j}(t)$ is the pheromone value of the ant in cluster centre j at ith iteration.

The steps followed in Map phase are as follows:

Map<key,value>
key: record number
value: gene vector

For each vector in DE population

- Find the active cluster centres using Equation 8.3.
- Calculate the distance between active cluster centres and gene vectors.
- Find the closest cluster centre for each gene vector.
- Apply mutation and crossover operators of DE using Equation 8.4 and Equation 8.5 respectively.
- Calculate the fitness of the target vector and mutant vector Equation 8.7.

Output<key',value'>
key': target vector || mutant vector
value': fitness(target vector)||fitness(mutant vector)

The input gene vectors are stored in HDFS as key–value pairs. The key is the record number of the gene vector and the value is the feature vector in the data set. The input is chunked and passed to the mappers. Each mapper takes the active cluster centres and calculates the distance between the active cluster centre and the feature vectors. Next, the feature vector is assigned to the cluster centre with which its distance is minimal. The first two steps of DE, such as mutation and crossover, are carried out in map phase. Finally, the output key' of the mapper is the target vector concatenated with the mutant vector and the value' is the fitness of target vector concatenated with the fitness of mutant vector.

The Reduce phase of parallel K-means clustering with DE and ACO is depicted as follows:

Reduce<key',value'>
key': target vector || mutant vector
value': fitness(target vector)||fitness(mutant vector)

For each key

- Perform selection operation of DE using Equation 8.6
- Update the probability value of the ant using Equation 8.8
- Update the pheromone value of the ant using Equation 8.9 and the activation threshold using Equation 8.10

key'': updated target vector
value'': updated probability || updated pheromone value

The key' and its value' of the mapper is passed to the reduce phase. The selection operator of DE is carried out and the best target vectors are obtained. Next, the probability of the ants and the pheromone value of ants are calculated. Finally, the output key'' of the reducer is the updated target vector and the value'' is the updated probability of ants concatenated with the updated pheromone value of ants.

8.6.2.1 Runtime Analysis of Parallel K-Means Clustering with DE and ACO

Let N be the number of gene vectors, K_{max} be the number of cluster centres in each vector of population size P and I_{max} denotes the number of iterations. The main operation involved is the calculation of distance between the K_{max} cluster centres and N feature vectors for a population size P. Let D be the number of data points in a HDFS block. The minimum number of map functions required is N/D. Considering m to be the number of maps required, the total running time of the input file is $\sum_{i=1}^{m} \frac{N \cdot K_{max}}{D}$. Each mapper calculates the distance between the target vector and

TABLE 8.1 Performance Evaluation of K-Means and K-Means with DE and ACO for Breast Cancer Gene Expression Data Set with Four Clusters

	K-means	K-means With DE & ACO
Cluster compactness	0.5631	0.1542
Cluster separation	0.3111	0.1312
Overall cluster quality	0.4371	0.1427

the feature vectors to perform K-means step of the algorithm, and also calculates the distance between the mutant vector and the gene vector to perform the differential operators such as mutation and crossover. Hence, the time taken by each mapper is $((P \cdot N \cdot K_{max}/D) + (P \cdot N \cdot K_{max}/D))$ which is $2 \cdot P \cdot N \cdot K_{max}/D$. The reducer performs the selection operation of DE, which takes constant time C to obtain the updated target vector. Next, it updates the probability of ants and pheromone values of ants. Hence, the time taken by each reducer is $2 \cdot K_{max}$. If CT is the communication time between the mapper and the reducer, then the total time taken by the proposed MapReduce based K-means clustering algorithm with DE and ACO is $I_{max} \cdot ((2 \cdot P \cdot N \cdot K_{max}/D) + CT + 2 \cdot K_{max})$. Thus, the time complexity of the proposed algorithm is $O(I_{max} \cdot P \cdot N \cdot K_{max}/D)$ since $CT << (2 \cdot P \cdot N \cdot K_{max}/D)$ and $2 \cdot K_{max} << (2 \cdot P \cdot N \cdot K_{max}/D)$.

8.6.3 Performance Evaluation of Clustering Algorithms

Performance evaluation measures like cluster compactness, cluster separation and overall cluster quality (He et al. 2004) were evaluated for the breast cancer data set. The results are tabulated in Table 8.1.

8.6.3.1 Cluster Compactness

The cluster compactness measure gives the global intra-cluster homogeneity of the clustering output and is based on the variance of a data set and is given in Equation 8.11.

$$Vax(X) = \sqrt{\frac{1}{N} \sum_{i=1}^{N} d^2(x, \bar{x})} \tag{8.11}$$

where $d(x,y)$ is the distance metric between two vectors x and y, N is the number of members in X and $\bar{x} = \frac{1}{N} \sum_i x_i$ is the mean of X. The Euclidean distance between two vectors of dimension d is given in Equation 8.12.

$$d(x, y) = \sqrt{\sum_{i=1}^{d} (y_i - x_i)^2} \tag{8.12}$$

The cluster compactness for the output clusters $c_1, \ldots, c_C$ is calculated using Equation 8.13.

$$Cmp = \frac{1}{C}\sum_i \frac{Var(c_i)}{Var(X)} \tag{8.13}$$

where C is the number of clusters generated on the data set X, $Var(c_i)$ is the variance of the cluster c_i, and $Var(X)$ is the variance of the data set X.

8.6.3.2 Cluster Separation

The cluster separation measure gives the global inter-cluster separation of the clustering output and is defined in Equation 8.14.

$$Sep = \frac{1}{C(C-1)}\sum_{i=1}^{C}\sum_{j=1,j\neq i}^{C} \exp\left(-\frac{d^2(x_{c_i}, x_{c_j})}{2\sigma^2}\right) \tag{8.14}$$

where σ is a Gaussian constant, C is the number of clusters, x_{c_i} is the centroid of the cluster c_i, and $d(x_{c_i}, x_{c_j})$ is the Euclidean distance between the centroid c_i and the centroid of c_j.

8.6.3.3 Overall Cluster Quality

The overall performance of a clustering system is determined by combining cluster compactness and cluster separation measures, and is given in Equation 8.15.

$$Ocq = \gamma.Cmp + (1-\gamma).Sep \tag{8.15}$$

where $\gamma \in [0,1]$ is the weight that balances measures cluster compactness and cluster separation. The smaller value for all three measures indicates a good output quality.

The results show that, on average, the overall cluster quality of the proposed K-means clustering with DE and ACO is improved by 0.2944 compared to the existing K-means clustering. These improvements are due to the mutation differential operator of the DE algorithm. Also, the fast convergence of K-means clustering with DE and ACO is due to improvements in solution quality based on pheromone update by the ants of the ACO.

8.7 CLASSIFICATION OF BREAST CANCER GENE EXPRESSION DATA

Data classification is the supervised process of assigning class labels to data objects whose class label is unknown (Cai et al. 2004, Caragea et al. 2012). Given a set of training data objects along with their class labels, classification aims at determining the class label for the unlabelled testing data object. That is, given a training data set of the form (x_i, y_i) where $x_i \in R^n$ be the ith data object and $y_i \in \{c_1, c_2, \ldots c_C)$ be the class label of the ith data object, a classifier aims at finding the learning model f such that $f(x_i) = y_i$. Classification is done in two phases such as

- *Training phase*: A classification algorithm analyses the training data objects and builds the classification model
- *Testing phase*: The model built in the training phase is used to classify the test data.

Classification approaches can be classified into eager learners and lazy learners. Given a training data set, the eager learners will construct a classification model before receiving the test data to classify. Lazy learners, when given a training data set, simply store it and wait for test data. Only when it receives the test data, does it perform classification of the test data based on its similarity to the stored training data set. Compared to eager learners, lazy learners are computationally expensive (Keller et al. 2000). The K-Nearest-Neighbour classifier is an example of a lazy learner.

The K-Nearest Neighbour (K-NN) classification requires little or no prior knowledge about the distribution of the data (Cunningham and Delany 2007, Halder et al. 2015). It classifies the data using the entire training data set. Given a data object with unknown class label, the K-NN algorithm will search through the training data set for the K-most similar data objects. The class label of the most similar data objects is returned as the class label for the unknown data object. For real-valued data, the Euclidean distance is used as the similarity measure most widely.

The algorithm is depicted as follows:

Input: Training data set (X_i,y_j) // $X_i \in R^n$; $y_j \in \{c_1,c_2,\ldots c_c)$
Unknown data sample UK_i // $UK_i \in R^n$
K //nearest neighbours
Output: y_j for UK_i // $y_j \in \{c_1,c_2,\ldots c_c)$; $UK_i \in R^n$

Algorithm

For i = 1 to T do

- Calculate the Euclidean distance between $(X_i)_k$ and UK_i.
  End for
- Compute set M containing data objects for the K smallest distances $d((X_i)_k, UK_i)$.
- Return the majority label of set M as y_j for UK_i

Given T training data set (X_i,y_j) where $X_i \in R^n$ is data objects with the attributes of n-dimensions; $y_j \in \{c_1,c_2, \ldots c_C)$ be the class labels of the data object, unknown data sample UK_i where $UK_i \in R^n$ and the parameter K, the K-NN first computes the distance between the unknown data sample and the N training data objects. Finally, the unknown data sample is classified by a majority vote of its K-neighbours.

The main advantages of a K-NN classifier are:

- K-NN is a simple classifier, which can be applied to the data from any distribution
- Its performance is good, if the number of samples is large.

The disadvantages of a K-NN classifier are:

- Choosing the parameter 'K' for the algorithm is difficult
- Since there is no training stage, the test stage is computationally expensive
- If there are equal votes from the neighbours, the classifier cannot predict the class label of the unknown sample correctly.

The computational intensive task in a K-NN classifier is the calculation of distance between the training sample and the test sample. So, in the proposed approach, the distance between the training sample and the test sample is calculated in a parallel manner using the MapReduce programming model. During the voting phase of the K-NN classifier, if there are equal votes from the neighbours, then the classifier cannot predict the class label of the unknown sample correctly. So, in the proposed approach, PSO is applied to optimize the distance between the neighbours.

8.7.1 Parallel K-NN Classifier with PSO

The flowchart of the proposed classification algorithm is shown in Figure 8.6. In the proposed MapReduce version of the K-NN classification algorithm improved by PSO, the input feature vectors are considered as the points in the n-dimensional pattern space. The data set is chunked and each mapper processes a chunk of data.

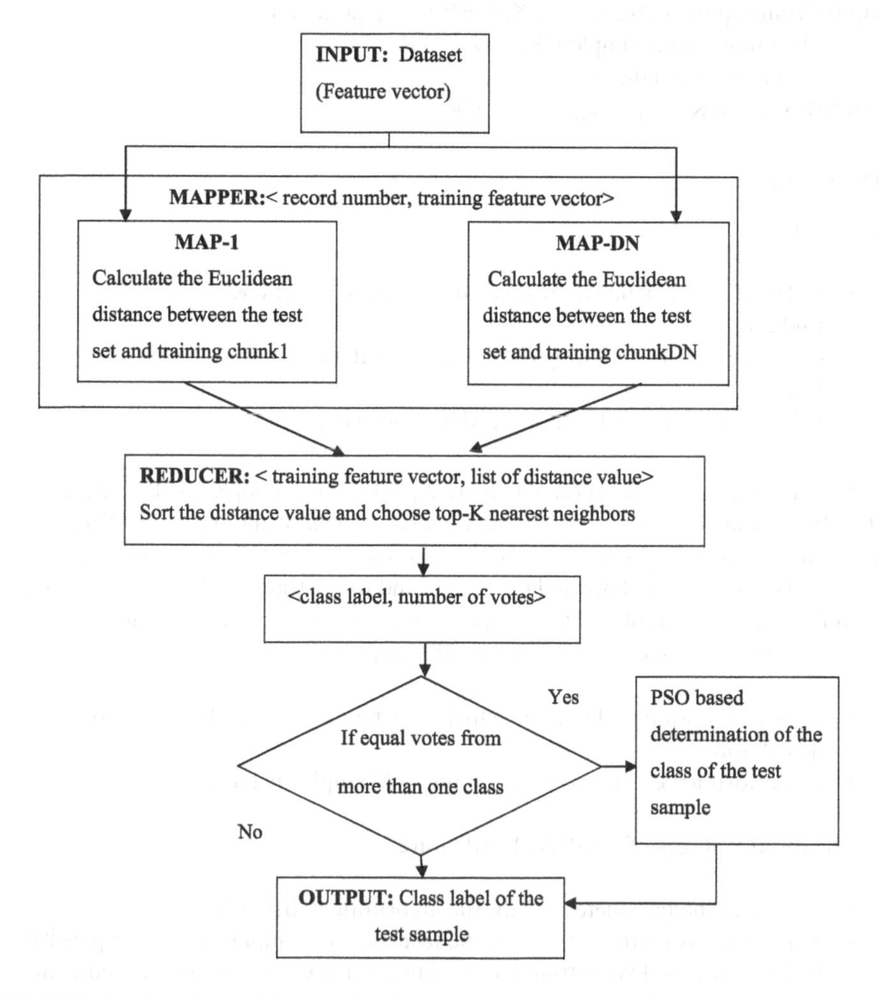

FIGURE 8.6 Flowchart of the parallel K-NN classification with PSO.

The mapper calculates the distance between the test sample and the training sample chunks in parallel. The reducer sorts the samples based on the distance value and chooses the top-2K neighbour samples. The top-2K neighbour samples are treated as particles in the PSO algorithm. Euclidean distance being the fitness function of the particles, the particle with the best value gives an optimum classification result. The test sample is assigned the class label of the optimum particle.

The steps followed in Map phase are as follows:

Map<key,value>
key: record number
value: training gene vector

For each key

- Calculate the Euclidean distance between the training chunk and the test sample.

Output<key',value'>
key': training feature vector
value': distance value

The input training sample file is stored in HDFS as key–value pairs. The key is the record number of the training sample and the value is the gene vector of the training sample. The training sample is chunked and passed to the mappers. Each mapper processes the value received by calculating the distance between the training chunk and the test data. Finally, the output key' of the mapper is the training gene vector and the value' is the distance between the training gene vector and the test gene vector.

The steps followed in the reduce phase are as follows:

Reduce<key',value'>
key': training gene vector
value': distance value

For each key

- Choose top K-neighbours based on the distance value.
- Count the number of votes for each class.

Output<key",value">
key": class label
value": number of votes

The output of the map phase is processed by the reduce phase. The reduce phase receives the training feature vector as the key' and distance value as its value' from the map phase. It then finds the K-nearest neighbours based on the distance value and then calculates the number of votes for each class from the K-neighbours. Finally, the output key" of the mapper is the class label and the value" is the number of votes for each class.

8.7.1.1 PSO for Class Determination

Particle swarm optimization (PSO) is a population-based stochastic optimization technique based on the social behaviour of birds flocking and fish schooling (Poli et al. 2007). The steps followed in PSO-based class determination are as follows:

Input: $m = 2K$ – the number of particles
c_1, c_2 – positive acceleration constants
Output: Class label of optimum solution

Algorithm

Initialize particles $P = \{P_1, P_2, \ldots, P_m\}$ with velocity $\{v_1, v_2, \ldots v_m\}$ and position $\{p_1, p_2, \ldots, p_m\}$
For $I = 1$ to I_{max}
For each particle P_i

- Calculate $f(P_i)$ using Euclidean distance.
- If $f(P_i) > pBest$ then $pBest = f(P_i)$.
- If $pBest > gBest$ then $gBest = pBest$.

For each particle P_i

- Calculate $v_i(t+1)$ using Equation 8.16.

$$v_i(t+1) = v_i(t) + c_1 r_1 (pBest(t) - p_i(t-1)) + c_2 r_2 (gBest(t) - p_i(t-1)) \quad (8.16)$$

where $pi(0) \sim U(xmin, xmax)$,
c_1 and c_2 – positive acceleration constants used to scale the contribution of the cognitive and social components respectively,
r_1 and r_2 – random values in the range [0,1], sampled from a uniform distribution

- Use gBest and v_i of the particle and position of the particle Equation 8.17.

$$p_i(t+1) = p_i(t) + v_i(t+1) \quad (8.17)$$

PSO-based determination of the class label for the test sample is followed if there occurs equal voting for each class. The top 2K-neighbours from the reduce phase are taken as particles for the PSO algorithm. The particles are initialized with its velocity v(0) and position p(0). During each iteration, the distance between the particle and the test sample is calculated as the fitness value. If the calculated fitness value is greater than the pBest, then it is assigned to pBest. Also, if the newly assigned pBest is greater than the gBest value, then it is assigned with the gBest. Next, the position of the particle and its velocity are updated based on its pBest and gBest values. The algorithm iterates I_{max} number of times. Finally, the class label of the particles that are much closer to the test sample are assigned as the class label for the test sample.

8.7.1.2 Run Time Analysis of Parallel K-NN with PSO

Let *t* denotes the number of test samples and *T* denotes the number of training samples. The time complexity of a simple K-NN algorithm is $O(tT)$, where distance calculation takes tT time, sorting operation takes TlogT time and voting takes constant time. To overcome this highest time complexity, MapReduce programming is followed, where each mapper processes one block of data which is 64MB in size. Let D be the number of data points in a block of data chunked by HDFS for processing using MapReduce programming. Then, the minimum number of map functions required is T/D maps. In our proposed algorithm, the distance calculation is done by the mapper and the reducer outputs the distance calculated in a sorted order. Considering m to be the number of maps required, then the total running time of the input file is $\sum_{i=1}^{m} \frac{tT}{D}$. The reducer takes constant time C to output the distance calculated and let CT be the communication time between the mapper and the reducer. Finally, the PSO algorithm takes 2KX time to evaluate the class of the test data. Here, 2K represents the number of particles taken from the K-NN algorithm and X is the number of iterations. Therefore, the total time taken by the proposed algorithm is (tT/D)+CT+C+2KX and the time complexity of the proposed algorithm is O(tT/D).

8.7.2 Performance Evaluation of Classification Algorithms

Performance evaluation measures like precision, recall and accuracy (Sokolova and Lapalme 2009) were evaluated for the breast cancer gene expression data set. Let TP_{Ci} be the set of records correctly classified into category C_i, TN_{C_i} be the set of records correctly classified other than C_i, FP_{C_i} be the set of records wrongly classified as C_i and FN_{Ci} be the set of records wrongly classified. The precision, recall and accuracy are defined as follows:

8.7.2.1 Precision

Precision measures the exactness of a classifier. Precision for a category C_i, measures the percentage of correct assignments among all the records assigned to C_i and is given in Equation 8.18.

$$\text{Precision} = \frac{\sum_{i=1}^{C} \frac{TP_{Ci}}{TP_{Ci} + FP_{Ci}}}{C} \tag{8.18}$$

8.7.2.2 Recall

Recall measures the completeness of a classifier. The Recall for C_i gives the percentage of correct assignments in C_i among all the records that should be assigned to C_i. It is given in Equation 8.19.

$$\text{Recall} = \frac{\sum_{i=1}^{c} \frac{TP_{Ci}}{TP_{Ci} + FN_{Ci}}}{C} \tag{8.19}$$

TABLE 8.2
Performance Evaluation of K-NN and K-NN with PSO for Breast Cancer Data Set with Four Classes

	K-NN	K-NN with PSO
Precision (%)	89.23	93.46
Recall (%)	90.42	96.22
Accuracy (%)	91.23	95.34

8.7.2.3 Accuracy

Accuracy is the percentage of all records that are correctly classified. It is estimated using the ratio between the total number of records correctly classified and the total number of records. It is given in Equation 8.20.

$$\text{Accuracy} = \frac{\sum_{i=1}^{C} \frac{TP_{Ci} + TN_{Ci}}{P_{Ci} + N_{Ci}}}{C} \tag{8.20}$$

From Table 8.2 it is observed that the accuracy of K-NN with PSO is 4.11% higher than K-NN alone. This means that K-NN with PSO classifies the protein sequences more accurately than simple K-NN. These improvements are due to the social and cognitive components of the PSO algorithm.

8.7.2.3.1 Speed-up Analysis of Parallel Algorithms

The computation time of the proposed parallel algorithms are tabulated in Table 8.3. The speed-up (Gebali 2011). is given in Figure 8.5. The time efficiency of a parallel algorithm is measured using speed-up which is calculated using Equation 8.21.

$$\text{Speed-up} = \frac{T_1}{T_p} \tag{8.21}$$

where p is the number of processors, T_1 is the execution time on one processor and T_p is the execution time of p processors.

TABLE 8.3
Computation Time of Clustering and Classification Algorithm

	Time in Minutes for different Number of Processors			
	1	2	4	8
Parallel K-means	254	251	73.3	35.8
Parallel K-means with DE and ACO	308.5	310	84.5	42.51
Parallel K-NN with PSO	50.8	52.2	14.2	7.31

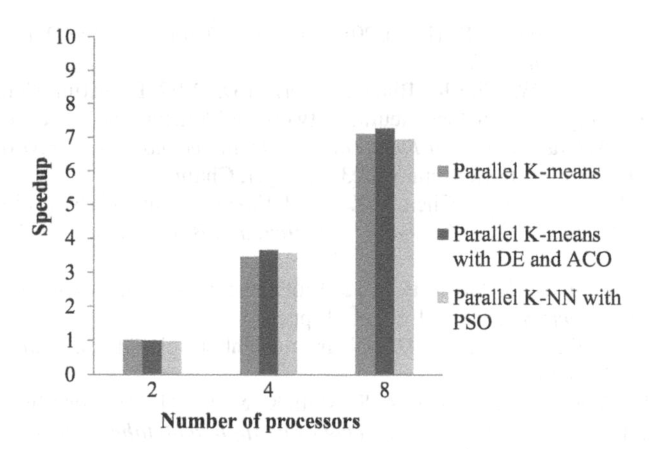

FIGURE 8.7 Speed-up graph.

From Figure 8.7, it is also observed that for $p = 2$, the scalability is not achieved because the mapper runs on the slave and the reducer runs on the master, which is same as $p = 1$ where the mapper and reducer run on the same machine. For the breast cancer gene expression data set, the mapper takes 23 runs on two processors, 6 runs on 4 processors and 3 runs on 8 processors to complete its task. Thus, the observation shows that MapReduce exhibits good scalability for a larger file size.

8.8 CONCLUSION

Nowadays, many researchers concentrate on processing biological data in a parallel manner. This chapter explains the prediction of breast cancer from gene expression data. The computationally intensive data mining tasks were parallelized using the MapReduce programming model. For processing 1.4GB of breast cancer gene expression data on 8 processors, the average speed-up obtained is 7.10. This shows a good scalability for processing larger file sizes.

REFERENCES

Acharjya, D., & Anitha, A. 2017. A comparative study of statistical and rough computing models in predictive data analysis, *International Journal of Ambient Computing and Intelligence (IJACI)*, 8(2), 32–51.

Ang, L. M., Seng, K. P., & Heng, T. Z., 2017. Information communication assistive technologies for visually impaired people. Smart Technologies: Breakthroughs in Research and Practice: Breakthroughs in Research and Practice, 17.

Baxevanis, A. D., & Ouellette, B. F., 2004. *Bioinformatics: A Practical Guide to the Analysis of Genes and Proteins*, John Wiley and Sons, New York.

Belarbi, M. A., Mahmoudi, S., & Belalem, G. 2017. PCA as dimensionality reduction for large-scale image retrieval systems, *International Journal of Ambient Computing and Intelligence (IJACI)*, 8(4), 45–58.

Benson, D. A., Cavanaugh, M., Clark, K., Karsch-Mizrachi, I., Lipman, D. J., Ostell, J., & Sayers E. W., 2013. GenBank, *Nucleic Acids Research*, vol. 41(D1), D36–D42.

Bhatt, C., Dey, N., & Ashour, A. S., (Eds.). 2017. *Internet of Things and Big Data Technologies for Next Generation Healthcare.*

Brito, R., Fong S., Song W., Cho K., Bhatt C., Korzun D., 2017. Detecting Unusual Human Activities Using GPU-Enabled Neural Network and Kinect Sensors. In: Bhatt C., Dey N., Ashour A. (eds) *Internet of Things and Big Data Technologies for Next Generation Healthcare.* Studies in Big Data, vol 23. Springer, Cham.

Cai, C. Z., Han, L. Y., Ji, Z. L., & Chen, Y. Z., 2004. Enzyme family classification by support vector machines, *Proteins: Structure, Function, and Bioinformatics*, vol. 55, no. 1, pp. 66–76.

Caragea, C., Silvescu, A., & Mitra, P., 2012. Protein sequence classification using feature hashing, *Proteome Science*, vol. 10, no. 1, pp. S14.

Cunningham, P., & Delany, S. J., 2007. k-Nearest neighbour classifiers, *Multiple Classifier Systems*, vol. 34, pp. 1–17.

Chakraborty, S., Chatterjee, S., Ashour, A. S., Mali, K., & Dey, N., 2017. Intelligent computing in medical imaging: A study, *Advancements in Applied Metaheuristic Computing*, 143. Doi: 10.4018/978-1-5225-4151-6.ch006

Daoudi, M., Hamena, S., Benmounah, Z., & Batouche, M., 2014. Parallel differential evolution clustering algorithm based on MapReduce, *Proceedings of the 6th International Conference on Soft Computing and Pattern Recognition*, pp. 337–341.

Das, S., Abraham, A., & Konar, A., 2008. Automatic clustering using an improved differential evolution algorithm, *IEEE Transactions on Systems, Man, and Cybernetics-Part A: Systems and Humans*, vol. 38, no. 1, pp. 218–237.

Deb, B., & Srirama, S. N., 2013. Parallel K-Means Clustering for Gene Expression Data on SNOW, *International Journal of Computer Applications*, vol. 71, no. 24, pp. 26–30.

Dembele, D., & Kastner, P., 2003. Fuzzy C-means method for clustering microarray data. *Bioinformatics*, vol. 19, no. 8, pp. 973–980.

Deng, Y., Kayarat, D., Elasri, M. O., & Brown, S. J., 2005. Microarray data clustering using particle swarm optimization K-means algorithm, *Proceedings of the eighth Joint Conference on Information Science*, pp. 1730–1734.

Dey, N., & Ashour, A. S., 2018. Computing in Medical Image Analysis. In *Soft Computing Based Medical Image Analysis* (pp. 3–11).

Dey, N., Ashour, A. S., & Borra, S. (Eds.). 2017a. *Classification in BioApps: Automation of Decision Making (Vol. 26).* Springer.

Dey, N., Ashour, A., Shi, F., Balas, V. E., 2018a. *Soft Computing Based Medical Image Analysi*, Academic Press.

Dey, N., Ashour, A. S., Shi, F., Fong, S. J., & Sherratt, R. S., 2017b. Developing residential wireless sensor networks for ECG healthcare monitoring. *IEEE Transactions on Consumer Electronics*, 63(4), 442–449.

Dey, N., Hassanien, A. E., Bhatt, C., Ashour, A., & Satapathy, S. C., (Eds.). 2018b. *Internet of Things and Big Data Analytics Toward Next-Generation Intelligence*, Springer.

Di Gesu, V., Giancarlo, R., Bosco, G. L., Raimondi, A., & Scaturro, D., 2005. GenClust: A genetic algorithm for clustering gene expression data, *BMC Bioinformatics*, vol. 6, no. 1, p. 289.

Dorigo, M., Birattari, M., & Stutzle, T., 2006. Ant colony optimization, *IEEE Computational Intelligence Magazine*, vol. 1, no. 4, pp. 28–39.

Du, Z., & Lin, F., 2005. A novel parallelization approach for hierarchical clustering, *Parallel Computing*, vol. 31, no. 5, pp. 523–527.

Elhayatmy, G., Dey, N., & Ashour, A. S., 2018. Internet of Things Based Wireless Body Area Network in Healthcare. In: *Internet of Things and Big Data Analytics Toward Next-Generation Intelligence* (pp. 3–20). Springer, Cham.

Fadlallah, S. A., Ashour, A. S., & Dey, N., 2016. Advanced titanium surfaces and its alloys for orthopedic and dental applications based on digital SEM imaging analysis. *Advanced Surface Engineering Materials*, 517–560. Doi: 10.1002/9781119314196.ch12

Fowdur, T. P., Beeharry, Y., Hurbungs, V., Bassoo, V., & Ramnarain-Seetohul, V., 2018. Big Data Analytics with Machine Learning Tools. In: Dey N., Hassanien A., Bhatt C., Ashour A., Satapathy S. (eds) *Internet of Things and Big Data Analytics Toward Next-Generation Intelligence*. Studies in Big Data, vol 30. Springer, Cham.

Gebali, F., 2011. *Algorithms and Parallel Computing*, John Wiley and Sons, New York.

Halder, A., Dey, S., & Kumar, A., 2015. Active Learning Using Fuzzy k-NN for Cancer Classification from Microarray Gene Expression Data, *Advances in Communication and Computing*, Springer, India, pp. 103–113.

Han, J., Pei, J., & Kamber, M., 2011. *Data Mining: Concepts and Techniques*, Elsevier, San Francisco.

Hartigan, J. A., & Wong, M. A., 1979. Algorithm AS 136: A k-means clustering algorithm, *Journal of the Royal Statistical Society, Series C (Applied Statistics)*, vol. 28, no. 1, pp. 100–108.

He, J., Tan, A. H., Tan, C. L., & Sung, S. Y., 2004. On Quantitative Evaluation of Clustering Systems, In: *Clustering and Information Retrieval*, Springer, US. pp. 105–133.

He, Y., & Hui, S. C., 2009. Exploring ant-based algorithms for gene expression data analysis, *Artificial Intelligence in Medicine*, vol. 47, no. 2, pp. 105–119.

Hore, S., Chakroborty, S., Ashour, A. S., Dey, N., Ashour, A. S., Sifaki-Pistolla, D., & Chaudhuri, S. R., 2015. Finding contours of hippocampus brain cell using microscopic image analysis. *Journal of Advanced Microscopy Research*, 10(2), 93–103.

Hulianytskyi, L., & Rudyk, V., 2014. Development and analysis of the parallel ant colony optimization algorithm for solving the protein tertiary structure prediction problem, *Information Theories and Applications*, vol. 21, no. 4, pp. 392–397.

Hung, C. L., & Lin, Y. L., 2013. Implementation of a parallel protein structure alignment service on cloud. *International Journal of Genomics*, vol. 2013, Article ID 439681, 7 pages.

Kalegari, D. H., & Lopes, H. S., 2013. An improved parallel differential evolution approach for protein structure prediction using both 2D and 3D off-lattice models, *Proceedings of IEEE Symposium on Differential Evolution*, pp. 143–150.

Kamal, S., Dey, N., Ashour, A. S., Ripon, S., Balas, V. E., & Kaysar, M. S., 2017a. FbMapping: An automated system for monitoring Facebook data. *Neural Network World*, 27(1), 27.

Kamal, M. S., Dey, N., & Ashour, A. S., 2017b. Large Scale Medical Data Mining for Accurate Diagnosis: A Blueprint. In: *Handbook of Large-Scale Distributed Computing in Smart Healthcare* (pp. 157–176). Springer, Cham.

Kamal, M. S., Parvin, S., Ashour, A. S., Shi, F., & Dey, N., 2017c. De-Bruijn graph with MapReduce framework towards metagenomic data classification. *International Journal of Information Technology*, 9(1), 59–75.

Kamal, M. S., Sarowar, M. G., Dey, N., Ashour, A. S., Ripon, S. H., Panigrahi, B. K., & Tavares, J. MR., 2017d. Self-organizing mapping based swarm intelligence for secondary and tertiary proteins classification. *International Journal of Machine Learning and Cybernetics*, 1–24. Doi: 10.1007/s13042-017-0710-8

Karaa, W. B. A., (Ed.). 2015. *Biomedical Image Analysis and Mining Techniques for Improved Health Outcomes*, IGI Global.

Karim, M. R., Bari, A. T. M. G., Jeong, B. S., & Choi, H. J., 2012. Cloud technology for mining association rules in Microarray Gene Expression Data sets, *International Journal of Database Theory and Application*, vol. 5, no. 2, pp. 61–74.

Keller, A. D., Schummer, M., Hood, L., & Ruzzo, W. L., 2000. *Bayesian Classification of DNA Array Expression Data*, Technical Report No. UW-CSE-2000-08-01, University of Washington.

Khan, A. A., Hassan, L., & Ullah, S. 2015. Open MP-based Parallel and Scalable Genetic Sequence Alignment, *Journal of Engineering and Applied Sciences*, vol. 34, no. 2, pp. 29–34.

Kraus, J. M., & Kestler, H. A., 2010. A highly efficient multi-core algorithm for clustering extremely large data sets, *BMC Bioinformatics*, vol. 11, no. 1, p. 169.

Kumar, M., Rath, N. K., & Rath, S. K., 2016. Analysis of microarray leukemia data using an efficient MapReduce-based K-nearest-neighbor classifier, *Journal of Biomedical Informatics*, vol. 60, pp. 395–409.

Leslie C. S., Eskin, E., & Noble W. S., 2002. The spectrum kernel: A string kernel for SVM protein classification, *Proceedings of the Pacific Symposium on Biocomputing*, pp. 566–575.

Marcais, G., & Kingsford, C., 2011. A fast, lock-free approach for efficient parallel counting of occurrences of k-mers, *Bioinformatics*, vol. 27, no. 6, pp. 764–770.

Marucci, E. A., Zafalon, G. F., Momente, J. C., Neves, L. A., Valencio, C. R., Pinto, A. R., Cansian, A. M., Souza, R. C., Shiyou, Y., & Machado, J. M., 2014. An efficient parallel algorithm for multiple sequence similarities calculation using a low complexity method, *Biomed Research International*, vol. 2014, pp. 563016–563016.

Matallah, H., Belalem, G., & Bouamrane, K., 2017. Towards a new model of storage and access to data in big data and cloud computing, *International Journal of Ambient Computing and Intelligence (IJACI)*, 8(4), 31–44.

Moraru, L., Moldovanu, S., Culea-Florescu, A. L., Bibicu, D., Ashour, A. S., & Dey, N., 2017a. Texture analysis of parasitological liver fibrosis images. *Microscopy Research and Technique*, 80(8), 862–869.

Moraru, L., Moldovanu, S., Dimitrievici, L. T., Shi, F., Ashour, A. S., & Dey, N., 2017b. Quantitative diffusion tensor magnetic resonance imaging signal characteristics in the human brain: A hemispheres analysis. *IEEE Sensors Journal*, 17(15), 4886–4893.

Mrozek, D., Brożek, M., & Małysiak-Mrozek, B., 2014. Parallel implementation of 3D protein structure similarity searches using a GPU and the CUDA, *Journal of Molecular Modeling*, vol. 20, no. 2, p. 2067.

Muggerud, A. A., Hallett, M., Johnsen, H., Kleivi, K., Zhou, W., Tahmasebpoor, S., Amini, R. M. et al., 2010. Molecular diversity in ductal carcinoma *in situ* (DCIS) and early invasive breast cancer, *Molecular Oncology*, vol. 4, no. 4, pp. 357–368.

Nandi, D., Ashour, A. S., Samanta, S., Chakraborty, S., Salem, M. A., & Dey, N., 2015. Principal component analysis in medical image processing: A study. *International Journal of Image Mining*, 1 (1), 65–86.

Needleman, S. B., & Wunsch, C. D., 1970. A general method applicable to the search for similarities in the amino acid sequence of two proteins, *Journal of Molecular Biology*, vol. 48, no. 3, pp. 443–453.

Ooi, C. H., & Tan, P., 2003. Genetic algorithms applied to multi-class prediction for the analysis of gene expression data, *Bioinformatics*, vol. 19, no. 1, pp. 37–44.

Othman, F., Abdullah, R., & Salam, R. A., 2004. Parallel k-means clustering algorithm on DNA data set, *Proceedings of the 5th International Conference on Parallel and Distributed Computing: Applications and Technologies*, pp. 248–251.

Poli, R., Kennedy, J., & Blackwell, T., 2007. Particle swarm optimization, *Swarm Intelligence*, vol. 1, no. 1, pp. 33–57.

Sen, S., Narasimhan, S., & Konar, A., 2007. Biological data mining for genomic clustering using unsupervised neural learning, *Engineering Letters*, vol. 14, no. 2, pp. 61–71.

Sokolova, M., & Lapalme, G., 2009. A systematic analysis of performance measures for classification tasks, *Information Processing and Management*, vol. 45, no. 4, pp. 427–437.

Storn, R., & Price, K., 1997. Differential evolution–a simple and efficient heuristic for global optimization over continuous spaces, *Journal of Global Optimization*, vol. 11, no. 4, pp. 341–359.

Sudha, S. G., & Baktavatchalam, G., 2009. A novel approach to multiple sequence alignment using hadoop data grids, *International Journal of Bioinformatics Research and Applications*, vol. 6, no. 5, pp. 472–483.

Tan, A. H., & Pan, H., 2005. Predictive neural networks for gene expression data analysis. *Neural Networks*, vol. 18, no. 3, pp. 297–306.

Wang, D., Li, Z., Cao, L., Balas, V. E., Dey, N., Ashour, A. S., & Shi, F., 2017. Image fusion incorporating parameter estimation optimized Gaussian mixture model and fuzzy weighted evaluation system: A case study in time-series plantar pressure data set. *IEEE Sensors Journal*, 17(5), 1407–1420.

Waterman, M. S., Smith, T. F., & Beyer, W. A., 1976. Some biological sequence metrics. *Advances in Mathematics*, vol. 20, no. 3, pp. 367–387.

Wei-Chun, Chang, Yu-Jung, Chen, Chien-Chih., Lee, Der-Tsai., & Ho, Jan-Ming. 2013. Optimizing a MapReduce module of preprocessing high-throughput DNA sequencing data, *Proceedings of the IEEE International Conference on Big Data*, pp. 1–6.

White, T. 2012. *Hadoop: The Definitive Guide*, O'Reilly Media Inc., Canada.

Wu, G., Zhang, J., Hu, X., Li, S., & Hao, S., 2013. A parallel clustering algorithm with MPI – MKmeans, *Journal of Computers*, vol. 8, no. 1, pp. 10–17.

Xiao, X., Dow, E. R., Eberhart, R., Miled, Z. B., & Oppelt, R. J., 2003. Gene clustering using self-organizing maps and particle swarm optimization, *Proceedings of the International Symposium on Parallel and Distributed Processing Symposium*, pp. 10.

Yin, L., Huang, C. H., & Rajasekaran, S., 2004. Parallel data mining of Bayesian networks from gene expression data, *Proceedings of the Eighth International Conference on Research in Computational Molecular Biology*, pp. 122–123.

Zou, Q., Li, X. B., Jiang, W. R., Lin, Z. Y., Li, G. L., & Chen, K., 2013. Survey of MapReduce frame operation in bioinformatics. *Briefings In Bioinformatics*, 15(4), pp. 637–647.

9 A MapReduce Approach to Analysing Healthcare Big Data

Md. Golam Sarowar, Shamim H. Ripon, and Md. Fahim Shahrier Rasel

CONTENTS

9.1 INTRODUCTION

In the present era of technological evolution, day-to-day human life depends increasingly on technologies. Therefore, every aspect of human life uses modern technologies. Consequently, recent improvements in high performance computing have made it possible to extract information from large volumes of higher dimensional data. However, the trend of human life with the use of modern technologies yields petabytes of data in every sector. When these data exceed certain speeds, velocities, volumes, dimensions, they are referred to as big data. In general, those data can be structured, unstructured and noisy and can be a negative factor when attempting to extract information from these data sets. Moreover, efficiency, lower use of energy, cost efficiency, easier manipulation is the main requirement of this era and technology sometimes paves the researchers, academies towards the way through which big data problems can be handled within a period of time along with efficiency. This big data set include obstacle while capturing, storing, visualizing, transferring, as well as analysing privacy and security. However, the main characteristics of big data are complex large volume sized data along with basic three dimensions, which are volume, velocity and verity. Recent studies [1–3] demonstrate that the world's technological data has doubled every 40 minutes since the 1980s. Additionally, since 2012 every day 2.5 exabytes of data are being generated. According to the prediction of the International Data Corporation, there will be 163 zettabytes of data available in 2025. Thus, big data can be referred as voluminous and complex data sets, which are difficult to pre-process using traditionally available approaches.

Almost all sectors of human life are now on the way to technological evolution and medical science is not an exception. Technological evolution has influenced medical science with computing technologies including computing mechanisms to analyse medical images [4], IoT-based wireless wearable components [5], quantitative diffusion tensor magnetic resonance [6], Gaussian mixture model and fuzzy weighted evaluation systems [7], large scale medical data computing techniques [8], wireless capsule gastrointestinal endoscopy based computing techniques [9] and so on. In general, medical science generates huge amounts of data from clinical trials, medicines, exercises and prescriptions [10,11]. Additionally, with the emergence of the IoT, further creation of big data in healthcare has emerged. In addition, emerging wireless sensor networks [12] monitoring human health continuously generate even more big data. Meanwhile, information including information of schedules of treatment, patient blood pressure monitoring, doctor's information, patient's, surgeons and nurses, and patient's insurance coverage need to be stored and processed to determine a patient's current condition and to decide on a further course of action. The exponential growth of those data is now a main concern for modern science. These massive data set cannot be manipulated using relational databases currently available in the literature and relational databases are able to work only with structured data sets; however, big data is the combination of structured as well as unstructured massive data. Emerging medical technologies including sensors, actuators, smart devices, cloud computing and smartwatches are generating huge numbers of data, which are almost impossible to parse and analyse using traditional approaches. Recent advancements in wearable technologies like smartwatches, Google glasses or smart sensors always remain active and continuously create huge amounts of data. Immediate exploration of these data is mandatory to determine a patient's current condition and for taking further action. Since medical data are generated from various sources as well and differs from patient to patient so it is conventional of appearing enormous irreverent as well as noisy data and those are the main obstacles for algorithms to work efficiently. Clinical decision-making is dependant on the analysis of these massive data sets. Thus, more emphasize on invention of feasible machine learning supervised as well as unsupervised approaches [13] along with compatible pre-processing mechanism need to be developed and discovered. Moreover, new pathways should be discovered to achieve more accuracy while extracting or predicting useful information from big medical data sets. Big data in medical science or human healthcare can be identified by three characteristics. These characteristics are represented in Figure 9.1.

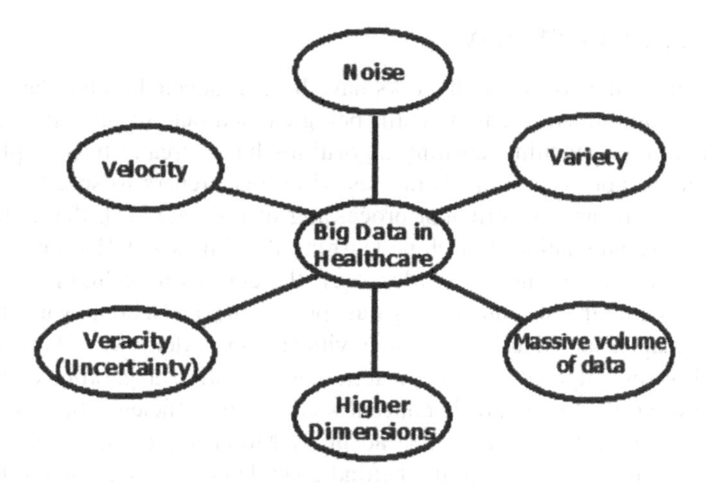

FIGURE 9.1 Basic level characteristics of big data in healthcare.

Basically, data extraction from big medical data sets that includes higher dimensions, high velocities, verity and speed is one of the top concerns now. Additionally, the variation within healthcare data is greater compared to others sciences because of the generation of data from various sources. Actually, big data is interrelated with analytics, which is a familiar concept. These analytics usually include many simulations, regression models and machine learning approaches. Realizing the worth of this recent big data trend in healthcare technology, we have assessed the impact of the MapReduce [14] approach in the context of medical big data sets while it is assembled with strong pre-processing mechanism. Moreover, rigorous comparisons among pre-processing based MapReduce, pre-processing less MapReduce, pre-processing based only mapping have been demonstrated. The overall infrastructure of our automated approach is illustrated in Figure 9.2.

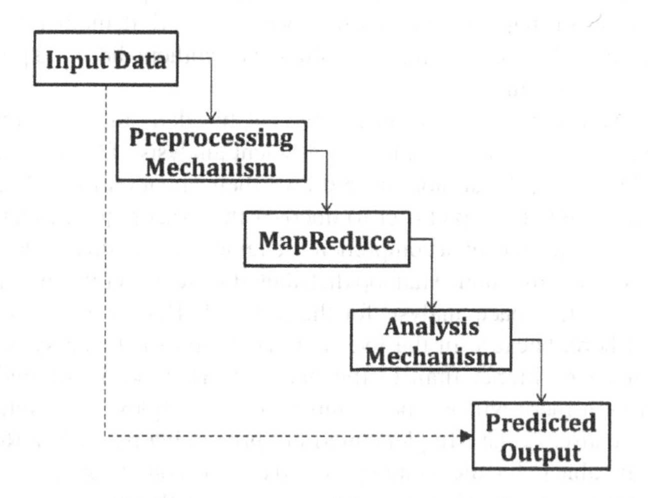

FIGURE 9.2 Overall working infrastructure of this contribution size.

9.2 LITERATURE REVIEW

Until now, thousands of research works have been conducted on big data analytics in healthcare and much research is still being carried out. To meet the challenges of big data, many machine-learning algorithms have already been explored and currently many approaches are being tested by researchers to speed up analysis and efficiency. To achieve efficient processing of big data [15], the authors have implemented a combination of machine-learning algorithms with Hadoop for efficient classification or processing tasks. Moreover, the authors have highly emphasized parallel computing in the context of big data processing, and a broadly used heuristic algorithm 'MapReduce' has been applied with Hadoop. Along with Hadoop various machine-learning algorithms are in action for creating a parallel computation set-up. However, this work mostly concentrated on time efficiency but also memory efficiency or space efficiency as well as accuracy. Moreover, for this work a multicore system has been considered lagging behind a single core system. In addition, both serial and parallel computing mechanisms have been implemented. In comparison with this, ours is a real-life application of the MapReduce algorithm combined with a strong pre-processing mechanism. MapReduce itself diminishes the total space size through mapping, as well as reducing duplicate information depending on the key value. Meanwhile, pre-processing mechanism pursuits efficient and undertakes less time for processing with proper efficiency as well as accuracy. Another work [16] on the MapReduce approach rigorously tests MapReduce for resolving various big data problems. However, nothing has been mentioned whether the data sets used have missing and irrelevant values. However, our work provides a pre-processing-based MapReduce approach, so there is no limitation of missing values. Another recent study [17] looked at finding dependency among granularity level, number of fuzzy variables, number of selected maps and so on. They have mostly concentrated on MapReduce for mapping as well as data reduction securing data privacy. In this work the authors have successfully proposed a method named chi-FRBCS, which seemed to be quite effective for data scattering, preserving good performance of data even when the number of maps increases. Moreover, this work demonstrates analysis of unstructured data while having missing values. The authors did avoid pre-processing to diminish missing values.

Using the MapReduce algorithm in medical big data analytics, reference [18] proposes a new automated system for the efficient analysis of big data in healthcare technology. Moreover, the authors have drawn their attention towards developing a five-node distributed Hadoop cluster to uncover the exact required features using a distributed system. Moreover, a comprehensive review within this work demonstrates the superiority of the five-node Hadoop distributed system over the one-node Hadoop system. However, the space analysis for the five-node Hadoop distributed system is not discussed here, because in the five-node Hadoop distributed system the space complexity must be higher than in the one-node Hadoop distributed system. In our consideration there will not be so much space complexity as well as memory complexity because we have implemented pre-processing based MapReduce which is exceptionally able to handle memory as well as time complexity in case of medical big data processing or any kind of feature detection with efficient manner. Another vision of the impact of big data in medical healthcare has been discussed from a clinical perspective in reference [19]. Additionally, the necessity of the analysis of medical big data sets for extracting required features is also discussed by the authors of this work. In addition, lots of real-time hospital-based heterogeneous healthcare big data analysis has been conducted recently. As one of the contributions to real-time analysis, the work of reference [20] is highly focused on the use of Hadoop for heterogeneous big data analysis. Although they did not mention any pre-processing tool used on missing values, it has been emphasized in our present work because it may slow the overall process. However, the authors of that contribution didn't headache on dimensions of healthcare big data. Unstructured data processing is just the similar as big data concept. Lots of work has been already conducted on controlling unstructured data. Contribution [21] negotiates with unstructured data by altering them into structured data. Thereafter, the authors have proposed a new dimensional approach for taking the processing forward to extract specific features from the Mapped and Reduced data sets. The authors also focused on pre-processing of data sets through dividing them into many of chunks or clusters. The motive reason of the work is to process big data efficiently even if with the presence of higher dimension, velocity, variety, volume etc. thus, they have propagated an automated with the combination of pre-processing, MapReduce, Sentiment analysis, collaborative filtering approach for sentiment analysis. Although accuracy level and efficiency are quite improved, space complexity can be the main obstacle for this automated system to work successfully. Technically, a pre-processing-based Mapreduce approach, which we have considered for this work, can be more effective when the data size is massive with higher dimension and higher velocity. Another feasible Mapreduce-based framework has been fabricated in the work of reference [22] for data scalability, efficiency and space reduction. Additionally, for healthcare data storage, patient's reports storage and analysis the authors have concentrated on Hadoop-based MapReduce and hive approaches to ensure fault tolerance. Moreover, a new perspective for a data management system has been proposed here which is quite trustworthy for efficient data upload, data query and data analysis. However, performance evolution of the algorithms has been conducted and it briefly illustrates the superiority and efficient manner of the proposed algorithms in terms of big data healthcare management. Big data has emerged as propagating a new paradigm of abundant data, massive volume, higher dimension, higher speed and so on. With the emergence of big data, processing, storing, analysing, transporting, mining and serving those data for feature detection or information extraction has been in the same trend of turning into almost impossible state. From computer science perspective to get rid of this type of situation lots of machine learning algorithms have been unfolded. Contribution [23] is a review article of the challenges, advantages, new approaches and observations of big data in healthcare. However, the principal motivation of this contribution is the explicit description of cloud computing and its impacts on the field of big data manipulation. In most cases, the authors have been motivated by the use of cloud computing for all the problems arising in big data. It has also been demonstrated that cloud computing assures optimal solution allowing less space complexity and less time consumption on which our current work is concerned massively.

Although available of thousands of algorithms to manipulate big data the researchers, students, academies are highly concerned with rule based, instance based, neural network based, support vector machine based approaches to be conceived and explored. Moreover, achieving higher accuracy, higher prediction accuracy, practical less space management and reduction in processing time is the principal goal of the big data world. Since decision-making task is the motive obligation for medical science, strong machine learning background infrastructure is the demand of time. Simply contribute to this portion the authors [24] in most cases drew their attention towards diagnosis based best suited classifier for disease detection. Technically, this work is the interrelation between three phases, which are big data management, diagnosis-based machine-learning approach and input/output details of the patients. This framework will highly confidential and satisfying for the doctors to be aided in their decision-making task through demonstrating higher and accurate measurement of disease. Thus, this can be evolving procedures for the detection of complex diseases with higher accuracy and efficiency. Basically, the most populous countries like China and India struggle with this recent big data trend. The number of hospitals, private healthcare organizations and most significantly the number of people is increasing in more or less exponential manner; therefore, it is hardest for those countries to manipulate the data for important decision-making with traditional database systems. Realizing the work of this contribution [25] focuses on imposing a new machine-learning approach through the use of built in tool Pig Latin script. Because of the tremendous improve in healthcare services the generation as well as quality of data has been increased a lot and lots of queries raised for big data from last decades which have not been explored for the constraint of big data efficiently manipulation technologies. Thus, in this study [25] the authors tried to resolve those queries from the perspective of big data and they have applied the Pig Latin script, which is similar to Hadoop for big data manipulation purposes. For ours we have concentrated on Hadoop because it is the most powerful and effective tool for big data analysis in the literature. Another study, reference [26], has been conducted on healthcare data analysis in India. This study bargains with lots of guidelines along with significant gap which must be noticed in the previous literature. Moreover, this systematic analysis of the literature has explored lots of useful and evolutionary contribution in this healthcare field of India. Since this era is the era of big data in the field of healthcare technology leading Exabyte of data every day, the authors of [27,28] raises their voice how manipulating big data better medical services and medication can be improved tremendously. This work is not only confined within medication and medical services, this is also aggregation of lots of work's drawbacks and effectiveness.

9.3 DATA COLLECTION

We have used a U.S. diabetes data set containing patients information from 130 hospital. It includes more than 30 features for each patient admitted in those hospitals. There are approximately 150,000 instances. There are more than 25 attributes. The data set attributes are depicted in Table 9.1.

TABLE 9.1
Details of Attributes Information of Diabetes 130 US hospitals for years 1999–2008

Feature Name	Description and Values	Missing (%)
Encounter ID	Unique identifier of an encounter	0
Patient number	Unique identifier of a patient	0
Race	Values: Caucasian, Asian, African American, Hispanic, and other	2
Gender	Values: male, female, and unknown/invalid	0
Age	Grouped in 10-year intervals: 0, 10), 10, 20), … , 90, 100)	0
Admission type	Integer identifier corresponding to 9 distinct values	0
Discharge disposition	Integer identifier corresponding to 29 distinct values	0
Admission source	Integer identifier corresponding to 21 distinct values	0
Number of lab procedures	Number of lab tests performed during the encounter	0
Number of procedures	Number of procedures (other than lab tests) performed during the encounter	0
Number of medications	Number of distinct generic names administered during the encounter	0
Number of outpatient visits	Number of outpatient visits of the patient in the year preceding the encounter	0
Number of emergency visits	Number of emergency visits of the patient in the year preceding the encounter	0
Number of inpatient visits	Number of inpatient visits of the patient in the year preceding the encounter	0
Diagnosis 1	The primary diagnosis (coded as first three digits of ICD9); 848 distinct values	0
Diagnosis 2	Secondary diagnosis (coded as first three digits of ICD9); 923 distinct values	0
A1c test result	Indicates the range of the result or if the test was not taken	0
Readmitted	Days to inpatient readmission. Values: "<30," ">30" & "No" for no record of readmission	0
Diagnosis 3	Additional secondary diagnosis	1

Source: B. Strack et al., *BioMed Research International*, vol. 2014, Article ID 781670, 2014, p. 11 [28].

9.4 METHODOLOGY

The aim of this work is to strengthen scalability, feasibility, availability and efficiency while processing and analysing healthcare big data. To resolve our problem we have focused on mapping based similar data reduction approach along with a flourished pre-processing mechanism. We implemented the overall infrastructure through the use of Hadoop; most successful and applied platform for implementation of MapReduce approach by the researchers and academia. We have accomplished our analysis and implementation on a personal computer running under Windows 8.1. The algorithm we applied here is described in details in Section 9.4.1.

9.4.1 MapReduce

Nowadays, lots of real life sectors including biology, chemistry, physics, mathematics, healthcare, information technology, bio-medical, molecular biology, computational biology, bio-data mining, pharmacy are propagating massive volume of energy. Moreover, they also propagate data with higher velocity, enormous variety, higher dimensions etc. [29]. processing as well as analysing such kind of data sets require powerful mechanism otherwise this may waste hours of time and this circumstance has been appeared as one of the major issue for IT sector to manipulate using available approaches [30–32]. In general, these types of big data has been analysed in data warehouse still exceptionally massive data sets can pave the overall process towards unsteady circumstances and unstable condition of such system may results in dead-end. Therefore, proper definitions of those types of data sets need to be propagated. The data sets which surpasse the storage capacity, dimensions, velocity of available approaches need to be evaluated with highest attention. Various recent studies have shown that conventional approaches available in the machine-learning field are not adequate to explore hidden features or extract decision-making information from these data sets. Thus, new techniques should be encouraged to develop efficient and low space consumption mechanisms to keep pace with exponentially increasing data sets [33]. Researchers and academia have defined the available classification, clustering and sampling algorithms into two categories: sampling and algorithmic [34]. However, approximately all the machine-learning approaches which outperforms big data problems have the ability to Map, fast and show efficiency even if in case of big data.

MapReduce [35–36] is a parallel programming model efficient at processing and analysing exabytes of data with higher efficiency and lower space consumption. This evolutionary algorithm is considered to be one of the best-suited programing models ever to explore big data problems. This algorithm works in a similar way to the divide and conquer mechanism. Initially, massive data sets are distributed into lots of sub-portions to apply parallel computing. Afterwards, all the sub-portions are aggregated together to reach the desired findings. Sometimes, it combines both parallel software as well as hardware to work on a distributed data set in subparts. In this case, the Hadoop Apache platform has been considered to implement pre-processing based on MapReduce. There are three basic phases to the MapReduce approach: mapping, shuffling and reducing.

In the first phase, the big data sets are mapped into several subparts depending on the actual data sets. The second phase aggregates all the similar data together within each subpart containing only similar data together in that region. In the last phase, all the similar data are reduced to just one along with the track of their occurrences within the data sets. The overall operations of MapReduce for healthcare data are illustrated in Figure 9.3.

9.4.1.1 Concepts and Advantages of MapReduce

MapReduce is a kind of motive programming language lying into the deep of apache Hadoop platform for manipulation and analysis of massive Hadoop clusters perceiving scalability, consistency and readability etc. This evolutionary paradigm

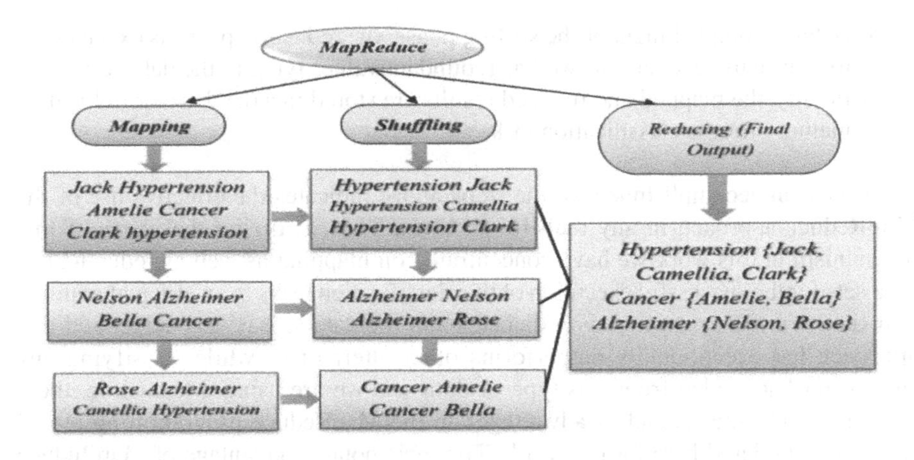

FIGURE 9.3 Overall operations of the MapReduce algorithm for healthcare data.

is primarily explored to negotiate with large volumes of data with thousands of clusters. In a simpler manner, MapReduce is one of the frameworks enabling the developers to implement thousands of functions capable of processing large volumes of data in a fast and efficient way. Through two basic phases this approach accomplishes overall manipulation process. These two phases are defined as the Map phase and the reduction phase where Map phase is capable of reducing dimension and rest reduction phase can be depicted as classification paradigm. Moreover, the job configuration supplies the map and reduce analysis functions and the Hadoop framework provides the scheduling, distribution and parallelization services.

The overall data analysis is accomplished by combining both the mapping and reducing approaches. The mapping phase can be considered as the most significant portion of the MapReduce approach since the reduction of complexity and higher dimensionality is effectively performed by mapping depending on specific key values in this phase. The key values may not be the same type for every node. However, a key value must be imposed for each and every assumption. Usually, both the mapping and the reducing cycles execute the following instructions and tasks.

1. Creates key value pairs after fetching input data depending on input data format
2. Afterwards apply scheduled Map function on each pair
3. Local sorting as well as aggregation is performed
4. If internal combiner within code is found then further aggregation is performed
5. Overall results are stored locally in the DRAM portion
6. Thereafter initially at reduction phase tasks resources are fetched from local DRAM
7. All the copies of key pair values are copied from the local memory and are copied

8. After accomplishment of the sorting phase, the reduction phase is executed to reduce unnecessary as well as redundant values lying in the data sets
9. Finally, the mapped and reduced results are stored in a database or in local memory for the classification task

Successful accomplishment of the steps above indicates the effective use of the MapReduce approach in any real-life problem. Thus, realizing the worth of this mechanism in this work we have concentrated on mapping as well as reducing our data sets for the successful execution of the classification tasks in an efficient manner. Our data sets are associated with diabetes and the data size is very large, and this approach has exceptionally outperforms other alternatives while classifying any significant knowledge from this type of data sets we are using. Meanwhile, there are lots of advantages and disadvantages in the MapReduce programming model we have considered here for this work. The most notable advantage of MapReduce is maintaining scalability and developing a bridge between traditional database management systems and recent technologies like Hadoop. For the sake of MapReduce, traditional database management system can be used effectively everywhere to extract specific knowledge. This combination of RDMS along with MapReduce allows user queries to be executed faster than before and can be represented efficiency even during implementing traditional systems.

Moreover, scalability, cost-effective function, flexibility, query speed along with secure authentication by removing redundancy and inconsistency are a few of the advantages of the approach. These advantages are achieved by applying parallel computing, distributed processing features, simple models of programming and availability provided by the MapReduce approach. Moreover, while implementing MapReduce programmers never need to worry about creating distributed computing mechanism inside overall computation since the parallel computing system has already been developed within the Apache Hadoop platform. These features greatly facilitate the implementation of the overall procedure. Additionally, unstructured data sets are a most prominent concern for large volumes of data. Since MapReduce works with simple key pairs, this mechanism is thus good enough to handle any type of data that fits within the model. Therefore, it becomes quite easy for the programmer to negotiate with MapReduce rather than Database. This approach is also able to tolerate fault occurred in any case. Instead of all those advantages it has also some disadvantages but quite negligible compared to advantages. Usually, one Reducer negotiates with one partitions which results in creating limitation in distributed computing paradigm. Moreover, if too many reducers are used then a lot small size of output files are generated which will hamper the HDFS system providing lots of pressure.

MapReduce is an efficient method for analysing large-scale databases. This algorithm is valuable because of its extraordinary ability to tackle big data problems through the use of parallel computing mechanisms. The advantages of the MapReduce approach are listed below:

- *Expandability*: The MapReducing mechanism lessens power consumption through the ability of parallel computing. It distributes portions of a data set to work in parallel, which saves time. Moreover, the reduction phase reduces data sets depending on similar parameters available within the data sets thus working efficiently with lower platform space requirements. Conventional approaches of big data analysis are unable to work as MapReduce does. Thus, MapReduce is a valuable programming platform for big data analysis in medicine;

- *Parallel processing*: One of the exceptional features of MapReduce is parallel processing. This feature divides data sets into chunks to achieve execution time reductions, requiring lower storage capacity and producing faster classifications;
- *Inexpensive solution*: MapReduce is an inexpensive solution for every sort of problem interpreting all the requirements. In conventional approaches, data storing and analysis becomes expensive for the users to keep tract or extract specific features or information. However, the MapReduce framework works on a Hadoop platform which is applicable to stored data. It diminishes data size as well as space complexity depending on the priority set by the implementer. Thus, this approach appears to be cost effective even though the data size is in terabytes.
- *Workability, privacy and security*: The MapReducing approach is able to work even with heterogeneous data sets, thus it is well suited for any source of data. Moreover, both structured and unstructured data sets can be analysed using MapReducing. Since healthcare data sets are heterogeneous data sets thus handling or accessing storage appears to be simpler than available approaches in healthcare big data analysis. Moreover, along with healthcare application fraud detection, easier way of data processing, market analysis can be demonstrated through the use of machine learning algorithms like MapReduce. Meanwhile privacy, authentication and security is a vital role for all sectors of human management and since MapReduce reduce searching time by mapping and reducing data redundancy, complexity and size, user authentication and privacy can be easily preserved. Even though the data size is terabytes or exabytes, user information can be easily extracted from the database and matched with the specific user and also no unstable condition seems to be happened.
- *Easy programming framework*: Since the MapReducing approach can be implemented easily using the Java programming model, this approach accommodates programmers to develop and redesign the mapping of input data set according their requirement and predicted output requirement. This mapping and reduction stage can easily handle multiple tasks through the use of parallel processing. Meanwhile, another benefit of MapReduce is identifying and detecting many illegal operations.

The above discussion interprets the outstanding performance for any sectors big data problems. This MapReducing algorithm is now being applied in lots of real-life situations including bioinformatics research, biodata mining, language analysis design, data warehouses and so on. This algorithm is a suitable and effective environment for gene sequence analysis. Map cluster is another new trend interrelated with the map reducing clustering approach which is distributed into different two phases defined as bottom up merging and top-down separation. This clustering behaviour is compatible with clustering biological data depending on their functionalities. Moreover, Hadoop [37] is the popular technique used as the best platform for distributed processing of big data for the MapReduce approach in the Google file system.

9.4.1.2 Mapping Phase

Basically, the map phase distributes the data set into smaller subparts so that they can be distributed to parallel processing nodes. However, the map stage is just mapping the whole data sets into subparts pair (p, q) for the shuffling phase. Here p is the key field whereas q is the key value, which has been conducted depending on the key field. Moreover, in the shuffling phase similar types of data – depending on their key fields and values – are aggregated together for diminishing redundancy and leading towards the next phase, which is the reduction phase. Among all the phases accomplished within the MapReduce mechanism, this mapping phase is the most complex because at a specific time only one pair can be constructed and operated within one phase. The shuffle stage is basically interpretation of merging similar data within one sets which is most effective for reduction stage to diminish the dimensions as well as density of data. However, after the shuffling phase the distributed subsets are transmitted towards a tachyon file system, which is a distributed file system. This file system distributes data sets towards the reduction stage.

9.4.1.3 Reducing Phase

The reduction phase is the correlation between various steps to reduce the merged data provided by the tachyon file system. The required steps are (i) data cleaning, (ii) rule purification, (iii) rule combination, (iv) rule conjugation and (v) updating rule. The steps are described below:

- *Data cleaning*: Data redundancy and inconsistency occurs while data are collected from various heterogeneous sources. Those data sets are contaminated with missing values, duplicate values and so on. However, for this work we have concentrated on eliminating those problems embedding pre-processing tool initially. Thus, invalid data, unintentionally mistaken data, unauthorized entry of data and data from ineffective sources were initially eliminated through a pre-processing tool; however, if somehow this couldn't be done, the data cleaning mechanism embedded in the MapReduce approach totally eliminated those data. The overall data cleaning process is characterized through the implementation of two phases defined as a heuristic approach and a optimistic heuristic approach. In most cases, those two approaches are implemented at the physical level. Since heuristic mechanisms are being implemented for logical purposes, its accomplishment largely depends on the programming model. The algorithm [31] for implementing an optimized heuristic mechanism is mentioned as follows:

Algorithm: Data Cleaning Algorithm

Start
 Add a key to every input record
 Extract attributes from input and output file
  Extract all sub-attribute properties from its attribute properties
 Examine all duplicate data from recorded data
 Aggregate all duplicate free data for every data attributes
End

From an implementation level perspective many logical operators, as well as effective algorithms, can be integrated together for the data cleaning programming model. For exploration of required features for data viewing, merging, storing, presenting cleaning mechanism is mandatory in case of heterogeneous big healthcare data.

- *Rule purification*: All the reduced data sets, after execution of the reduction phase, are combined together again through the use of easier statistical function and all the data sets are aggregated into only one set. This one set can be described as follows,

$$AS = AS_1 \times AS_2 \times AS_3 \times \cdots \times AS_n \tag{9.1}$$

Here AS is an aggregated data set consisting of different similar sets. However, this combination never ensures the noise free condition of the overall new data set. Therefore, the concern regarding noise and irrelevant data still exits. Here rule conjugation is the solution for this concern. The rule conjugation process is described below:

- *Rule conjugation*: In general, rule conjugation is a pre-processing system embedded with a filtering approach. This mechanism processes the AS set and removes the irrelevant as well as noisy data when extracting information. Recent studies found in the literature on this issue indicate that lots of researchers have concentrated on the heuristic technique for this rule conjugation process. For our work, we have embedded this pre-processing or filtering approach, which divides the data sets in two portions, initially depending on the relation between those two attributes. Further steps complete the execution by dividing those sets into another two sets. Thus, the process continues until the relation between attributes stops. This approach eliminates lots of noisy, as well as unmatched, data from the sets which is able to make a faster view of overall process.
- *Updating rule*: Data redundancy is one of the obstacles for making processes faster and efficient. This approach basically deals with data redundancy by eliminating redundant data within the input data sets. It is, however, part of the overall reduce phase in the MapReducing framework. For the overall process to happen successfully, a fusion-clustering algorithm (defined as clustering ensemble) has been applied. This fusion-clustering algorithm basically aggregates multiple portions of a data set into a single portion

depending on the similar value inside those data sets. Moreover, for this fusion clustering K-means clustering is used. If we consider that the input data sets are distributed into N partitions, then the Euclidean distance function is implemented as a distance calculation, thus this is the most useful way to calculate distances among various types of data. Thus K-means clustering can be illustrated mathematically as follows

$$D\left(C_1,C_2,C_3\ldots\ C_n = \sum_{k=1}^{n}\left[(1/n)\sum_{i=1}^{n} m_i \parallel x_i - C_{\Re}\right]\right) \quad (9.2)$$

where $m_i = 1$, if x_i belongs to k cluster, otherwise 0. The notation ǁ denotes the absolute form. This is how the filtering or elimination of data redundancy is implemented though.

9.4.2 Pre-Processing Mechanism

The massive growth in all aspects of human existence has been observed recently which is the factor of obstacle for technological evolution and strengthened the threat towards the data analysis procedure. Nevertheless, available approaches and improvement of recent technological as well as machine learning approaches make it possible for the big data feature detection using various sources of recent information technology [38–40]. Addressing big data sets is most challenging and non-traditional part of this recent big data research trend.

Pre-processing trend is usually a combination of various tasks to be executed during pre-processing mechanism implementation. The tasks are defined as data cleaning, data normalization, data transformation, missing values imputation, data integration, noise identification, merging data, reducing redundancy and so on. All of these tasks are illustrated in Figure 9.4.

However, for this work we have approached pre-processing by dividing our overall input data sets into two parts depending on the interrelation between two attributes. We have embedded the pre-processing mechanism within the MapReduce approach,

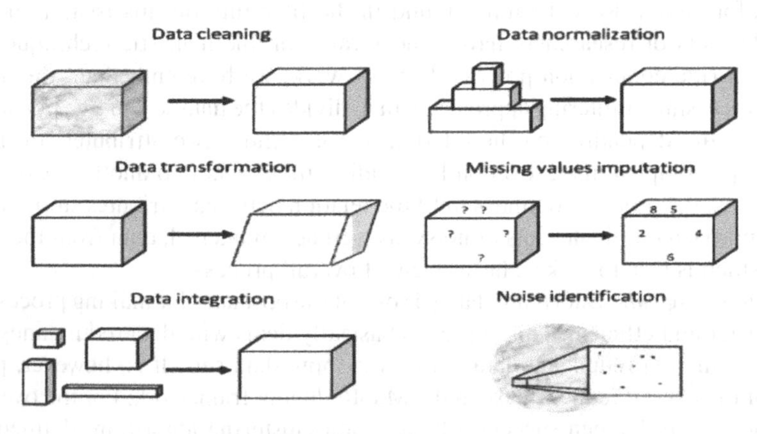

FIGURE 9.4 Tasks of pre-processing mechanism. (From S. García et al. *Big Data Analytics*, vol. 1, no. 9, 2016 [46].)

which initially reduces irrelevant data from the considered data sets. This is a type of real-time pre-processing. When one instance of data is considered then we check whether it has any missing data and act accordingly. For better clarification, the whole data sets is divided into two subsets depending on the interrelation between the attributes 'RACE' and 'Gender'. While accessing RACE, if the corresponding value of gender for that particular race is missing then the data is removed, and the entire row is reduced as well. A similar process is applied to the Gender attribute. The division of data sets will be continued until male and female file not created from whole data sets. Once those two files are created containing male and female frequency along with their corresponding A1C, insulin level and hospital readmission rate the others can be measured through the use of those information we extracted so far.

9.5 RESULTS AND FINDINGS

The results of applying this proposed algorithm to the data sets are presented in tabular format in this section. The measurement of accuracy, specificity, and rate of reduction of our implemented MapReducing algorithm has been thoroughly analysed. Due to its parallel computing, the MapReduce approach produces better and more accurate results than the traditional approaches. Additionally, when the data sets are considerably large the working principle of MapReduce appears to be more effective. Thus, when the number of tasks is increased, our proposed approach responds with faster and more efficient manipulation. Table 9.2 illustrates the total number of male and female patients considering their races using our proposed technique as well as in WEKA.

Table 9.2 shows that rate of male patients is greater than female. However, our approach detects 99,492 patients combining both male and female, but the actual number is 101,763. Thus, the accuracy measurement of pre-processing based on a MapReduce approach is:

$$\begin{aligned}\text{Accuracy} &= \frac{\text{Total findings}}{\text{Total number of patients}} * 100\\ &= \frac{99{,}492}{101{,}763} * 100\\ &= 97.76\%\end{aligned}$$

TABLE 9.2
Total Number of Patients Detected After Imposing Pre-Processing Based MapReduce

Race	Male	Female
African–American	7482	11,728
Caucasian	36,410	39,689
Hispanic	945	1092
other	757	748
Asian	323	318
Total detected	45,917	53,575
Total (male + female)	99,492	
Total (in WEKA tool)	101,763	

We can determine the reduction rate by considering *Sizeof(Classified_group)* and *Sizeof(total_instances)*. Here 9 attributes are considered as classified groups and total size of instances are approximately 101,763. Thus, the reduction rate for pre-processing based on the MapReduce approach can be determined by the following equation:

$$\text{Reduction\_rate} = \frac{Sizeof\,(Classified\_groups)}{Sizeof\,(total\_instances)} * 100\%$$
$$= \frac{9}{101,763} * 100\%$$
$$= 0.0088\%$$

Table 9.3 illustrates the number of male and female patients without performing pre-processing on the data set.

The accuracy of reduction rate of pre-processing without MapReduce can be estimated as follows:

$$\text{Accuracy} = \frac{\text{Total findings}}{\text{Total number of patients}} * 100$$
$$= \frac{95941}{101,763} * 100$$
$$= 94.27\%$$

The reduction rate of this algorithm can be represented in a similar way as before considering *sizeof(classified_groups)* and *sizeof(total_instances)*. For this case only 3 groups are considered as classified groups internally while implementing the MapReduce algorithm itself. Thus, the reduction rate can be calculated as,

TABLE 9.3
Total Number of Patients Detected Without Imposing Pre-Processing in MapReduce

Race	Male	Female
African–American	7123	10,923
Caucasian	35,320	38,900
Hispanic	745	982
Asian	302	312
Other	710	624
Total detected	44,200	51,741
Total (male + female)	95,941	
Total (in WEKA tool)	101,763	

$$\text{Reduction\_rate} = \frac{Sizeof\,(Classified\_groups)}{Sizeof\,(total\_instances)} * 100\%$$
$$= \frac{3}{101{,}763} * 100\%$$
$$= 0.00294\%$$

Pre-processing based on MapReduce seems to be more effective for big data analysis as it represents a higher level of accuracy and reduction rate of nodes within the internal process of the MapReduce algorithm.

We have evaluated the relationship between various attributes in the diabetes data set. Table 9.4 represents the relationship between insulin level and effect of HbA1c for diabetes detection and prediction of patients. Moreover, the result clearly shows that when the value of HbA1c is greater than 8, the number of people affected by diabetes is more than that of other values for all insulin levels. Thus, the patients who have HbA1c greater than 8 are at a high risk and should be treated immediately. The graphical representation of the relationship between insulin level and HbA1c for Caucasian is illustrated in Figure 9.5.

The graph in Figure 9.5 is a 3D representation regarding the effect of HbA1c on diabetes. This graph specifically demonstrates that a value of HbA1c > 8 is high enough interrelated to be affected by diabetes. Technically, here the x-axis represents the value of HbA1c, the y-axis is the number of patients and the z-axis indicates glucose level. From Table 9.4, it can also be estimated that the value for HbA1c is greater than 8 within those people who adopt steady insulin level. However, the patients who take down as well as up level of insulin in Caucasian race are found to be approximately similar in size. A similar result can be observed for pre-processing without the MapReduce approach, which is illustrated in Table 9.5.

A slight difference is observed for the number of patients for each level of insulin. If we consider only one race and one insulin level for clarification of our comparison between the results of our proposed approach with pre-processing less mechanism, then the difference can be estimated. Table 9.6 shows the comparison depending on Caucasian race steady insulin level patients' HbA1c value.

A remarkable difference is found between the results of pre-processing with and without MapReduce for our data sets. The overall approach is best suited for all kinds of big data manipulation.

Another interrelationship is shown in Table 9.7 between insulin level and the number of readmissions of patients. It is found that the patients who are readmitted in more than 30 days adopt high level of insulin dose per day. This trend has been observed for every race according to our findings.

Table 9.7 shows a combination of insulin level and value of readmission of patients of different race to predict the patients who are most cases prescribed to meet consultant in more than 30 days and also considered as less risky than UP level insulin user. Figure 9.6 shows the comparative graph of the same analysis for Caucasian.

TABLE 9.4
Relation Between Insulin Level and HbA1c Using Pre-Processing Based MapReduce

Race	Insulin Level	Number of People Based on Value of HbA1c		Total
Asian	Up	HbA1c(>7)	3	20
		HbA1c(>8)	14	
		HbA1c(Normal)	3	
	Steady	HbA1c(>7)	5	39
		HbA1c(>8)	21	
		HbA1c(Normal)	13	
	Down	HbA1c(>7)	6	22
		HbA1c(>8)	10	
		HbA1c(Normal)	6	
African–American	Up	HbA1c(>7)	53	550
		HbA1c(>8)	360	
		HbA1c(Normal)	137	
	Steady	HbA1c(>7)	209	1276
		HbA1c(>8)	711	
		HbA1c(Normal)	356	
	Down	HbA1c(>7)	61	3959
		HbA1c(>8)	319	
		HbA1c(Normal)	179	
Caucasian	Up	HbA1c(>7)	392	1910
		HbA1c(>8)	1164	
		HbA1c(Normal)	354	
	Steady	HbA1c(>7)	909	3786
		HbA1c(>8)	1775	
		HbA1c(Normal)	1102	
	Down	HbA1c(>7)	344	1872
		HbA1c(>8)	1152	
		HbA1c(Normal)	376	
Hispanic	Up	HbA1c(>7)	15	78
		HbA1c(>8)	54	
		HbA1c(Normal)	9	
	Steady	HbA1c(>7)	32	150
		HbA1c(>8)	78	
		HbA1c(Normal)	40	
	Down	HbA1c(>7)	10	75
		HbA1c(>8)	48	
		HbA1c(Normal)	17	

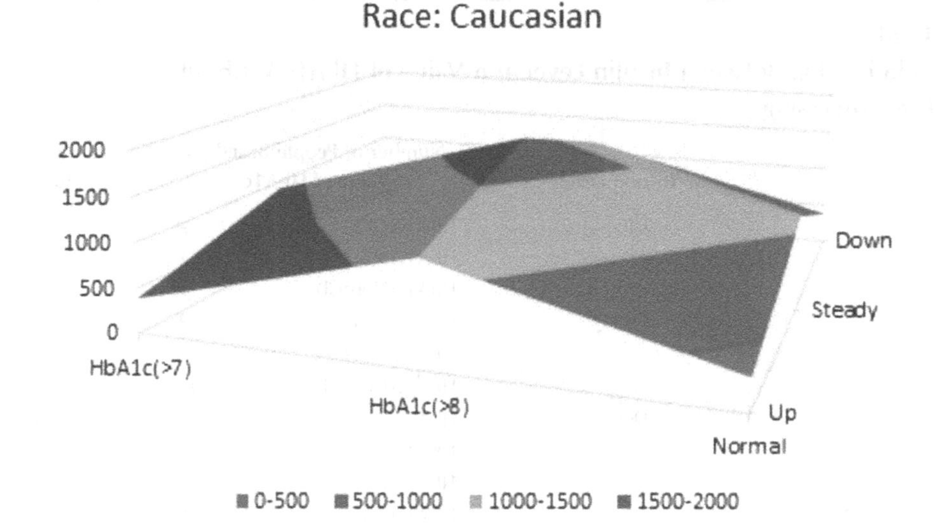

FIGURE 9.5 Impact of HbA1c on diabetes for the Caucasian race.

Figure 9.6 shows the relationship between insulin level and readmission of patients in the hospital for treatment purpose. The graph also illustrates that readmission >30 days are higher for every level of insulin user and among them the rate is higher for those of steady levels. For pre-processing without the MapReduce approach, a similar table has been created and shown in Table 9.8.

The main difference between the findings in Tables 9.7 and 9.8 is whether MapReduce was used in pre-processing or not. Moreover, pre-processing ensures better classification and more efficient clustering along with less memory consumption, which is noticeable in Table 9.7 compared to Table 9.8.

9.6 DISCUSSION

With the emergence of computing technologies in medical science and human health monitoring, diagnosing systems now need to be automated. Because of automated manner, in every single unit of time this sort of automated system needs to interact with server or database to analyse, manipulate, and update data and this immediate interaction with efficient and faster way is mandatory to create harmless human care. Consequently, various recent trends invented in modern technologies are considered to propagate this task and taking further decision. However, with the emergence of these technologies, data created from these types of sources are becoming larger in volume, greater in velocity, higher in dimension and creates bottlenecks for the analysis and extraction of significant information. Meanwhile it is important to think of ways to easily manipulate these data sets. In order to resolve these problems we have focused on the MapReduce approach over flourished preprocessing tool. We have thought of a large volume of U.S. healthcare data sets for efficient extraction of necessary information instantly. Surprisingly, the impact of the pre-processing mechanism on unstructured or missing data sets seems to be quite significant when

TABLE 9.5
Relationship Between Insulin Level and Value of HbA1c Without Pre-Processing

Race	Insulin Level	Number of People Based on Value of HbA1c		Total
Asian	Up	HbA1c(>7)	2	14
		HbA1c(>8)	11	
		HbA1c(Normal)	1	
	Steady	HbA1c(>7)	4	35
		HbA1c(>8)	20	
		HbA1c(Normal)	11	
	Down	HbA1c(>7)	5	16
		HbA1c(>8)	9	
		HbA1c(Normal)	2	
African–American	Up	HbA1c(>7)	50	517
		HbA1c(>8)	340	
		HbA1c(Normal)	127	
	Steady	HbA1c(>7)	203	1259
		HbA1c(>8)	701	
		HbA1c(Normal)	355	
	Down	HbA1c(>7)	60	547
		HbA1c(>8)	315	
		HbA1c(Normal)	172	
Caucasian	Up	HbA1c(>7)	342	1720
		HbA1c(>8)	1124	
		HbA1c(Normal)	254	
	Steady	HbA1c(>7)	904	3621
		HbA1c(>8)	1715	
		HbA1c(Normal)	1002	
	Down	HbA1c(>7)	304	1812
		HbA1c(>8)	1132	
		HbA1c(Normal)	376	
Hispanic	Up	HbA1c(>7)	11	70
		HbA1c(>8)	52	
		HbA1c(Normal)	7	
	Steady	HbA1c(>7)	32	150
		HbA1c(>8)	78	
		HbA1c(Normal)	40	
	Down	HbA1c(>7)	10	69
		HbA1c(>8)	42	
		HbA1c(Normal)	17	

TABLE 9.6
Comparison of With and Without Pre-Processing MapReducing for HbA1c and Insulin Level

Race	Pre-Processing Based	Without Pre-Processing
Asian	21	11
Caucasian	1775	1715
African–American	711	701
Hispanic	78	78

TABLE 9.7
Relation Between Insulin Level and Patients' Readmission

Race	Insulin Level	Number of Readmitted Patients		Total
Asian	Up	Readmitted (>30)	16	27
		Readmitted (<30)	11	
	Steady	Readmitted (>30)	24	44
		Readmitted (<30)	20	
	Down	Readmitted (>30)	25	34
		Readmitted (<30)	9	
African–American	Up	Readmitted (>30)	124	13
		Readmitted (<30)	40	
	Steady	Readmitted (>30)	227	293
		Readmitted (<30)	66	
	Down	Readmitted (>30)	109	141
		Readmitted (<30)	32	
Caucasian	Up	Readmitted (>30)	499	637
		Readmitted (<30)	138	
	Steady	Readmitted (>30)	573	722
		Readmitted (<30)	149	
	Down	Readmitted (>30)	433	580
		Readmitted (<30)	147	
Hispanic	Up	Readmitted (>30)	21	26
		Readmitted (<30)	5	
	Steady	Readmitted (>30)	24	29
		Readmitted (<30)	5	
	Down	Readmitted (>30)	10	16
		Readmitted (<30)	6	

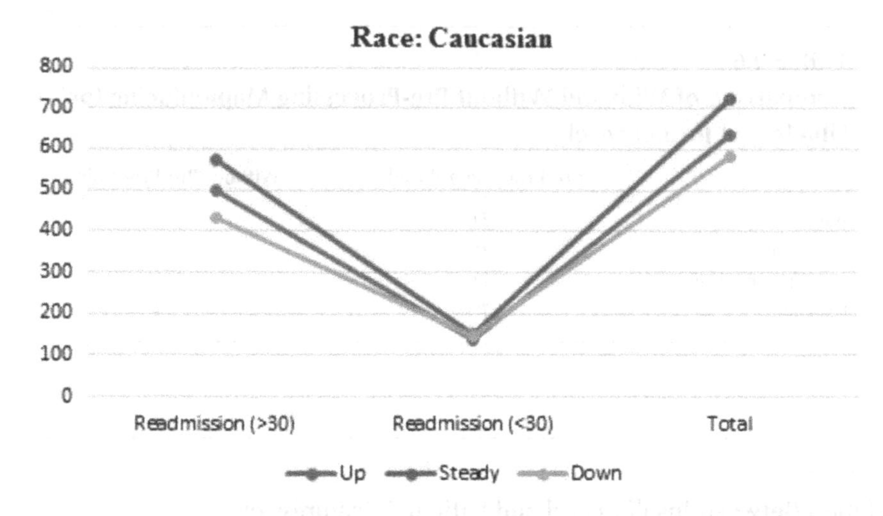

FIGURE 9.6 Pre-processing with and without MapReducing result for Caucasian race.

processing big data sets. We have successfully and efficiently extracted important information, which is demonstrated theoretically as well as graphically in Section 9.5.

9.6.1 Limitations

In spite of these successful accomplishments there are some drawbacks which might create poor performance with the MapReduce approach. While exploring massive healthcare data sets, the lack of high-level database language like SQL can be a cause of problems for efficient manipulation. Moreover, it appears to be challenging to implement an iterative algorithm in support of ad-hoc data exploration as well as stream processing. Additionally, while processing data online by using MapReduce, the performance as well as the latency seems to be poor in action. Sometimes, optimization job inside MapReduce creates problem that occurs while integration of MapReduce, distributed file system, RDBMSs and NoSQL stores. While considering extended large volume data sets, various statistical and computational challenges appeared which could be neglected and removed with the help of a pre-processing mechanism. To resolve these kinds of problems there are various approaches including Pregel [41], Giraph [42], Spark [43], HaLoop [44] and Twister [45].

9.6.2 Challenges

There are some challenges while working on this contribution appeared in case of heterogeneity and high dimensionality which are discussed here

- Noise accumulation can be a problem for the programmers, and it may slow down the overall potential of the MapReduce approach
- Sometimes, false interrelationships can be appear
- Inconsistencies and false discoveries are conducted sometimes

- While focusing on fault tolerance, results of mapping phase are stored in local files before sending to reducers. Meanwhile, reducers store the final result to a high-overhead distributed file system, which eventually adds significant latency to the overall processing mechanism
- The security system is very poor and HDFS needs to be protected from vulnerabilities and breaches

There is a lot of research work currently underway to overcome these limitations. However, in the present work we have minimized some of the major limitations by embedding a *pre-processing* tool using the MapReduce mechanism. In spite of this, inconsistency as well as redundancy might appear while data size is considerably large to manipulate and in that case optimization approach over MapReduce could be a possible solution. Moreover, classification algorithms might pave overall findings towards a positive peak.

TABLE 9.8
Insulin Level vs. Value of Readmission Without Applying Pre-Processing on MapReducing

Race	Insulin Level	Number Readmitted Patients		Total
Asian	Up	Readmitted (>30)	10	16
		Readmitted (<30)	6	
	Steady	Readmitted (>30)	22	38
		Readmitted (<30)	16	
	Down	Readmitted (>30)	12	17
		Readmitted (<30)	5	
African–American	Up	Readmitted (>30)	101	136
		Readmitted (<30)	35	
	Steady	Readmitted (>30)	201	264
		Readmitted (<30)	63	
	Down	Readmitted (>30)	100	128
		Readmitted (<30)	28	
Caucasian	Up	Readmitted (>30)	422	544
		Readmitted (<30)	111	
	Steady	Readmitted (>30)	525	666
		Readmitted (<30)	141	
	Down	Readmitted (>30)	411	523
		Readmitted (<30)	112	
Hispanic	Up	Readmitted (>30)	20	22
		Readmitted (<30)	2	
	Steady	Readmitted (>30)	20	23
		Readmitted (<30)	3	
	Down	Readmitted (>30)	4	7
		Readmitted (<30)	3	

9.7 CONCLUDING REMARKS

The present world is challenged by the recent trends and advancements of big data. Proper management of this big data plays a key role in mining required as well as unforeseen information from this huge volume of data. The present work demonstrates how the combination of a proper pre-processing mechanism with MapReduce can effectively handle this challenge of extracting knowledge lying inside the data. This work shows how to manage and lower space requirements and time consumption for big data analysis. Comparative analyses are demonstrated for a clear vision of this contribution. Bunch of comparisons interpret that our considered approach is flourished enough to be remark as big data manipulator. So far we have analysed male and female diabetes patient data sets collected from one American hospital. We have also illustrated the effects of HbA1c value along with insulin level for the prediction of diabetes. The interrelationship between insulin level and readmission of patients has also been shown. Accuracy, specificity as well as reduction rate of nodes have also been conducted to measure performance of our adopted approach. Future directions of our work include optimization and normalization techniques to demonstrate more accuracy and efficiency.

9.8 SCOPE AND RELATION

The present situation is characterized by the revolution of modern technologies as well as the digitalization of information, communication, management and retrieval systems. Moreover, the emergence of wearable devices strengthens the potential for developing enormously effective and efficient prediction capacities as well as analysis mechanisms in the field of medical healthcare. IoT along with recent technologies plays a dominating role while resolving any issues associated with human healthcare where DNA, cells, tissues, organisms arc the primary concern. These organisms create a situation which makes us concerned about data processing and analysing. That is basically known as big data. The aggregation of medical big data with IoT may encourage the development of new automated machine-learning approaches for the prediction of disease, diagnosis and treatment. Therefore, from the perspective of this context it is immediately estimated that machine learning related algorithms, tools, and infrastructures can be a helpful paradigm to model as well as prediction of overall management system. However, in the field of medicine and human healthcare it is quite difficult to mine, analyse, process, characterize, classify and cluster a wide range of medical data while preserving privacy, security, establishment of standards, scalability, availability, user friendliness and the continuous improvement of the tools and technologies. Thus, it is the demand of the present world to develop and implement machine-learning approaches to analyse, process and manipulate big data sets.

This book is titled *Medical Big Data and Internet of Medical Things: Advances, Challenges, and Applications* and deals with recent trends and advances for mining, learning, predicting and analysing medical big data, in case of both real time and off-line systems. This book also explores a variety of data formats, data types and complex features of data sets. Furthermore, this book also represents distributive, dynamic, data management, optimization techniques and governance of medical big data collected from multiple heterogeneous sources. Also, trust and privacy issues in healthcare big data analytics have been discussed. Thus, above discussion makes our realization towards the scope and contents of this book and so this book demonstrates recent advancement, challenges, applications in healthcare embedded with internet of things. However, advancements refer to the development side of various machine-learning approaches and their applications in big data analysis. Moreover, challenges while analysing these sorts of data are also indicated here. Meanwhile, our chapter depicts the applications and challenges of machine-learning approaches for medical healthcare big data. Moreover, our work covers one of the recent most successful trends in the manipulation of healthcare big data. The title of the book illustrates advances, challenges and applications of various machine learning approaches to analysis big data through the use of various wearable medical IoT devices and corresponding to that our proposed title demonstrates how the data generated from the use of wearable medical internet of things devices can be classified and analysed efficiently and immediately to take further decision using MapReduce along with flourished pre-processing mechanism.

REFERENCES

1. The World's Technological Capacity to Store, Communicate, and Compute Information. MartinHilbert.net. Retrieved 13 April 2016.
2. D. Boyd, and K. Crawford, Six provocations for big data (September 21, 2011). *A Decade in Internet Time: Symposium on the Dynamics of the Internet and Society*, September 2011, http://dx.doi.org/10.2139/ssrn.1926431.
3. Data, data everywhere. *The Economist*. 25 February 2010. Retrieved 9 December 2012.
4. N. Dey, and A. S. Ashour, Computing in medical image analysis. In *Soft, Computing Based Medical Image Analysis* (pp. 3–11), 2018.
5. G. Elhayatmy, N. Dey, and A. S. Ashour, 2018. Internet of things based wireless body area network in Healthcare. In *Internet of Things and Big Data Analytics toward Next-Generation Intelligence* (pp. 3–20). Springer, Cham.
6. L. Moraru, S. Moldovanu, L. T. Dimitrievici, F. Shi, A. S. Ashour, and N. Dey, Quantitative diffusion tensor magnetic resonance imaging signal characteristics in the human brain: A hemispheres analysis, *IEEE Sensors Journal*, vol. 17, no. 15, 2017, 4886–4893.
7. D. Wang, Z. Li, L. Cao, V. E. Balas, N. Dey, A. S. Ashour, and F. Shi, Image fusion incorporating parameter estimation optimized Gaussian mixture model and fuzzy weighted evaluation system: A case study in time-series plantar pressure data set, *IEEE Sensors Journal*, vol. 17, no. 5, 2017, pp. 1407–1420.
8. M. S. Kamal, N. Dey, and A. S. Ashour, Large scale medical data mining for accurate diagnosis: A blueprint. In *Handbook of Large-Scale Distributed Computing in Smart Healthcare* (pp. 157–176). Springer, Cham, 2017.
9. N. Dey, A. S. Ashour, F. Shi, and R. S. Sherratt, Wireless capsule gastrointestinal endoscopy: Direction-of-arrival estimation based localization survey, *IEEE Reviews In Biomedical Engineering*, vol. 10, 2017. pp. 2–11.
10. C. Bhatt, N. Dey, A. Ashour, *Internet of Things and Big Data Technologies for Next Generation Healthcare*, vol. 23, Series ISSN 2197-6503, doi: 10.1007/978-3-319-49736-5, Springer, 2017.
11. S. Chakraborty, S. Chatterjee, A. S. Ashour, K. Mali, and N. Dey, Intelligent computing in medical imaging: A study, *Advancements In Applied Metaheuristic Computing*, vol. 143, 2017, pp. 143–163.

12. N. Dey, A. S. Ashour, F. Shi, S. J. Fong, and R. S. Sherratt, Developing residential wireless sensor networks for ECG healthcare monitoring, *IEEE Transactions on Consumer Electronics*, vol. 63, no. 4, 2017, pp. 442–449.
13. N. Dey, A. S. Ashour, and S. Borra, (Eds.). *Classification in BioApps: Automation of Decision Making*, vol. 26, Springer, 2017.
14. N. Singh, and S. Agrawal, A review of research on MapReduce scheduling algorithms in Hadoop, *International Conference on Computing, Communication & Automation*, Noida, 2015, pp. 637–642, doi: 10.1109/CCAA.2015.7148451.
15. J. V. N. Lakshmi, and A. Sheshasaayee, Machine learning approaches on map reduce for big data analytics, *2015 International Conference on Green Computing and Internet of Things (ICGCIoT)*, Noida, 2015, pp. 480–484.
16. K. Shim, MapReduce algorithms for big data analysis, *Proceedings of VLDB Endowment*, Vol. 5, no. 12, August 2012, pp. 2016–2017, http://dx.doi.org/10.14778/2367502.2367563.
17. A. Fernandez, S. Río, and F. Herrera, Fuzzy rule based classification systems for big data with MapReduce: Granularity analysis, *Advances in Data Analysis and Classification*, Vol. 11, 2017, pp. 711–730, doi: 10.1007/s11634-016-0260-z.
18. Q. Yao, Y. Tian, P. F. Li et al., Design and development of a medical big data processing system based on Hadoop, *The Journal of Medical Systems*, vol. 39, no. 23, 2015, https://doi.org/10.1007/s10916-015-0220-8.
19. Ho Ting Wong, Qian Yin, Ying Qi Guo, Kristen Murray, Dong Hau Zhou, and Diana Slade, Big data as a new approach in emergency medicine research, *Journal of Acute Disease*, vol. 4, no. 3, 2015, pp. 178–179, ISSN 2221-6189, https://doi.org/10.1016/j.joad.2015.04.003.
20. M. Ojha, and D. K. Mathur, Proposed application of big data analytics in healthcare at Maharaja Yeshwantrao Hospital, *3rd MEC International Conference on Big Data and Smart City*, 2016.
21. V. Subramaniyaswamy, V. Vijayakumar, R. Logesh, and V. Indragandhi, Unstructured data analysis on big data using map reduce, *Procedia Computer Science*, vol. 50, 2015, pp. 456–465, ISSN 1877-0509, https://doi.org/10.1016/j.procs.2015.04.015.
22. H. Yu, and D. Wang, Research and implementation of massive health care data management and analysis based on Hadoop, *2012 Fourth International Conference on Computational and Information Sciences, Chongqing*, 2012, pp. 514–517.
23. C. Yang, Q. Huang, Z. Li, K. Liu, and F. Hu, Big Data and cloud computing: Innovation opportunities and challenges, *International Journal of Digital Earth*, vol. 10, no. 1, 2017, pp. 13–53.
24. E. P. Ephzibah, R. Sujatha, Big data management with machine learning inscribed by domain knowledge for health care, *International Journal of Engineering & Technology*, vol. 6, no. 4, 2017, pp. 98–102, doi: 10.14419/ijet.v6i4.8214.
25. M. Dayal, and N. Singh, Indian health care analysis using big data programming tool, *Procedia Computer Science*, vol. 89, 2016, pp. 521–527, ISSN 1877-0509, https://doi.org/10.1016/j.procs.2016.06.101.
26. M. A. Alkhatib, A. Talaei-Khoei, and A. H. Ghapanchi, Analysis of research in healthcare data analytics, *Australasian Conference on Information Systems*, Sydney, 2015, pp 1–16.
27. S. Sonnati, Improving healthcare using big data analytics, *International Journal of Scintific and Technology Research*, vol 6, no. 03, March 2017, pp. 142–146.
28. B. Strack, J. P. DeShazo, C. Gennings, J. L. Olmo, S. Ventura, K. J. Cios, and J. N. Clore, Impact of HbA1c measurement on hospital readmission rates: Analysis of 70,000 clinical database patient records, *Biomed Research International*, vol. 2014, Article ID 781670, 2014, pp. 1–11.

29. S. Mahmoudi, M. A. Belarbi, and G. Belalem, 'PCA as Dimensionality Reduction for Large-Scale Image Retrieval Systems,' *International Journal of Ambient Computing and Intelligence*, vol. 8, no. 4, October 2017, pp. 45–58.
30. W. Han, Y. Kang, Y. Chen, and X. Zhang, A MapReduce approach for SIFT feature extraction, *2013 International Conference on Cloud Computing and Big Data*, Fuzhou, 2013, pp. 465–469, doi: 10.1109/CLOUDCOM-ASIA.2013.22.
31. S. Kamal, S. H. Ripon, N. Dey, A. S. Ashour, and V. Santhi, A MapReduce approach to diminish imbalance parameters for big deoxyribonucleic acid data set, *Computer Methods and Programs in Biomedicine*, vol. 131, 2016, pp. 191–206, ISSN 0169-2607, https://doi.org/10.1016/j.cmpb.2016.04.005.
32. D. A. A. Raj, and T. Mala, Cloud Press: A next generation news retrieval system on the cloud, *2012 International Conference on Recent Advances in Computing and Software Systems*, Chennai, 2012, pp. 299–304, doi: 10.1109/RACSS.2012.6212684.
33. V. López, S. del Río, J. Benítez, and F. Herrera, Cost-sensitive linguistic fuzzy rule based classification systems under the MapReduce framework for imbalanced big data, *Fuzzy Sets Systems*, vol. 258, 2015, pp. 5–38.
34. R. Batuwita, and V. Palade, Adjusted geometric-mean: A novel performance measure for imbalanced bioinformatics data sets learning, *Journal of Bioinformatics and Computational Biology*, vol. 10, no. 4, 2012, pp. 1–23.
35. D. Miner, and A. Shook, *MapReduce Design Patterns: Building Effective Algorithms and Analytics for Hadoop and Other Systems*, O'Reilly Media, 2012.
36. S. G. Manikandan, and S. Ravi, Big Data analysis using Apache Hadoop, *2014 International Conference on IT Convergence and Security (ICITCS)*, Beijing, 2014, pp. 1–4, doi: 10.1109/ICITCS.2014.7021746.
37. X. Yang, J. Zola, and S. Aluru, Parallel metagenomic sequence clustering via sketching and maximal quasi clique enumeration on map-reduce clouds, *Parallel & Distributed Processing Symposium (IPDPS), 2011 IEEE International, IEEE*, 2011, pp. 1223–1233.
38. N. Dey, A. E. Hassanien, C. Bhatt, A. S. Ashour, and S. C. Satapathy, *Internet of things and big data analytics toward next-generation intelligence*, vol. 30, ISSN 2197-6503, doi: 10.1007/978-3-319-60435-0, 2018.
39. M. S. Kamal, M. G. Sarowar, N. Dey et al., *International Journal of Machine Learning and Cybernetics*, 2017, https://doi.org/10.1007/s13042-017-0710-8.
40. D. Acharjya, and A. Anitha, 'A comparative study of statistical and rough computing models in predictive data analysis,' *International Journal of Ambient Computing And Intelligence*, vol. 8, no. 2, April 2017, pp. 32–51, doi: https://doi.org/10.4018/IJACI.2017040103.
41. G. Malewicz, M. H. Austern, A. J. C. Bik, J. C. Dehnert, I. Horn, N. Leiser, and G. Czajkowski, Pregel: A system for large-scale graph processing, *Proceedings of the 2010 ACM SIGMOD International Conference on Management of Data*, 2010.
42. Apache Giraph, https://giraph.apache.org/.
43. Apache Spark, https://spark.incubator.apache.org/.
44. Y. Bu, B. Howe, M. Balazinska, and M. D. Ernst, HaLoop: Efficient iterative data processing on large clusters, *Proceedings of The VLDB Endowment*, vol. 3, no. 1–2, 2010, pp. 285–296.
45. J. Ekanayake, H. Li, B. Zhang, T. Gunarathne, S. Bae, J. Qiu, and G. Fox, Twister: A runtime for iterative MapReduce, *Proceedings of the 19th ACM International Symposium on High Performance Distributed Computing*, 2010.
46. S. García, S. Ramírez-Gallego, J. Luengo, J. Manuel Benítez, and F. Herrera, Big data pre-processing: Methods and prospects, *Big Data Analytics*, vol. 1, no. 9, 2016, https://doi.org/10.1186/s41044-016-0014-0.

10 IoT and Robotics in Healthcare

Varnita Verma, Vinay Chowdary, Mukul Kumar Gupta, and Amit Kumar Mondal

CONTENTS

10.1 INTRODUCTION

Robotics is one of the emerging and innovative fields of research in universities since the last two decades. The robot was first introduced in 1920 by Karl Capek. *There is no fixed definition of the robot but as per the Robot Institute of America (RIA): a robot is a programmable multifunctional manipulator designed to move material, parts, tools or specialized devices through variable programmed motions, for the performance of a variety of tasks.*

Robotics is the engineering science and technology that has applications mainly in military applications, intelligent home applications, industrial automation, health services and outer space applications. Robotic unmanned spacecrafts are used as the key to exploring the stars, planets and so on.

With the development of electronic technology like the very large scale of integration (VLSI), ultra large scale of integration (ULSI) and the giga scale of integration (GSI), the new sensor robots are being used in new fields. Robotics now play a vital role in the area like:

a. Aerial robotics
b. Assistive living
c. Bioenergy and self-sustainable
d. Biomimetic and neuro-robotics
e. Medical robotics
f. Non-linear robotics
g. Robot vision
h. Safe human–robot interaction
i. Self-repairing robotics systems
j. Smart automation
k. Soft robotics
l. Swarm robotics
m. Unconventional computation etc.

Since the last decade robotics has played a tremendous role in the field of healthcare and rehabilitation. Robotics in healthcare focus on the provision of a grade of autonomy, or a service to people who have permanently or temporally lost partial autonomy owing to physical or mental abilities. If the current trend continues, the global shortage of healthcare workers is projected to be more than 14 million in 2030 according to a world trade organization estimation. The coming times will be of robotics, as can be verified from the fact that in 2011 the global medical robotic systems market was worth $5.48 billion, and was expected to more than double to $13.6 billion in 2018. IoT start-ups are finding new implementations within healthcare and are leveraging connected sensors for better diagnoses, monitoring and guiding patient treatment. The major advantages of the IoT-driven healthcare are the following:

a. Reduces costs of medical treatment
b. Betters outcomes of treatment
c. Better diseases management and reduced errors
d. Improves patient experience
e. Upgrades management of drugs

The robots in the healthcare industry are often used as service robots, and economic considerations may be the driving factor for their application. In healthcare, teleoperations play an important role.

Teleoperated system are such that they have a replica of the receiver end mimicking the functions on the specialist side, which stands in for the operating system remotely [1,2], thus human beings can control the teleoperated robots. The control signal sent by humans can be wired (through cables) or wireless. The need for teleoperated systems came from the urge to operate devices able to perform controlled functions from a distance, and in that respect teleoperations are different from telepresence. Telepresence refers to a category of technology that allows users to sense and feel stimuli from another individual. Telerobotics is the combination of these two fields, telepresence and teleoperation. A fully autonomous device is a robot. Teleoperators are also called telemanipulators, but devices that perform autonomous work are called telerobots. A teleoperated robot consists of a central processing unit, which sends data from the input device (generally haptic devices) to a microcontroller through Wi-Fi or cable. Microcontroller Wi-Fi modules are connected to others that read and received data, and then actuators are energized—commonly used actuators are motors. The microcontroller decides on the direction of operation of the motor, that is, clockwise or anti-clockwise.

Teleoperated robot systems have been developed for operators to perform various tasks via a remote control system. The teleoperation method, which is most frequently used in typical teleoperated robotic systems, involves a master device that collects target task commands from an operator and sends them to a slave robot to carry out them out. For example, Heikkila et al. proposed the functional design of a manufacturing robot cell [3]. Yamada et al. introduced a construction telerobot system using virtual reality [4]. Zhao et al. developed a construction telerobotic system that has wide applications in restoration work in stricken areas [5]. Kwon et al. developed a microsurgical telerobot system [6]. Geerinck et al. introduced the operability of an advanced demonstration platform incorporating reflexive teleoperated control concepts developed on a mobile robot system [7]. A comparative study of different robotic solutions can be seen in Table 10.1.

Teleoperations find their applications in many fields such as:

a. Medicine
b. Robotic surgery
c. Space exploration
d. Large machinery management
e. Disaster management
f. Elderly care
g. Bomb disposal
h. Remotely operated vehicles
i. Handling radioactive materials

Other related healthcare benefits from robotics are robots used in rehabilitation, whose aim is to provide people with a tool that (slightly) compensates their disabilities.

10.2 APPLICATION OF ROBOTICS IN HEALTHCARE

The synergy between healthcare and robotics is deeply intertwined in the field of surgery, providing great aid to the surgeons' hands. Total integration of surgical care includes simulation training, preoperative planning and rehearsal, intraoperative navigation, post-operative assessment, open surgery, minimally invasive and remote surgery. Robotic surgery reduces surgical trauma and loss of healthy tissue, leading to the fast recovery of the patient, as these systems operate without tremors and make precise micro-motions with prespecified micro-forces.

The robotic assistive surgery (RAS) includes a motorized system to control the motion of the intervention, and in computer assisted surgery (CAS) the computer

TABLE 10.1
Comparative Study of Different Robotic Solutions

#	Robotic Solution	Brief About the Robotic Solution	Specialty	Year	Category
1.	PROBOT	Is a floor robot used for teaching advanced control technique	Prostatecttomy	2004	Stand alone device
2.	ROBODOC (Integrated Surgical Supplies Ltd. from Sacramento, CA, USA) [8,9]	To create robotic surgery system that would redefine precision in joint replacement procedures	Hip replacement surgery	2006	Hands-on compliant control
3.	FIPS Endoarm [10]	4 DoF, maintain an invariant point of constraint motion, FIST Endoarm was remotely controlled with a finger ring, which was clipped to surgeon's instruments			Tele Robotic
4.	Endoassist (Armstrong Healthcare Ltd.) [11,12]	Robotic camera holder, allows surgeon to control its movement with the surgeons head movement			Tele Robotic
5.	Gagner et al. [13,14]	6 DoF prototyped robotic surgical assistant, controlled via joystick by a surgeon in a remote room	Laparoscopic surgery	1993	Tele Presence
6.	da Vinci Surgical System (Intuitive Surgical Inc.)	It is designed to facilitate complex surgery using minimal invasive approaches and is controlled by a surgeon from a consol	Urological, gynecological and gastronomical surgery [15–17], cardiac surgery [18–20], abdominal surgery [21–23]	1999	Tele Robotic
7.	da Vinci Surgical System S model	Improved robotic arm movements, console display and simpler set ups		2006	
8.	da Vinci Si	Offers dual consoles for simultaneous operations of two individuals		2009	

(*Continued*)

TABLE 10.1 (*Continued*)
Comparative Study of Different Robotic Solutions

#	Robotic Solution	Brief About the Robotic Solution	Specialty	Year	Category
9.	ZEUS (Computer Motion Inc., Goleta, CA) [24]	Operates in master–slave mode, helps surgeon (master) to operate a console, which controls a robot (slave)	Cardiac surgery [25–27]	1994	Tele Robotic
10.	Automated endoscope system for optimal positioning (AESOP) (Computer Motion, Santa Barbara, CA, USA) [28–31]	The robotic system has an end effector that is adaptive to hold surgical instruments such as an endoscope	Urological [32], laparoscopic surgery	1994	Voice Controlled/ Tele Robotic
11.	SGRCCS [33]	Modified version of AESOP, to work on colour tracking mechanism		1999	Tele Robotic
12.	Advanced robotic tele manipulator for minimally invasive surgery (ARTEMIS) [34,35]	6 DoF. Voice controlled, finger-ring joy stick. It has force-feedback		1999	Tele Robotic
13.	Laprotek (Brock-Rogers, Boston, USA) [36]	It provides 6 DoF and system is moved by mechanical metal cables. Control of end-effectors is realized by electronic data gloves		2000	Tele Robotic

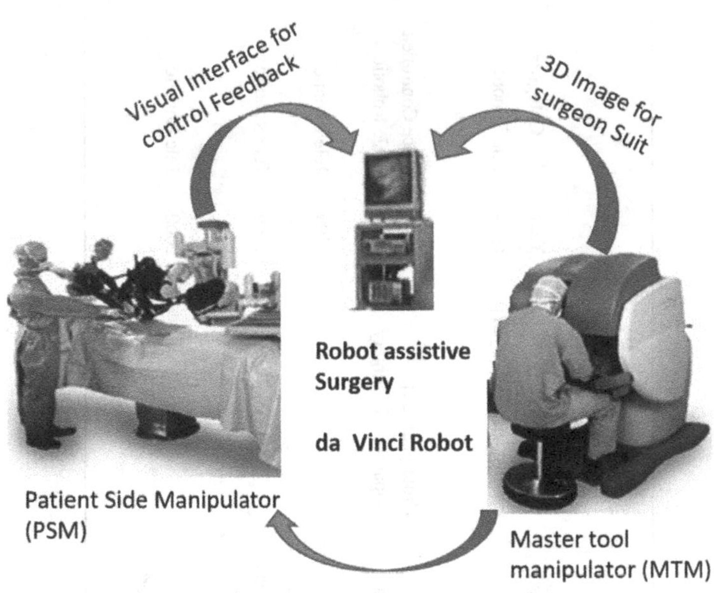

FIGURE 10.1 Robot assistive surgery. (Modified from Okamura, A.M. et al. *IEEE Robotics & Automation Magazine*, 2010. 17(3): p. 26–37. [83])

interface provides assistance to the doctor to control the intervention suite manually. CAS is generally used for the detection of tools, motion planning, identification of cuts, cancer cells and so on. RAS consists of a robot at the patient side manipulator (PSM) that can be autonomously or telemetry controlled by the master tool manipulator (MTM), which gave manipulability capabilities to the doctor, as can be seen in Figure 10.1.

RAS provides human dexterity and ergonomics to make the system more reliable with fewer tremors and more resistance to human errors. The upcoming technology not only focuses on the enhancement of minimally invasive surgery (MIS), but also on the localization of foreign bodies. Different assistive robots are used to provide aid to elderly or handicapped people, such as Cody, Hybrid Assistive Limb (HAL) and many more. The pros and cons of assistive robots can be consulted in Table 10.2.

10.2.1 Mechanical Design Consideration

The specific application(s) for a robot guides its design and structure. The design of robots for surgical tasks needs to feature high precision, high stiffness and limited dexterity, which are suitable for needle placement in a sensitive organ like the eyes, heart, brain and so on, as well as for orthopaedic bone shaping, tumour removal and many other applications. The robots used for surgery usually have high gear ration, low speed and low back-drivability. On the other hand, robots used for MIS and complex surgeries on soft tissues need to have high dexterity, compactness and responsiveness. The robots usually have high speed, high back-drivable mechanisms and low stiffness.

TABLE 10.2
Pros and Cons of Different Assistive Devices

#	Assistive Device	Salient pts.	Limitations
1.	SENsor-Aided intelligent wheelchaiR navigatIOn (SENARIO) [37]	a. High-level navigation to wheelchair b. Works in semi autonomous with the help of joystick or voice commands c. In autonomous modes, follows recorded paths	a. In autonomous mode, navigation is limited to defined paths b. Voice commands need proper filtering of noise, due to the shivering voice of the elderly people
2.	VAHIM [38]	a. Works in manual mode with anti-collision system b. In assisted manual mode, it uses wall following or obstacle detection c. In automatic mode, it uses globally planned paths	
3.	Wheelesley [39]	a. Equipped with graphical user interface in place of joystick	For people with limited upper body mobility
4.	Navchai [40,41], SIAMO, Rolland [41]	a. Automated navigation system	
5.	Kuno et al. [42,43]	a. Caregiver following wheel chair	
6.	Electrooculography (EOG) [44,45]	a. These wheelchairs works based on the retina movement of the user b. Electrodes are placed around the eye to capture the eye movement	
7.	Direction gazing [46,47]	a. Works on the movement of camera (placed in active or passive way) mounted with the wheel chair or user's head	

(Continued)

TABLE 10.2 (*Continued*)
Pros and Cons of Different Assistive Devices

#	Assistive Device	Salient pts.	Limitations
8.	Brain actuated [48,49]	a. Works on brain–computer interface b. EEG signals via passive electrodes are used to capture the brain signals	a. Aged people may find it difficult to use [50]
9.	ROAD [51]	a. Assists users in standing up, sitting down and locomotion b. Robot is connected via ceiling and provides locomotion over a predefined path	a. Most of the operations are addressed for lower limb rehabilitation b. Infrastructural upgradation required c. Navigation is limited to defined paths
10.	Robot for Interactive Body Assistance (RIBA) (developed by RIKEN-TRI Collaboration Centre for Human-Interactive Robot Research) [52]	a. Strong and versatile to lift up patients b. Could be used for modifying lifting trajectories, caregiver's instructions, voice commands etc.	a. Highly sophisticated b. Risk of slippage from robotic arms (user safety)
11.	Intelligent Sweet Home (ISH) [53,54]	a. Smart house consist of various assistive robotic subsystems such as intelligent bed, robotic hoist, intelligent wheel chair etc. b. All the devices are connected	a. Highly sophisticated b. Need modifications to infrastructures
12.	Home Lift, Position and Rehabilitation (HLPR) (developed by National Institute of Standards and Technology, USA) [55–58]	a. Multipurpose assistive chair for locomotion, transfer to toilet or bed, lift assistance and rehabilitation b. Autonomous navigation	a. Highly sophisticated b. Dimensions of HLPR are constrained for operations in small spaces

Previously, industrial robots were modified and adapted to medical implementations. This helped researchers reduce cost and time for the development of robots, and increase rapid prototyping with high reliability through modifications to assure safety and sterility. For these specialized end-effector, tools are required for tool holding, positioning, sniping, suturing and so on, which will decrease the task of assisting individuals and provide satisfactory control over surgery. Different control techniques are used to interface between the surgeon and the surgical tools like joysticks, visual tracking and voice recognition, thereby enabling surgeons to operate more intuitively. The perfect example of a foot-actuated joystick and voice recognition system is the Aesop endoscopic positioner. Aesop and Zeus [59] robots are capable of pivoting along the insertion points, whereas the rotation of the tool end-effector at its distal point provides more tranquillity to the surgeon. The above techniques are used by the DaVinci robot, which is commercially very successful.

A variety of designs have been proposed by numerous research groups who focus on the development of robots able to provide high performance in constrained environments. This has led to MIS systems operating through quarter inch incisions to reduce the loss of healthy tissues. In the construction of instruments for MIS systems, researchers have investigated different technologies such as cable-actuated wrists, shape memory alloy actuators [60], micro-hydraulic systems [61] and electroactive polymers [62].

The placement of a robot also plays an important role during surgery. Usually, robots are mounted on the operation theatre floor, ceiling or bedside of patients. The mounting of robots is necessary to ensure that even if a patient moves the relative motion of the patient and robot would not change. Robotic systems are embedded with different sensory technologies to helps in estimating the pose of the end-effector, imaging environments and different other constraints. Robots should be able to manoeuvre along the patients' contour. The workspaces of robots should be free from geometrical constraints. Also, the images captured for processing should not interfere with the robots' structures or actuators.

10.2.2 Control Paradigms

The controls of surgical robots play a crucial role, as without them robots cannot be considered safe and may cause harm. Different sensors, dynamics tools and vision systems have been mounted on robots, and been controlled by the doctor w.r.t patient. Usually, the control of a robot can be achieved through the three ways listed below:

a. *Hands-on Compliant Control*: Figure 10.2 below shows a robotic tool or control handle grasped by the surgeon's hand to locate the position of the robotic end-effector. It is totally dependent on the surgeon's eye–hand coordination. The force sensors provide the intuitive feedback to the surgeon, enabling visual or sensory feedback. Robodoc [63], EndoBot [64] and many others are used for this purpose. The coordinates of end-effectors can be continuously sent to servers for tracking an operation.
b. *Teleoperative Control*: The surgeon directs the desired motion of a robot through an interface device where PSM controls MTM. Through visual

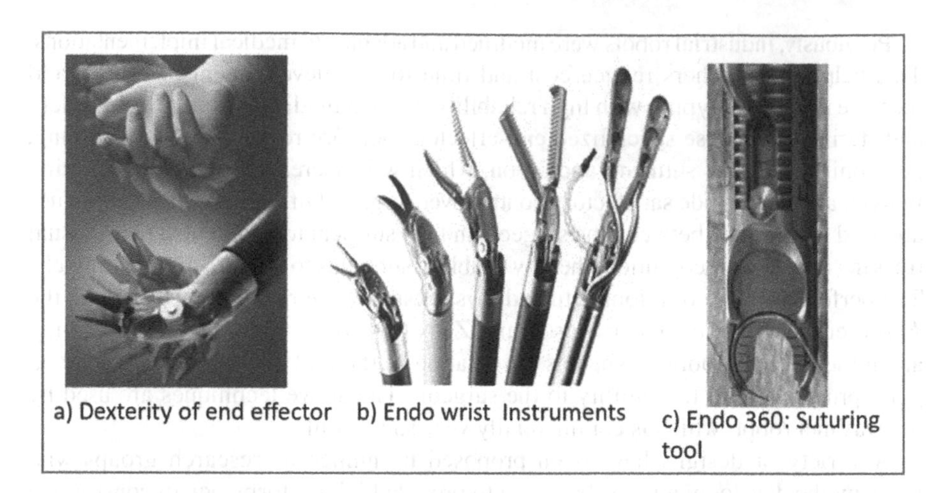

a) Dexterity of end effector
b) Endo wrist Instruments
c) Endo 360: Suturing tool

FIGURE 10.2 Different types of end-effector tools. (Adapted from intuitive surgical and Leonard, S. et al., *IEEE Transactions on Biomedical Engineering*, 2014. 61(4): p. 1305–1317. [65])

interfaces surgeons can have an inside view of patients, and the robot motion-interfacing device replicates the movement of the surgeon's hand. Force feedback is provided through a haptic interface to make surgery more intuitive for surgeons. Such examples include the most widely used teleoperated robot, that is, the daVinci robot.

c. *Preprogrammed, Semi-Autonomous Control*: Here, the extensive details of the robot's movements are fed and saved into the control system, which then guides the robot's movements. Generally, the parameters are fetched from the image and processed further according to requirements. The use of dynamic systems is robust to disturbance parameters. Examples shown in Figure 10.3 include the smart tissue anastomosis robot (STAR) [65], the self-actuating flexible needle system [66], Acrobot [66] and many more in which the insertion point, cutting path, suturing points are preprogrammed and predefined. The robot is guided with the help of visual feedback, which makes the system real-time operated with minimal error.

10.3 SYSTEMS AND APPLICATION

10.3.1 Surgical Simulation

Simulation tools provide added advantages to surgeons to analyse the work environments of surgeries and even rehearse before performing complex surgeries. For training young surgeons in simulation environments, they help prepare them to perform in real-world scenarios without any hesitations. Robotic simulators, when integrated/fused/in collaboration with virtual reality (VR) technologies provide an eminent platform for treating human conscious and subconscious psychology affecting directly physiological health. Some people's fears affect their psychology and different biological activities deeply, resulting in early age heart failures,

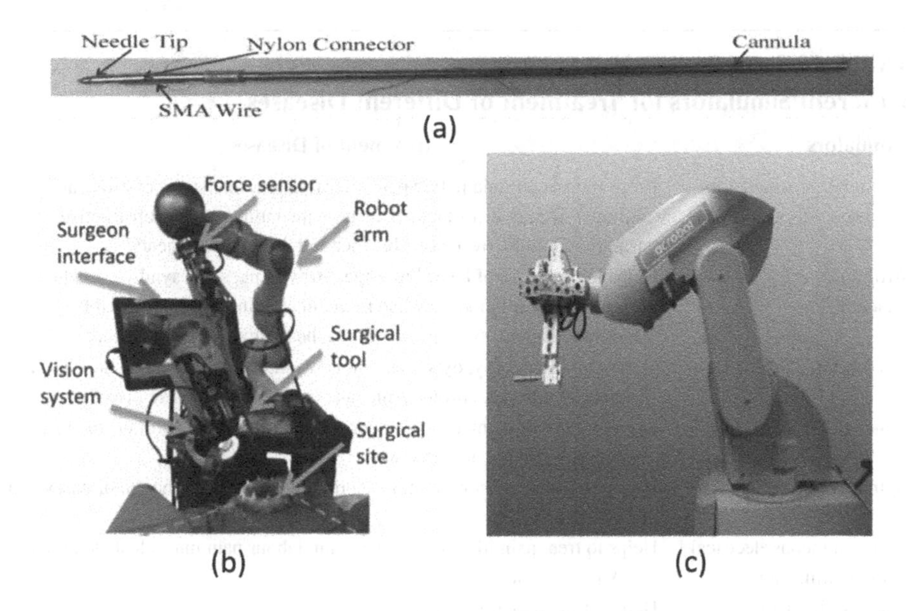

FIGURE 10.3 Preprogrammed or semi-autonomous robot. (a) Self-actuating flexible needle system. (b) STAR. (c) Acrobot. (Modified from Leonard, S. et al., *IEEE Transactions on Biomedical Engineering*, 2014. 61(4): p. 1305–1317. [65]; Joseph, F.O.M. et al., *Journal of Automation and Control Engineering*, 2015. 3(5): p. 428–434. [60])

disturbances in hormonal activities and mental instabilities. Therefore, doctors can take advantage of the integration of robotic simulators with VR, and patients are asked to sit, stand or laydown on the simulator wearing the VR glasses. This can help patients face their fears and overcome them. During this process different biological parameters are measured and updated on patients' profiles directly through IoT for doctors and patients to monitor, as well as to provide records for future use.

Table 10.3 above describes the different simulators used to treat different diseases and monitor, record and treat different real-time biological parameters.

10.3.2 Nanites

Taking a leap ahead for complex operations on the human body such as non-invasive or minimally invasive surgery is a challenging task, leading to the evolution of micromechanics to produce well-equipped microrobots to nanorobots. Nanorobots are also named nanobots, nanoids, nanites or nanomites and can be used for exploration, weaponry, cleaning and stunts placements. These nanites can easily traverse the human body and estimate the exact position and intensity of tissue damage, the presence of cancer cells, the presence of cysts and can localize foreign bodies.

Current theories suggest that oxygen, glucose or natural body sugars may be used to initiate the propulsion of nanobots inside the body. Bidirectional communication channels with nanobots are necessary, and theory suggests that robots can receive power or be reprogrammed from external sources through acoustic signals. Nanites, entering through natural body orifices can treat cancer cells, tumours and cysts with

TABLE 10.3
Different Simulators for Treatment of Different Diseases

Simulators	Treatment of Diseases
Sacral nerve stimulators	Helps to treat chronic pelvic pain, constipation, interstitial cystitis and different urinary disorder such as urge incontinence, non-obstructive urinary retention, overactive bladder and faecal incontinence
Artificial cardiac pacemakers	Helps in treatment of heart blockage, arrhythmia, sinus syndrome, which causes irregular blood flow due to disfunctioning of heart, possibly leading to angina, syncope, dizziness, heart attack or heart failure
Deep brain stimulators	Treatment of epilepsy, dystonia, brain essential tremors, Parkinson's disease, Tourette syndrome, chronic pain and obsessive compulsive disorder
Vagus nerve stimulators	Used to treat inflammatory disorder of joint diseases, gastrointestinal tract, epilepsy, heart and depression
Spinal cord stimulators	Helps to restore sensory, motor functioning, synapse, ischemic diseases and failed back surgery
Transcutaneous electrical nerve stimulators	Helps to treat pain like: lower back pain, labour pain musculoskeletal pain, neck pain and used in dentistry
Obesity simulator	Help to treat bariatric patient

little or no loss of healthy tissues. Using acoustic communication channels, doctors would not only be able to instruct but also monitor the in vivo progress of the healing stages. After the completion of tasks, nanobots could either be flushed from the body or made to biodegrade in due course.

Different hardware architectures of nanobots have been developed to test the behaviour of robot software like Nanorobot Control Design (NCD), as shown in Figure 10.4. The integration of IoT with nanites helps not only to operate and monitor

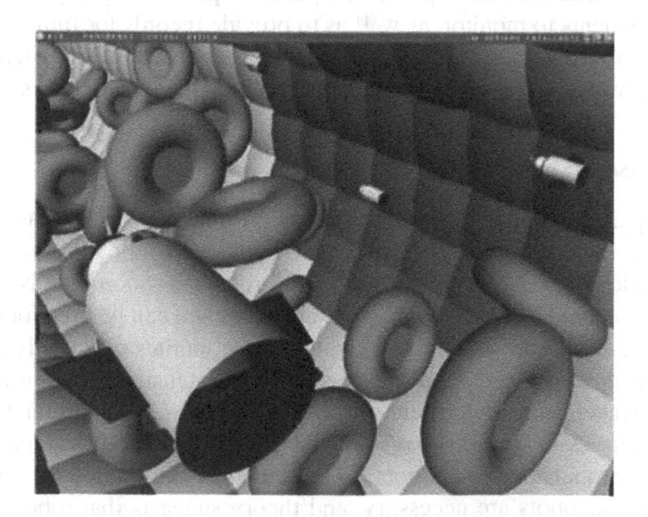

FIGURE 10.4 Detection of alpha-NAGA signals within blood stream using NCD software. (From Thangavel, K. et al. *Nanotechnology*, 2014. 2(5): p. 525–528. [84])

severe diseases but also to establish a structural healing diagram of the human body, which contributes significantly to the medical profession.

10.3.3 BIOTEX (Bio Sensing Textile)

The development of smart textiles for the continuous evaluation of physiological parameters has made a contribution towards this progression in healthcare technology. The collection of body fluids and multiple parameters in a non-invasive setting has given this technique a unique edge applicable to people of all ages. The smart textile used is embedded with multiple sensors to measure ECG [67], sweat rate, respiratory volume and oxygenated blood. The sweat test analysis helps in diagnosing cystic fibrosis (CF). The excessive loss of body fluids may cause dehydration, electrolyte unbalances and can lead to death.

Various types of sensors can be used for measuring health parameters such as conduct metric biosensors used for dialysis. Electrochemical biosensors are basically known as potentiometric and amperometric, and are significantly used for detecting the various chemical compositions of different body fluids. Optical biosensors, piezoelectric biosensors, calorimetric biosensor and so on are used for the detection of various foreign bodies like parasites, bacteria, viruses and so on. Blood glucose biosensors are used to measure levels of glucose in blood. Fiber-optic biosensor measure changes in oxygen concentration levels.

10.3.4 RFID Technology

RFID has emerged as an important technology, which has brought automation in healthcare management for tracking assessment, remote monitoring, real-time locating and managing of healthcare databases. The three primary components required for an RFID system integration are as follows:

1. RFID tag
2. RFID reader
3. Host computer

The RFID tag has a unique identification code that distinguishes one system from another. The reader captures the unique identification code and forwards the information to a server, which locates the RFID position and provides valuable information to the server. There are two types of RFID tags: active and passive tags. Active tags have an inbuilt embedded power supply and have a long readability range, while passive tags don't have a power supply and can only be read at a shorter range requiring passive tag readers. Figure 10.5 below shows the total integration of RFID in healthcare.

The RFID covers a wide range of implementations, which are growing exponentially daily due to fast and more accurate results. The major advantages of RFIDs include more automated readings, writing capability, greater data handling capability and more precise and integrated information, which helps to make more accurate decision-making. Figure 10.6 below clearly describes the advantages of RFID technology for meeting healthcare challenges.

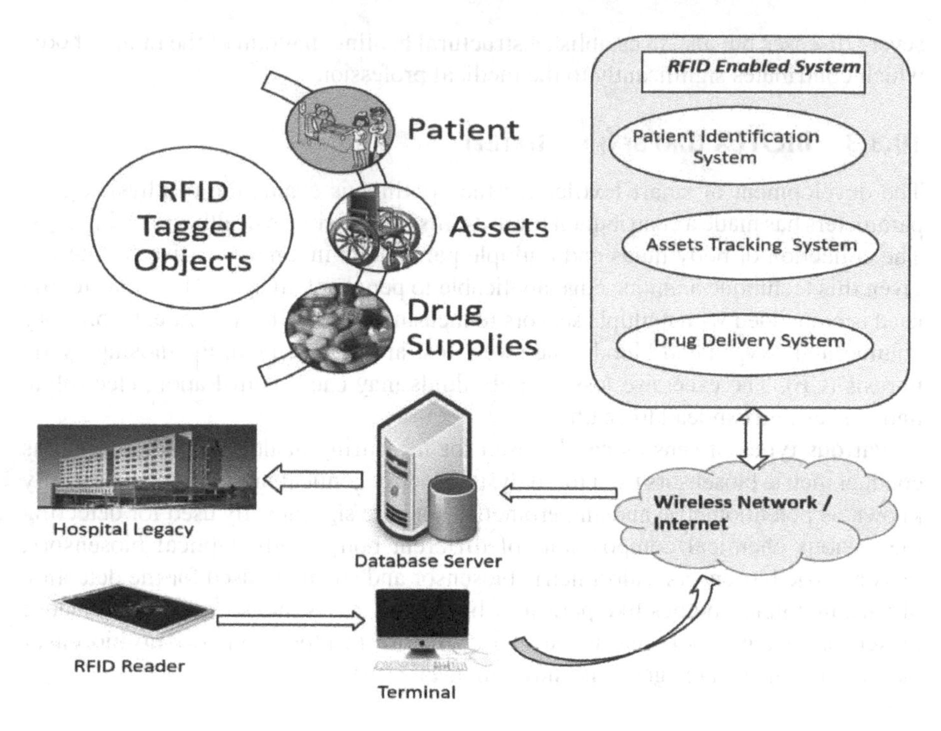

FIGURE 10.5 RFID integration in healthcare.

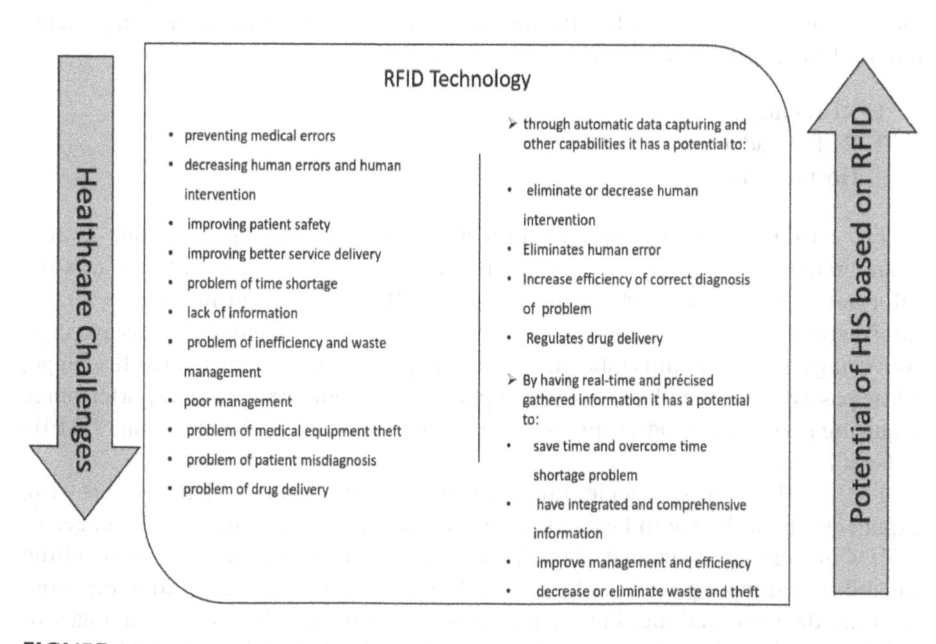

FIGURE 10.6 Potential solution for healthcare challenges based on RFID technology.

10.4 INTERNET OF THINGS (IoT)

Healthcare is but one among the numerous applications of IoT. A very simple straightforward definition of IoT is 'anything, anytime, anywhere'. IoT has played a prominent role in the redesigning of present healthcare systems with the current advancements in the field of technology associated with IoT. IoT-based solutions guarantee health monitoring not only in hospitals [68–73] and house environment [74–78], but also in outdoor environments. Now, when it comes to the situation of implementing IoT along with robotics in healthcare, a challenging task lies in the integration of both. Therefore, this section discusses the integration of robotics and IoT in healthcare. The use IoT-based robotics in healthcare should mainly focus on assisting doctors for local and remote monitoring of patients. The area of IoT-based robotics has a very wide coverage, and ranges from establishing short-range communication for local monitoring to a wide-area communication network for remote monitoring (where a wireless body area network is often set up), as well as to designing a protocol suite and implementing and testing it.

Applications in healthcare with respect to robots and IoT has been tested in the recent past but their integration has yet to be studied in depth.

As per the census of 2011, there are 100+ million people aged > 60+, and as per the latest report of the United Nations, the number of elderly people will grow to 170+ million by 2025. This means that presently the number of senior citizens is at an all-time high. As a result, there is a need for at least two individuals to take care of someone in his/her 60–70 s. Also, as per the 2014 data of the World Bank, the amount of public expenditures towards healthcare accounts for 1.4% of Indian GDP. Taking into account all these mentioned statistics, robots capable of sensing, monitoring and communicating, both locally and remotely form effective alternatives.

10.4.1 IoT REQUIREMENTS

The range of IoT devices available starts from a small tiny device that can be controlled by Bluetooth, and extends to a complex device to remotely monitor healthcare equipment. Therefore, with respect to remote healthcare monitoring systems a list of requirement for IoT-based robotics is essential. These requirements are summarized into main components, as shown below:

- *Source of power and its management*: Every IoT device needs a power source to operate. The type of power source can be directly from a line source or it can be through battery with energy harvesting. When batteries are used power management becomes essential as batteries provide limited power supply, which may run out of power anytime.
- *Sensors and Actuators*: Robotics-based IoT healthcare is practically possible only with the help of sensors continuously monitoring the vital parameters of the human body.
- *Microprocessor/Microcontroller*: Processors or controllers are the heart of any embedded system, and robotics-based remote health monitoring is an advanced embedded system, therefore, a high-end processor or controller is always desirable.

- *Communication*: Wireless communication is required for remote monitoring, therefore, selection of an appropriate communication methodology, and the types of transceivers to be used depend on the type of application.

10.4.2 Use of IoT and Robotics in Healthcare

IoT platforms along with robots used in today's healthcare world should be capable of operating in heterogeneous networks where huge amounts of data will be generated. These heterogeneous networks may have systems like medical equipments, monitoring systems (both local and remote), smart sensors fitted onto the patient's body and systems for tracking operations. The data generated by these systems have to be uploaded in real time round the clock. It means that the data rate the bandwidth requires for such operations should always be on the higher side. Figure 10.7 given below describes the global scenario of IoT-based robotics.

As per [79], IoT-based robotics in healthcare will include:

- Providing assistance to patients being monitored
- Providing aid for persons with disabilities to help them during standing from seating positions and, vice-versa
- Detecting drugs, medical equipment's automatically based on some predefined input criteria and also their movement accordingly
- Assisting medical staff in their activities

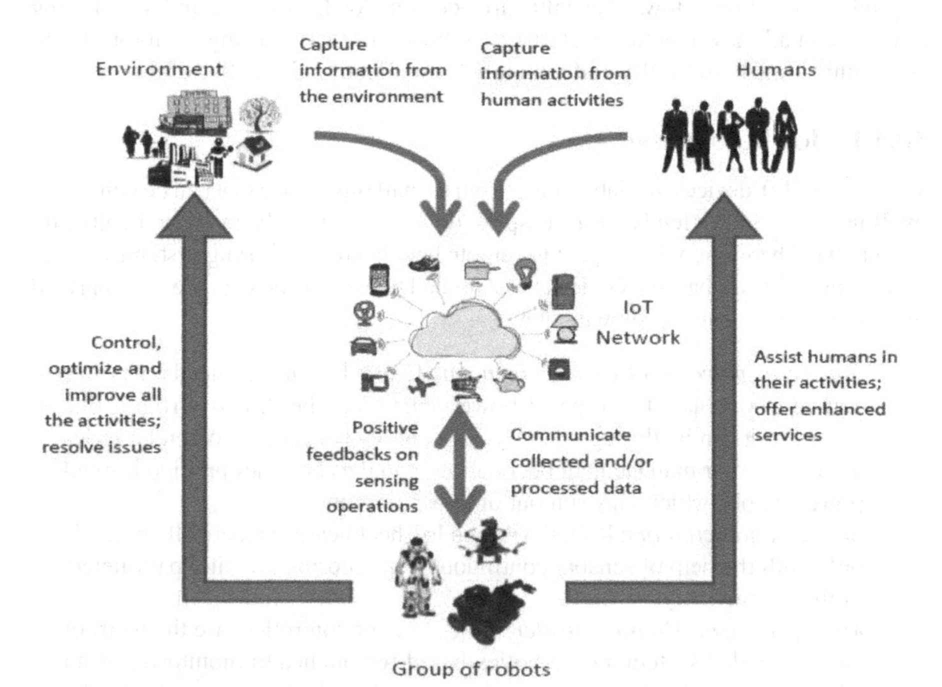

FIGURE 10.7 IoT-based robotics scenario.

- Assisting patients in panic situations
- Reminding patients in case they miss their medications and alerting, as the drug is about to be over

10.4.3 Open Source IoT Platforms

An IoT platform is required to collect data generated by Internet-enabled devices, then to further extract some meaningful information from it and relay it back to decision makers. Managing and analysing this data is very important. In this section, a comparison of some of the best open-source IoT platforms is performed. This comparison is done with respect to some key metrics, which are shown in Table 10.4 and are defined below:

- *Device Management*: Devices used in IoT are of many types, for example sensors of different range, microcontrollers acting as sink devices and gateways used for routing data to different groups. As the network grows, the variation in types of devices will also grow. Regardless of this variation, device management provides efficient methods to cater to the diverse nature of devices.
- *Data Integration*: IoT is all about data integration. As the number of devices connected with IoT increases, the data generated by these devices will also increase and therefore the rate at which this data is generated will also be high. In order to make sure networks do not get too crowded by these huge amounts of data, one should implement data integration techniques. These techniques guarantee that only meaningful data will be forwarded to the next hop whereas redundant data will be suppressed.
- *Encryption/Authentication*: Security is a prime concern in data communication. To avoid any intrusion or usage of data by unknown means, link security should be implemented in IoT networks. Authentication is needed to make sure that data arrives from intended sources and will be delivered only to intended destinations.
- *HTTP/MQTT protocol*: The most common protocols used for messaging in IoT are Hyper Text Transfer Protocol (HTTP) and Message Queue Telemetry Transport (MQTT). One major difference between the two is that MQTT is a lighter weight message protocol than HTTP. That means MQTT uses fewer bits to represent data than HTTP. Security of HTTP is stronger than MQTT.
- *Real-Time Data Analytics*: Data analytics in IoT is used to inspect data sets of varying sizes and of varying properties. This inspection helps in extracting some meaningful information from the data sets. Real time is also a very important metric in IoT robotics-based healthcare in order to deal with emergency and panic situations.
- *Round the Clock Availability*: Availability of the network 24 × 7 is highly desirable when IoT and robotics are implemented in healthcare industry. Monitoring and tracking of patients is of prime importance, therefore the assurance that the IoT network will be available round the clock is highly desirable.

TABLE 10.4
Comparison of Different IoT Platforms

	Kaa IoT Platform	Site Where	Things Speak	Device Hive	Zetta	Distributed Services Architecture	Thingsboard.io	Thinger.io	WSo2
Device Management	√	√	×	√	×	×	√	√	√
Data Integration	√	√	√	√	√	√	√	√	√
Encryption/ authentication	Link Encryption	Link Encryption	Basic Authentication	Basic Authentication	Basic Authentication	Basic Authentication	Basic Authentication	Link Encryption	Link Encryption
HTTP/MQTT protocol	Both	MQTT	HTTP	MQTT	HTTP	HTTP	Both	Both	Both
Real time data analytics	√	√	√	√	√	×	√	√	√
Round the clock availability	√	√	√	√	√	√	√	√	√
Scalability	√	√	×	√	×	×	√	√	√
Data storage	√	√	×	√	×	×	√	√	√

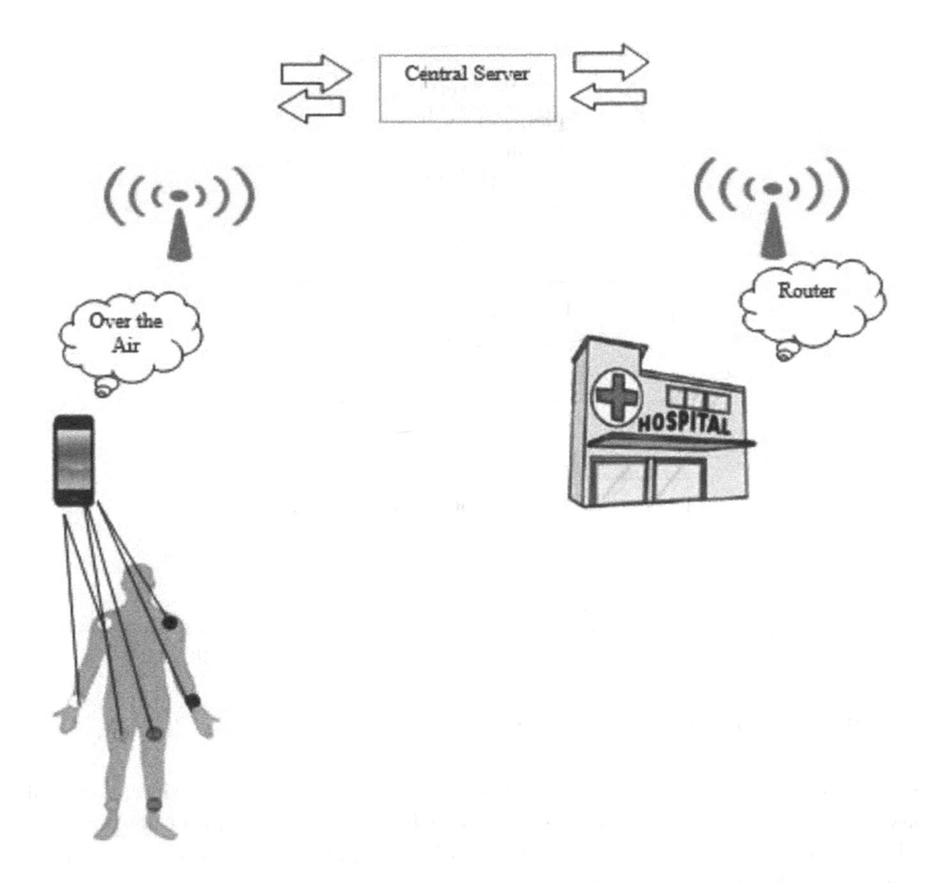

FIGURE 10.8 Architecture of nano-sensors and its communication module in healthcare.

- *Scalability*: Adding devices to the existing network of devices is always recommended. If the network allows the addition of new devices without any major modifications then it is said to be scalable.
- *Data Storage*: Data, when measured by any sensor and uploaded on IoT network, should be stored so that it gives an opportunity for the purposes of comparison with the instantaneous data.

10.4.4 Internet of NanoTechnology (IoNT)

Nanotechnology has gained a lot of attention in recent times due to rapid increases in the field of technology, especially towards miniaturization of embedded hardware devices. Due to these advancements, the sensors, their interfacing hardware and the size of software are becoming smaller and smaller.

A nano-sensor in the medical field can be defined as any sensor capable of detecting nanoparticles. Nano-sensors are also capable of monitoring the physical and biological parameters at nano scale. Nano-sensors can be used in the field of healthcare for remote monitoring of any individual who has wearable sensors capable of detecting and monitoring vital parameters. Figure 10.8 above shows the architecture

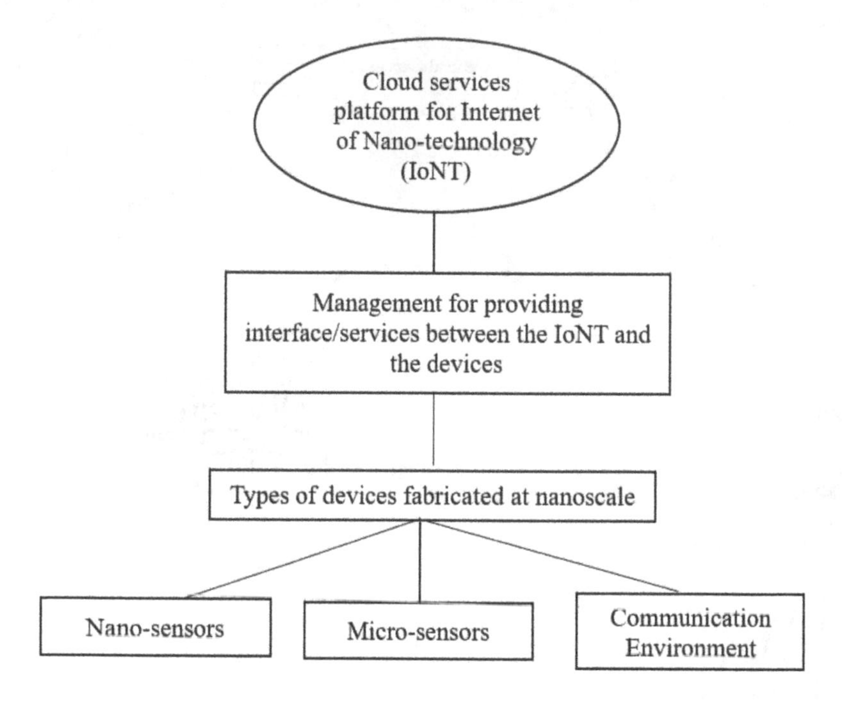

FIGURE 10.9 IoT architecture for nanotechnology.

of one such system that can be implemented and tested in real time. Here, a patient has more than one wearable nano-sensor, which are placed in appropriate positions (placement of these nano-sensors will depend on the parameter they are detecting or monitoring). Each of these nano-sensors will communicate with a smartphone over the air either through a low energy Bluetooth or any other RF communicative device (or even it may be a wired connection). A central server can be established governing, storing and controlling the complete architecture of the system. Communication to the central server is with a smartphone on the patient side, and with the router on the hospital side, which will be completely wireless. Here a central server will also have an online portal displaying the complete profile of the data measured/detected or monitored by the nano-sensors, and it will also have the feedback/prescription or suggestion given by the medical representative from the hospital side. The IoT will fulfil all the requirements needed for the establishment, control and supervision required for the online portal.

Figure 10.9 illustrates the IoT architecture for nanotechnology, which integrates it with IoT using a cloud service platform (IoNT). The effective management interface was to develop to access the data from server that was real time monitored due to different types of sensors implanted at a nanoscale. The fabrication of different nanoscale healthcare devices is possible due to advanced enhancements in nanoelectronic solid state materials to develop fast communication channels, large measured parameters, increased response times and so on. To establish a robust network for communicating between different nanites, IoNT has been developed that depends on different communication channels.

10.5 DISCUSSION

As discussed in the previous sections, assistive/robotic devices perform selective tasks with the help of multiple sub-devices, assistive devices mentioned like ROAD and so on require additional arrangements for navigation purpose. Without modifying the arm of RIBA, it is not possible to use it for applications related to transfer. Psychological and behavioural tendencies create hurdles for the adoption of these assistive devices for people, or perhaps these devices need improvements in terms of ergonomics and aesthetics for easily adoption by people.

Multi-cooperative assistance is one of the areas addressing many issues pertaining to aligning and docking of different assistive devices together to cater to one problem. Usual issues like toppling, jerking and so on also need to be addressed. Apart from hardware issues like battery life and battery-charging time, power consumption by circuit components also greatly affects the acceptance of assistive devices. Implementation of intelligent monitoring systems minimizes the risk of failures.

Assistive devices are sophisticated, expensive and mostly designed as per the standards of developed countries. Hence, research areas in the domain of assistive devices is wide open and researchers have miles to go.

Robotic devices generally face two issues: imaging and limitation of DoF. Although imaging solutions have been resolved to a greater extent by training surgeons and applying good imaging techniques [80–82]; however, increasing the DoF causes other limitations like backlash, singularity and so on. Be it telerobotic, telepresence or telementoring all are connected via wired or wireless solutions, and in order to perfectly replicate the actual motion of the surgeon, sensor data from the host station needs to be fed to the user station immediately. This can only be achieved by uninterrupted high-speed Internet connections.

10.6 CONCLUSION

The meshing of IoT and robotics has yielded great breakthroughs in healthcare by:

- Improving the quality of interventions by ensuring safety, accuracy and consistency of operations
- Providing continuous track of patient healthcare and assessments
- Enhanced skill assessment by the use of simulators, which provide quantitative data analysis and record the clinical operations while training
- Effective use of patient information, which enables doctors to treat individual patients and provides easier ways for other doctors to be informed of case histories
- Eliminating human errors and limitations in treating patients

The rapid development of IoT in the healthcare and medical sector needs the latest technology of telemetry, which provides proper connection between PSM and MTM to attain the objectives of IoT-driven manipulators. Medical data mining, caretaker arrangements, privacy and security of data, frequent updating and notification of different healthcare parameters and the ethics of physicians are few of the challenges that remain to be addressed.

REFERENCES

1. Hokayem, P.F. and M.W. Spong, Bilateral teleoperation: An historical survey. *Automatica*, 2006. 42(12): p. 2035–2057.
2. Sheridan, T.B., Teleoperation, telerobotics and telepresence: A progress report. *Control Engineering Practice*, 1995. 3(2): p. 205–214.
3. Heikkilä, T., M. Järviluoma, and T. Juntunen, Holonic control for manufacturing systems: Functional design of a manufacturing robot cell. *Integrated Computer-Aided Engineering*, 1997. 4(3): p. 202–218.
4. Yamada, H., N. Tao, and Z. Dingxuan, Construction tele-robot system with virtual reality. In *Robotics, Automation and Mechatronics, 2008 IEEE Conference on*. 2008. IEEE.
5. Zhao, D. et al., 6 DOF presentation of realistic motion in operating a construction tele-robot system. In *Proceedings of the JFPS International Symposium on Fluid Power*. 2002. The Japan Fluid Power System Society.
6. Kwon, D.-S. et al., Microsurgical telerobot system. In *Intelligent Robots and Systems, 1998. Proceedings., 1998 IEEE/RSJ International Conference on*. 1998. IEEE.
7. Geerinck, T. et al., Tele-robot with shared autonomy: Distributed navigation development framework. *Integrated Computer-Aided Engineering*, 2006. 13(4): p. 329–345.
8. Paul, H.A. et al., Development of a surgical robot for cementless total hip arthroplasty. *Clinical Orthopaedics and Related Research*, 1992. 285: p. 57–66.
9. Taylor, R.H. et al., Computer-integrated revision total hip replacement surgery: Concept and preliminary results. *Medical Image Analysis*, 1999. 3(3): p. 301–319.
10. Arezzo, A. et al., Experimental trial on solo surgery for minimally invasive therapy. *Surgical Endoscopy*, 2000. 14(10): p. 955–959.
11. Yavuz, Y. et al., A comparative experimental study evaluating the performance of surgical robots aesop and endosista. *Surgical Laparoscopy Endoscopy & Percutaneous Techniques*, 2000. 10(3): p. 163–167.
12. Arezzo, A. et al., Positioning systems for endoscopic solo surgery. *Minerva Chirurgica*, 2000. 55(9): p. 635–641.
13. Gagner, M. et al., Robotic interactive laparoscopic cholecystectomy. *The Lancet*, 1994. 343(8897): p. 596–597.
14. Begin, E. et al., A robotic camera for laparoscopic surgery: Conception and experimental results. *Surgical Laparoscopy Endoscopy & Percutaneous Techniques*, 1995. 5(1): p. 6–11.
15. Anderson, J.E. et al., The first national examination of outcomes and trends in robotic surgery in the United States. *Journal of the American College of Surgeons*, 2012. 215(1): p. 107–114.
16. Barbash, G.I. and S.A. Glied, New technology and healthcare costs—The case of robot-assisted surgery. *New England Journal of Medicine*, 2010. 363(8): p. 701–704.
17. Bharathan, R., R. Aggarwal, and A. Darzi, Operating room of the future. *Best Practice & Research Clinical Obstetrics & Gynaecology*, 2013. 27(3): p. 311–322.
18. Falk, V. et al., Developments in robotic cardiac surgery. *Current Opinion in Cardiology*, 2000. 15(6): p. 378–387.
19. Carpentier, A. et al., Computer-assisted cardiac surgery. *The Lancet*, 1999. 353(9150): p. 379–380.
20. Kappert, U. et al., Robotic-enhanced Dresden technique for minimally invasive bilateral internal mammary artery grafting. In *Heart Surgery Forum*. 2000. Forum Multimedia Publishing.
21. Himpens, J., G. Leman, and G. Cadiere, Telesurgical laparoscopic cholecystectomy. *Surgical Endoscopy*, 1998. 12(8): p. 1091.
22. Cadiere, G. et al., Nissen fundoplication done by remotely controlled robotic technique. In *Annales de Chirurgie*. 1999.

23. Cadiere, G. et al., Evaluation of telesurgical (robotic) NISSEN fundoplication. *Surgical Endoscopy*, 2001. 15(9): p. 918–923.
24. Isgro, F. et al., Robotic surgery using Zeus™ MicroWrist™ technology. *Journal of Cardiac Surgery*, 2003. 18(1): p. 1–5.
25. Kiaii, B. et al., Robot-assisted computer enhanced closed-chest coronary surgery: Preliminary experience using a Harmonic Scalpel® and ZEUS™. In *Heart Surgery Forum*. 2000. Forum Multimedia Publishing.
26. Ducko, C.T. et al., Robotically-assisted coronary artery bypass surgery: Moving toward a completely endoscopic procedure. In *The Heart Surgery Forum*. 1999.
27. Gulbins, H. et al., 3D-visualization improves the dry-lab coronary anastomoses using the Zeus robotic system. In *The Heart Surgery Forum*. 1999.
28. Sackier, J.M. and Y. Wang, Robotically assisted laparoscopic surgery. *Surgical Endoscopy*, 1994. 8(1): p. 63–66.
29. Jacobs, L., V. Shayani, and J. Sackier, Determination of the learning curve of the AESOP robot. *Surgical Endoscopy*, 1997. 11(1): p. 54–55.
30. Johanet, H. Voice-controlled robot: A new surgical aide? Thoughts of a user. In *Annales de chirurgie*. 1998.
31. Allaf, M. et al., Laparoscopic visual field. *Surgical Endoscopy*, 1998. 12(12): p. 1415–1418.
32. Partin, A.W. et al., Complete robot-assisted laparoscopic urologic surgery: A preliminary report. *Journal of the American College of Surgeons*, 1995. 181(6): p. 552–557.
33. Omote, K. et al., Self-guided robotic camera control for laparoscopic surgery compared with human camera control. *The American Journal of Surgery*, 1999. 177(4): p. 321–324.
34. Schurr, M.O., A. Arezzo, and G.F. Buess, Robotics and systems technology for advanced endoscopic procedures: Experiences in general surgery. *European Journal of Cardio-Thoracic Surgery*, 1999. 16(Supplement 2): p. S97–S105.
35. Schurr, M. et al., Robotics and telemanipulation technologies for endoscopic surgery. *Surgical Endoscopy*, 2000. 14(4): p. 375–381.
36. Birkett, D.H., Electromechanical instruments for endoscopic surgery. *Minimally Invasive Therapy & Allied Technologies*, 2001. 10(6): p. 271–274.
37. Katevas, N.I. et al., The autonomous mobile robot SENARIO: A sensor aided intelligent navigation system for powered wheelchairs. *IEEE Robotics & Automation Magazine*, 1997. 4(4): p. 60–70.
38. Bourhis, G. and Y. Agostini, The vahm robotized wheelchair: System architecture and human-machine interaction. *Journal of Intelligent and Robotic Systems*, 1998. 22(1): p. 39–50.
39. Yanco, H.A., Wheelesley: A robotic wheelchair system: Indoor navigation and user interface. In *Assistive Technology and Artificial Intelligence*. 1998, Springer. p. 256–268.
40. Levine, S.P. et al., The NavChair assistive wheelchair navigation system. *IEEE transactions on Rehabilitation Engineering*, 1999. 7(4): p. 443–451.
41. Lankenau, A. and T. Rofer, Mobile robot self-localization in large-scale environments. In *Robotics and Automation, 2002. Proceedings. ICRA'02. IEEE International Conference on*. 2002. IEEE.
42. Kinpara, Y. et al., Situation-driven control of a robotic wheelchair to follow a caregiver. In *Frontiers of Computer Vision (FCV), 2011 17th Korea-Japan Joint Workshop on*. 2011. IEEE.
43. Iwase, T., R. Zhang, and Y. Kuno, Robotic wheelchair moving with the caregiver. In *SICE-ICASE, 2006. International Joint Conference*. 2006. IEEE.
44. Rokonuzzaman, M. et al., Design of an autonomous mobile wheel chair for disabled using electrooculogram (EOG) signals. In *Mechatronics* 2011, Springer. p. 41–53.
45. Barea, R. et al., Wheelchair guidance strategies using EOG. *Journal of Intelligent & Robotic Systems*, 2002. 34(3): p. 279–299.

46. Tomari, M.R.M., Y. Kobayashi, and Y. Kuno, Development of smart wheelchair system for a user with severe motor impairment. *Procedia Engineering*, 2012. 41: p. 538–546.
47. Halawani, A. et al., Active vision for controlling an electric wheelchair. *Intelligent Service Robotics*, 2012. 5(2): p. 89–98.
48. Lopes, A.C., G. Pires, and U. Nunes, Assisted navigation for a brain-actuated intelligent wheelchair. *Robotics and Autonomous Systems*, 2013. 61(3): p. 245–258.
49. Perrin, X. et al., Brain-coupled interaction for semi-autonomous navigation of an assistive robot. *Robotics and Autonomous Systems*, 2010. 58(12): p. 1246–1255.
50. Tapus, A., C. Ţăpuş, and M.J. Matarić, User—Robot personality matching and assistive robot behavior adaptation for post-stroke rehabilitation therapy. *Intelligent Service Robotics*, 2008. 1(2): p. 169–183.
51. Carrera, I. et al., ROAD: Domestic assistant and rehabilitation robot. *Medical & Biological Engineering & Computing*, 2011. 49(10): p. 1201.
52. Mukai, T. et al., Realization and safety measures of patient transfer by nursing-care assistant robot RIBA with tactile sensors. *Journal of Robotics and Mechatronics*, 2011. 23(3): p. 360–369.
53. Park, K.-H. et al., Robotic smart house to assist people with movement disabilities. *Autonomous Robots*, 2007. 22(2): p. 183–198.
54. Jung, J.-W. et al., Advanced robotic residence for the elderly/the handicapped: Realization and user evaluation. In *Rehabilitation Robotics, 2005. ICORR 2005. 9th International Conference on*. 2005. IEEE.
55. Bostelman, R. and J. Albus, Robotic patient transfer and rehabilitation device for patient care facilities or the home. *Advanced Robotics*, 2008. 22(12): p. 1287–1307.
56. Bostelman, R. and J. Albus. A multipurpose robotic wheelchair and rehabilitation device for the home. In *Intelligent Robots and Systems, 2007. IROS 2007. IEEE/RSJ International Conference on*. 2007. IEEE.
57. Bostelman, R. and J. Albus, Robotic patient lift and transfer. In *Service Robot Applications*. 2008. InTech.
58. Bostelman, R. and J. Albus. Sensor experiments to facilitate robot use in assistive environments. In *Proceedings of the 1st international conference on PErvasive Technologies Related to Assistive Environments*. 2008. ACM.
59. Reichenspurner, H. et al., Use of the voice-controlled and computer-assisted surgical system ZEUS for endoscopic coronary artery bypass grafting. *The Journal of Thoracic and Cardiovascular Surgery*, 1999. 118(1): p. 11–16.
60. Joseph, F.O.M. et al., Control of shape memory alloy actuated flexible needle using multimodal sensory feedbacks. *Journal of Automation and Control Engineering*, 2015. 3(5): p. 428–434.
61. Ikuta, K. et al., Multi-degree of freedom hydraulic pressure driven safety active catheter. In *Robotics and Automation, 2006. ICRA 2006. Proceedings 2006 IEEE International Conference on*. 2006. IEEE.
62. Scrosati, B., *Applications of electroactive polymers*. Vol. 75. 1993: Springer.
63. Bargar, W.L., A. Bauer, and M. Börner, Primary and revision total hip replacement using the Robodoc (R) System. *Clinical Orthopaedics and Related Research*, 1998. 354: p. 82–91.
64. Kang, H. and J.T. Wen. Endobot: A robotic assistant in minimally invasive surgeries. In *Robotics and Automation, 2001. Proceedings 2001 ICRA. IEEE International Conference on*. 2001. IEEE.
65. Leonard, S. et al., Smart tissue anastomosis robot (STAR): A vision-guided robotics system for laparoscopic suturing. *IEEE Transactions on Biomedical Engineering*, 2014. 61(4): p. 1305–1317.
66. Joseph, F.M. et al., Development of self-actuating flexible needle system for surgical procedures. *Journal of Medical Devices*, 2015. 9(2): p. 020945.

67. Dey, N. et al., Developing residential wireless sensor networks for ECG healthcare monitoring. *IEEE Transactions on Consumer Electronics*, 2017. 63(4): p. 442–449.
68. Ren, Y. et al., Monitoring patients via a secure and mobile healthcare system. *IEEE Wireless Communications*, 2010. 17(1): p. 59–65.
69. Mansor, M.N.B. et al., Patient monitoring in ICU under unstructured lighting condition. In *Industrial Electronics & Applications (ISIEA), 2010 IEEE Symposium on*. 2010. IEEE.
70. Wong, W.K. et al., Wireless webcam based omnidirectional healthcare surveillance system. In *Computer Research and Development, 2010 Second International Conference on*. 2010. IEEE.
71. Ko, J. et al., Wireless sensing systems in clinical environments: Improving the efficiency of the patient monitoring process. *IEEE Engineering in Medicine and Biology Magazine*, 2010. 29(2): p. 103–109.
72. Dey, N. et al., *Internet of Things and Big Data Analytics Toward Next-Generation Intelligence*. 2018: Springer.
73. Elhayatmy, G., N. Dey, and A.S. Ashour, Internet of things based wireless body area network in healthcare. In *Internet of Things and Big Data Analytics Toward Next-Generation Intelligence*. 2018. Springer. p. 3–20.
74. Taleb, T. et al., Angelah: A framework for assisting elders at home. *IEEE Journal on Selected Areas in Communications*, 2009. 27(4): p. 480–494.
75. Xu, X. et al., Outdoor wireless healthcare monitoring system for hospital patients based on ZigBee. In *Industrial Electronics and Applications (ICIEA), 2010 the 5th IEEE Conference on*. 2010. IEEE.
76. Junnila, S. et al., Wireless, multipurpose in-home health monitoring platform: two case trials. *IEEE Transactions on Information Technology in Biomedicine*, 2010. 14(2): p. 447–455.
77. Rosati, R.J., Evaluation of remote monitoring in home healthcare. In *eHealth, Telemedicine, and Social Medicine, 2009. eTELEMED'09. International Conference on*. 2009. IEEE.
78. Yang, Y., X. Liu, and Y. Wang, Immediate communication system for remote medical monitoring based on internet. In *Automation Congress, 2008. WAC 2008. World*. 2008. IEEE.
79. Grieco, L.A. et al., IoT-aided robotics applications: Technological implications, target domains and open issues. *Computer Communications*, 2014. 54: p. 32–47.
80. Chakraborty, S. et al., Intelligent computing in medical imaging: A study. *Advancements in Applied Metaheuristic Computing*, 2017: p. 143.
81. Wang, D. et al., Image fusion incorporating parameter estimation optimized Gaussian mixture model and fuzzy weighted evaluation system: A case study in time-series plantar pressure data set. *IEEE Sensors Journal*, 2017. 17(5): p. 1407–1420.
82. Dey, N. and A.S. Ashour, Computing in medical image analysis. In *Soft Computing Based Medical Image Analysis*. 2018. Elsevier. p. 3–11.
83. Okamura, A.M., M.J. Mataric, and H.I. Christensen, Medical and health-care robotics. *IEEE Robotics & Automation Magazine* 2010. 17(3): p. 26–37.
84. Thangavel, K. et al., A survey on nano-robotics in nano-medicine. *Nanotechnology* 2014. 2(5): p. 525–528.

11 Internet of Medical Things

Remote Healthcare and Health Monitoring Perspective

Sitaramanjaneya Reddy Guntur, Rajani Reddy Gorrepati, and Vijaya R. Dirisala

CONTENTS

11.1 INTRODUCTION

The latest emerging trends and advances in communication technology continue to sweep the global healthcare industry. Recent advances in innovative design and development of medical devices enhance the quality of patient care. New technological trends, from physical devices to smart systems, are transmitting essential information in real time enabling specialists, healthcare providers and patients to interface in new ways and recognize life-threatening situations [1]. The vision of medical services at 'anytime, anywhere and anything' is changing to reach the patient expectations and is motivating the next generation of innovations [2]. Presently, advances in smartphone frameworks are helpful and comfortable enough to enable specialists or doctors with consultations for medical assistance. IoT regions interfaces with frameworks related to enormous information, security and protection, which possess a serious challenge. In addition, it allows people to upload, retrieve, store and collect information, which ultimately forms big data. IT enables individuals to transfer, recover, store and accumulate the data.

In addition to this, it examines the investigation of continuous large flows of information to infer significant knowledge in big data applications in a few areas. It investigates conceivable computerized arrangements in everyday life including structures and mechanized frameworks of IoT innovation, and additionally medicinal services frameworks that oversee a lot of information to enhance clinical decisions [3,4]. In this chapter, we highlight the innovative technological advances with major implications in the field of IoMT and remote medical services.

11.1.1 Overview of Internet of Medical Things (IoMT)

The IoMT is playing a pivotal role in remote healthcare and monitoring to increase the efficacy of medical devices, and the speed and accessibility of medical services. The IoMT can be used to collect remote patient health data utilizing wearable sensors and devices connected to Internet-based health monitoring systems. IoMT is processed by connecting and communicating machine-to-machine (M to M) through medical devices equipped with Wi-Fi. The received healthcare data from the IoMT devices are stored in the cloud server database, which is linked to cloud platforms and then analysed.

Remote health monitoring (RHM) refers to a continuous process of monitoring a person's health by closely following the course of exercises and contrast to ascertain what is going on and what is relied upon to happen. Clinical services are not particularly incorporated into RHM, in spite of that it may certainly lead to the provision of such services. It may include monitoring physiological parameters such as the heart rate, pulse rate, blood pressure and temperature in addition to other parameters such as medication and diet monitoring to help patients. Remote healthcare is a process that periodically collects and reviews health data on program implementation and provides elementary analysis so as to enable progress of expectations. RHM may be used in conjunction with healthcare, which allows a patient to use a smartphone or wearable device to perform a routine test and send the test data to a healthcare professional in real time. RHM and healthcare services have improved the physicians ability to monitor and manage patients in non-traditional healthcare settings. RHM utilizes advanced methods to gather the health information from people in a single area, such as a patient's home, and transmit the data to healthcare providers in a different area for analysis and recommendations by home nurses, or disease management programs.

The concept of IoMT is the integration of healthcare devices with computer networks through the web, which receives information in real time, and also allows interaction with patients [5–7]. Basically, IoMT associates living and non-living things through the web [8]. Another concept of IoMT is 'things oriented', particularly day-to-day objects via smart systems utilizing smart interfaces such as ZigBee, Bluetooth, RFID, LAN, Wi-Fi or by other working frameworks to interface and communicate in social, natural and client contexts [9–11]. The object-oriented IoMT module considers a smart physiological thing that allows communicate through the internet technologies virtually [12]. IoMT collects and communicates the patient's physical parameters at any time, any location, about anything by using the ideal service in any path or network, as shown in Figure 11.1. The IoMT is able to remotely connect people about chronic diseases by using the patients' and hospitals' locations and tracking medication orders and wearable health devices [13]. The large amount of health data generated by health devices or sensors, including blood pressure, heart rate, body temperature, respiratory rate, pulse rate and so on is sent to the healthcare providers [14]. At present, hospital beds and analytics, dashboards are being connected with

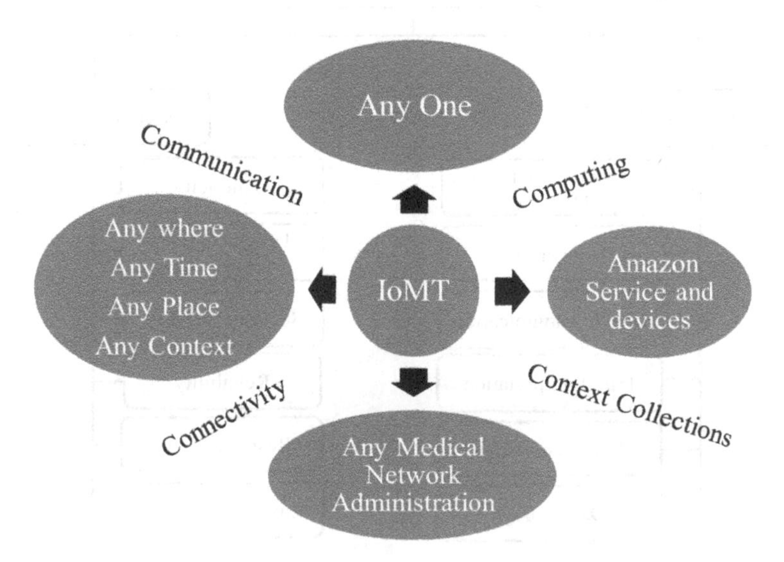

FIGURE 11.1 IoMT in various environments.

a good number of wearable sensors for measurement of a patient's physiological parameters in real-time, and the medical devices convert or deploy the data with the IoMT technology [15].

A large volume of generated data (often called big data) is reccived from sensors, actuators and embedded systems; however, these are unable to process it readily by utilizing conventional data processing methods and applications. Besides this, numerous database groups and additional resources are required to store the data [16]. However, storage and recovery is not the only problem but it is also of paramount importance to obtain a meaningful pattern relevant to patient diagnostic information [17]. IoMT enables faster diagnosis of disease and decision-making by compiling numerous medical data (i.e., big data) on time coupled with wise investigation [18]. IoMT includes in-depth analysis of patient monitoring with network and communication technology. A novel health monitoring healthcare framework and computational models that handles large volume of data driven by patient information such as big data and along with the available tools, mechanisms and algorithms to deal with those problems as well as a case study was presented and addressed the security and privacy issues as well as the big data challenges [19–21].

11.1.2 The Requirements of the IoMT Systems

The requirements for remote healthcare and health monitoring capture the data required to build IoMT functionalities by a framework. These practical prerequisites of IoMT requirements are classified into two categories: functional and non-functional, as shown in Figure 11.2.

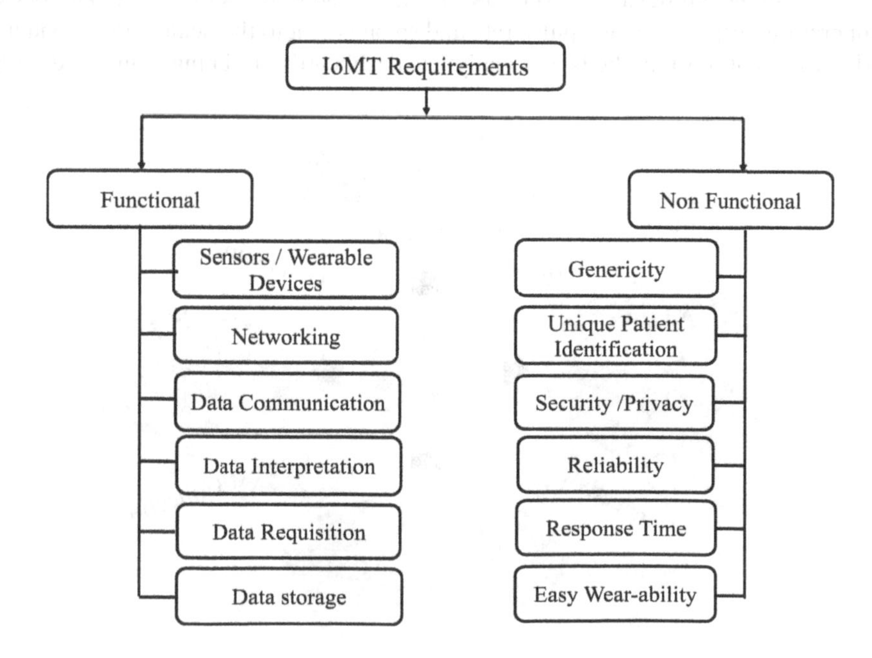

FIGURE 11.2 Requirement of IoMT.

Sensors/wearable devices: Sensors and wearable devices are prerequisites for obtaining patient health data. Non-invasive, non-intrusive sensors are prime components of mobile and long-term health check-up frameworks. Wearable sensors are more comfortable and less obstructive, while suitable for checking a person's health without intruding on their day-by-day activities. The sensors, placed on various locations of the body, can measure physiological parameters as well as the activity and movement of a person. Different sensors have been developed for an assorted scope of health information specific for healthcare services. Such sensors are prospective enough to transfer the data to similar service through the IoT. On the other hand, wearable devices can accompany an arrangement of highlights for proper the IoT design along these lines, the joining of previously mentioned sensors into wearable items is obvious. Such sensors are sufficiently imminent to exchange the information comparative administrations through the IoT. Then again, wearable gadgets can go with a game plan of features for appropriate the IoT outline. Thusly, the joining of beforehand specified sensors into wearable things is self-evident.

Network: The framework should allow for an interface between the patient and specialist. The IoMT refers to both the network and the computing platform, and the service layer focuses on patient healthcare. This structure demonstrates that an efficient and precise hierarchical model of caregivers or operators can access various databases from the application layer with the assistance of a supporting layer.

Data communication: Data communication is the process of transferring the received health information from wearable sensors. First, the acquired health data is transferred to the IoMT in real time. The correspondence between the sensor and health monitoring system is called intra-correspondence, and data should to be transferred to the cloud server.

Data interpretation/differentiation: The system should be able to analyse the basic health parameters from the sensors. Biosignals must be effectively deciphered by remote patient monitoring systems and cloud servers. RPHM establishes the connections between various physiological parameters and life styles.

Data requisition: In order to analyse the patient's health condition, clinicians require the current heath information together with the past records from the database. The data-storing servers are able to provide relevant data on request.

Data storage: IoMT-based remote health monitoring systems consist of a large volume of information from various patients or patient groups. The data storage severs should have enough space and memory and be able to accommodate large data quantities.

Non-Functional Requirements:

Genericity: The IoMT system is able to adjust with the patients' monitoring requirements and should not be concise with a particular disease, group or group of individuals.

Patient identification: Patients must be furnished with a unique patient ID.

Security/privacy: The connection between the biometric sensor layer, IoMT base unit and server should be secure and verified.

Reliability: IoMT systems play a vital role in the healthcare industry for increasing the accuracy, reliability and productivity of electronic devices. They should be able to perform consistently and produce health information precisely.

Response time: IoMT systems are always fast enough to provide services for emergencies and also fast enough to provide patient information during crisis situations to avoid unnecessary confusions and provide quick assistance to patients.

Easy wear-capacity: The sensors used to measure on the body area network are portable and sensitive, so that the sensors are simple, easy to use and convenient for the patient.

11.1.3 Remote Patient Healthcare and Health Monitoring System

Currently, people are busy with their work schedules and have limited time to visit specialists for routine health examinations. Because of this, medical problems continue to increase and people suffer from various avoidable diseases. Likewise, a majority of elderly people suffer from various health ailments and are unable to visit hospitals routinely. People are not prepared to wait for a long time for consultations and medical examinations. Once in a while, the treatment may not be accessible immediately nearby even when a person is suffering from a major health problem, and the patient needs to travel a long way for the treatment. Based on the assistance of the RHM framework, health parameters can be self-assessed by sitting at home and the information can be shared with a qualified doctor who is far away. If the patient is suffering from a major medical ailment, and the specialist treating the patient is unable to help, these parameters can be remotely sent to a doctor who can help him/her from any location. In the worst case, even if the treatment isn't accessible in his/her country, he/she can communicate and continue the treatment with a specialist from a technically advanced country. Thus, mortality rates can be decreased and the quality of care can be enhanced using RHM with IoMT [22]. The RHM is a small and compact system and can be carried by the patient whenever and wherever it is needed; these devices are available at very low cost and provide long-term service [23].

Remote patient monitoring is also called self-testing or monitoring of health parameters. This enables remote monitoring of patient's physiological parameters using different medical equipment by the physician. This method is useful for monitoring the patient's physiological status as well as diagnosing diseases and prescribing medications according to data analysis feedback [24]. In remote monitoring, the wearable sensors capture data pertaining to body temperature, heartbeat, pulse rate and so on and transmit the data to a specialist either in real time, or it is stored and then forwarded. RHM includes home-based surveillance of ECGs [25], wireless gastrointestinal capsules [26], dialysis, multi-parameter ICUs, telehealth and diseases [27]. Hence, the RHM system is used to continuously monitor the physiological parameters of a patient with the help of sensors. These parameters are tracked and sent to the physician and in case of any abnormality, the problem can be rectified. The existing patient monitoring systems are fixed monitoring systems, which are only available in hospital ICUs, as shown in Figure 11.3. These systems are huge in size and useful to monitor patients in hospital beds only. No automatic system exists able to regularly provide important data about the patient when he or she is mobile [28]. Medical assistance, health monitoring, and rehabilitation for the older and disabled people is an imminent challenge, as it requires an ideal network between people, medical equipment and health service providers. For this reason,

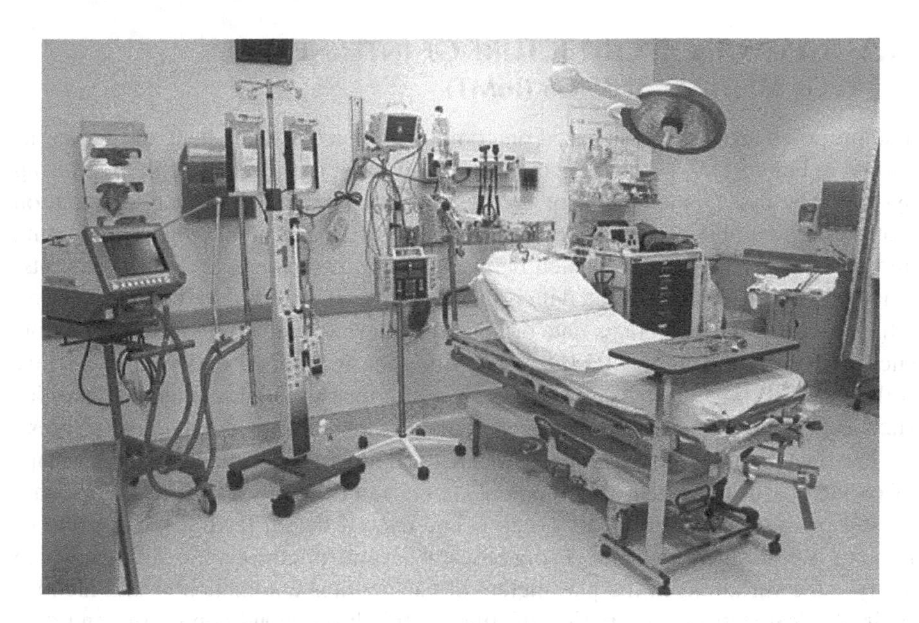

FIGURE 11.3 Existing patient monitoring system.

reliable and cost-effective wearable devices with low-power are needed to enhance their quality of life. The IoMT framework offers promising innovative advances to accomplish before the specified patient care services, and enhance further the health service systems [29].

The IoMT platform is useful to collect important health information form clients and wirelessly communicate the information to the cloud server for evaluating the previous record of the client [30]. Such a network with this equipment and these processes will assist in preventive measures, or for providing immediate care. Recently, a few IoMT frameworks have been created for remote medical care services and assisted living applications. An excellent IoT system, with capable medical devices and standard communication, was produced by Xu et al. [31]. A resource-based data receiving method is better suited for medical care intensive data applications [32]. A medical support system based on IoT was proposed and implemented by Kolici et al. Moreover, several experiments were performed to evaluate the system [33]. Ang et al. developed a smart guide portable system with a handheld device which helps visually impaired and low vision individuals to move around using a camera sensor system to improve observing capacities for an additional level of security and reliability with mobile association innovation [34]. IoMT-based remote patient monitoring data sets consist of large volumes of information from various groups, which is difficult to separate the data sets between groups and within the groups. In order to arrange datasets, a Fisher's discrimination criterion was applied to clean the raw datasets in row, columns and time stamps [35,36]. This chapter is mainly focused on proposing a new remote patient healthcare monitoring system, and starts with a review of the related work, monitoring and examines the proposed architecture, available advantages, limitations, and the challenges addressed to enhance the practice of medicine.

11.2 NETWORK ARCHITECTURE OF INTERNET OF MEDICAL THINGS (IoMT)

The proposed architecture is based on communication between client and server, as shown in Figure 11.4, which depicts the main layers of the remote health monitoring system. The first biometric sensor layer is data sensing and collecting information from smart wearable sensors from the whole network. Local and intelligent computing and processing units are connected between the hardware devices and the patients' skin surface [37]. The collected data is processed and transferred to the service layer, which consists of data storage, data organization devices and wire/wireless protocols such as Bluetooth, Zigbee, RFID, WiFi, Ethernet and 3G/4G networks; these are linked to the IoMT layer for communicating the measured parameters to facilities such as hospitals, emergency centres, ambulances, care takers and medical centres.

The physicians have easy access to the medical histories of patients or large groups of patients as well as their physiological status, and the analyses of suspicious data (blood pressure, respiration, ECG, etc). The overall health testing for a group of patients is performed by doctors using efficient hardware and software devices, which analyse the variations in the parameters of each patient automatically and identify their physical status over a period of time. In this layer, cloud computing and data protection as well as patient privacy services are provided. The fourth application layer uses the interface between doctors and remote patients to easily provide information status on demand or on a regular basis. The second layer has a significant importance for processing the sensor data online. A cloud-based information storage framework is used for arranging and recovering lots of information and provides highly stable, high speed, and cheap storage. S_3 serves as a raw sensor data storage and sorting server for image, audio, and video data in IoMT systems. Amazon offers commercial web services that permit designing MySQL, Oracle or Microsoft SQL servers on the cloud.

Amazon Kinesis is a web service that permits real-time processing of information, and it deals with automatically scaling large amounts of streaming data originating from a large number of sources. Amazon SQS offers an exceptionally adaptable and predictable hosted queue for stores, storing and release messages in a scalable manner between distinct components of applications. Amazon EMR is a web service that utilizes the Hadoop framework and permits processing large-scale data, consequential, and is suitable for IoMT applications that produce large amounts of data that need to be analysed. Matallah et al. [38] addressed the service metadata improvement by proposing a blended arrangement amongst centralization and circulation of metadata to upgrade the execution and versatility of the model utilizing Hadoop Distributed file (HDFS) framework.

The IoMT can be seen in three patterns based on sensors oriented acquisition things, internet-oriented middleware monitoring, and knowledge-oriented action systems (Figure 11.4). However, the IoMT is useful for the development of applications based on the three patterns. As the purpose of the hardware layer is to interconnect sensors and the device components such as storage, processing, and internal parts. Physiological parameters such as temperature, ECG, blood pressure and so on are measured using wearable sensors, which must be very precise and small in size.

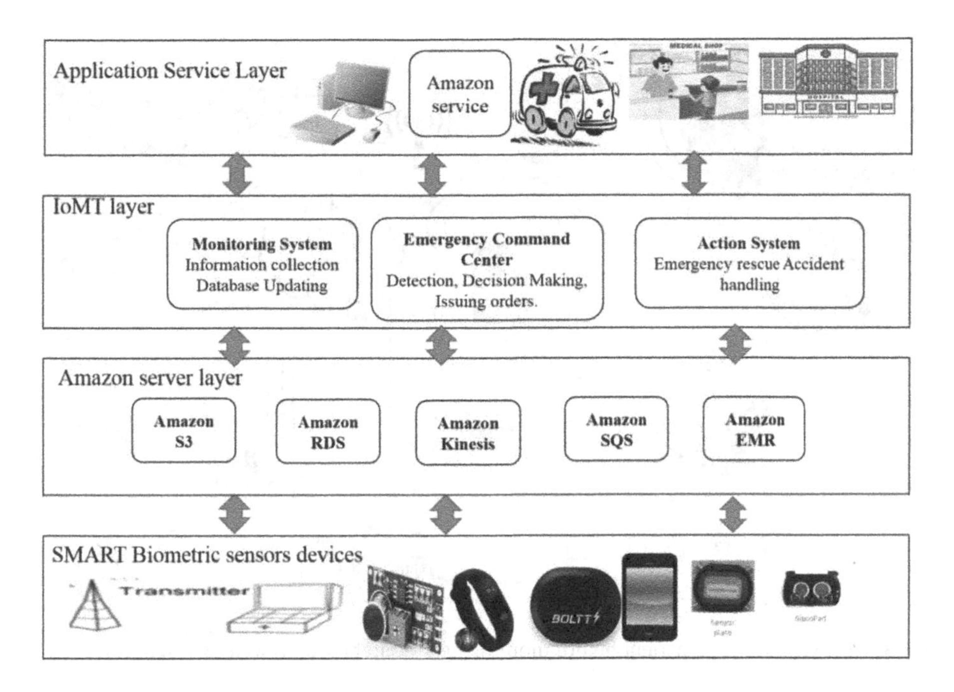

FIGURE 11.4 Architecture of remote healthcare monitoring system.

The communication layer is useful to connect many devices to the network and transmit the information to the application layer. The application layer presents the required services such as hospital, ambulance, emergence, medication to the remote patient by receiving the information from the IoMT layer. Moreover, it allows the possibility of obtaining, processing and recommending valuable information to patients and improving their way of life. Moreover, it is necessary to exploit the advantages of innovative technologies such as IoT for communicating medical data from machine-to-machine (M2M), and person-to-person (P2P) to improve healthcare.

11.3 REAL-TIME ANALYSIS REMOTE PATIENT HEALTH MONITORING

A remote patient monitoring system based on wearable sensors with IoMT for obtaining chronic patient conditions is shown in Figure 11.5. Patients purchase biosensors from commercial vendors and install network services on their mobiles. These services can be connected with the help of GPS and BPO services for location-tracking of patients at low cost. The patients' physiological parameters and patients' health status can be monitored with a smartphone and the information can be transferred to the IoMT through Zigbee, Bluetooth or Wi Fi. IoMT can be used to store and process the data received from the sensors in a cloud server and provide required medical assistance on the basis of sending application service to the hospital, physician, medication centre, home and so on in real time. IoMT technology is implemented using biometric sensors in hospital settings or smartwatches that enables the patients to self-monitor

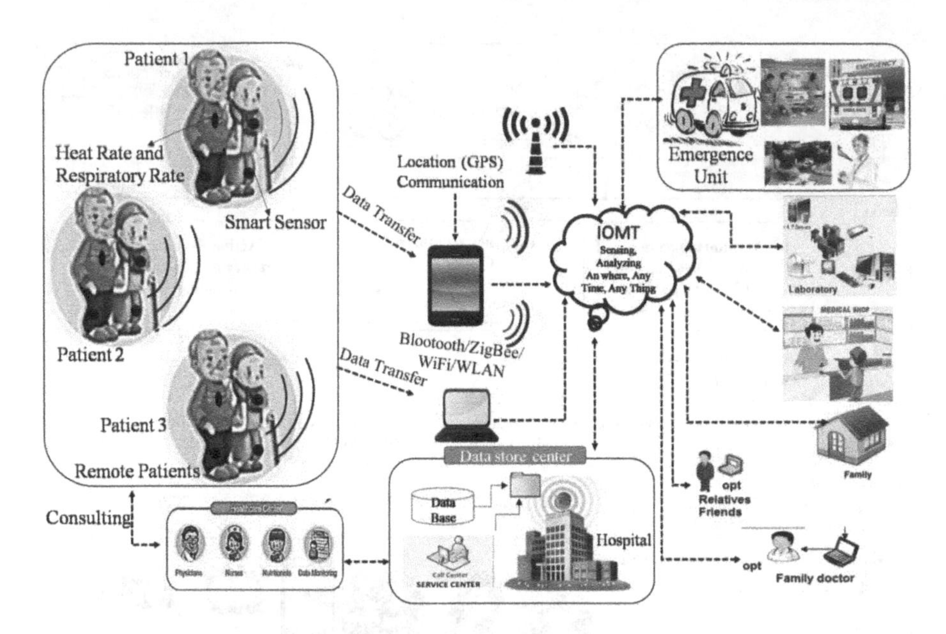

FIGURE 11.5 Remote patient health monitoring system based on wearable sensors.

and collect sensory information. Built-in smartwatch sensors are able to monitor heartbeat, pulse rate, blood pressure, temperature or respiration and even interface with remotely located medical devices. In case of medical emergencies, these smart devices can be used to contact the physician at any time, or an opened pharmacy in order to save a life. Blood pressure measurements help in explaining patterns of blood stream variations. IoMT-based blood pressure sensing is done in real-time by sensors attached to patients and connected to relevant medical services. An IoMT network collects specific information about blood pressure using communication devices, which includes a blood pressure sensors, smartphones and processors.

11.4 METHODOLOGY AND ANALYSIS

Patient physiological information is collected by wearable sensors for body temperature, pulse, respiration, and ECG [39]. The sensors are connected to the network through data accumulators in the region of the patient [40]. The patients' physiological data are recorded and transmitted in real time (by using the components of the system from any remote location) to the data centre with assured security and privacy [41].

Typically, data acquired from wearable sensors is transferred to the accumulator through Zigbee or Bluetooth. Collected information is further transmitted to a cloud using web connectivity on the accumulator, typically via a cellular phone data connection or Wi-Fi [42]. Each sensors' data can be accessed through the Internet via the accumulator based on the IoMT architecture. Most often, data storage and processing devices are a mobile client, or local cloud storage, and a processing unit (PC), which are directly accessible by the accumulator through Wi-Fi [43,44].

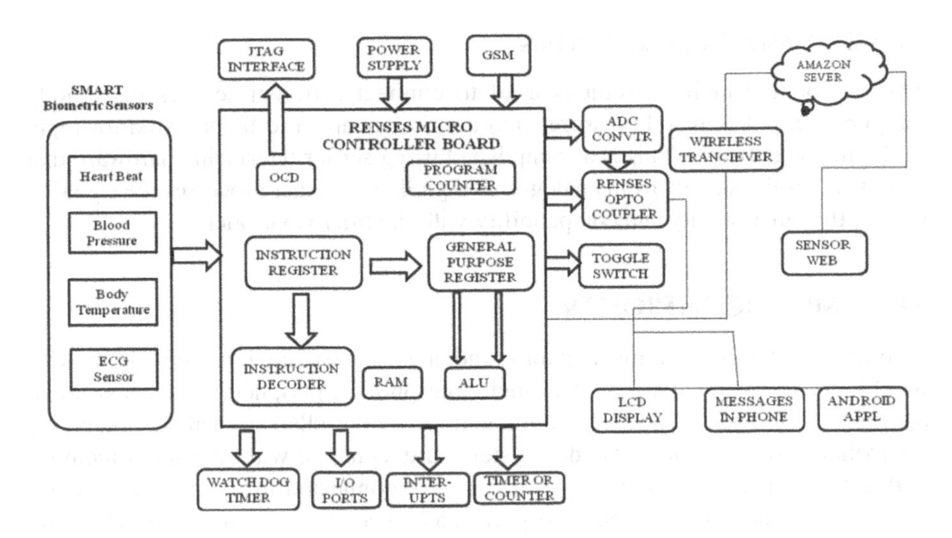

FIGURE 11.6 Master block diagram of a remote health monitoring and healthcare system.

Cloud servers are used to partially store the data before communication and are also used to concentrate the patient's data when mobile devices have some limitations [45]. A block diagram of the proposed remote health monitoring and healthcare system is shown in Figure 11.6; the design consists of a Renses microcontroller, smart biometric sensors, GSM/GPRS, power supply, and so on. Biosensors such as heartbeat, temperature, respiration, pulse and ECG sensors are connected to the microcontroller though port pins and the output signals are analogue; thus, the output pins are connected to the analogue digital converter pins of the microcontroller. Microcontrollers read and process the data and transfer it to the next level through a Wi-Fi. The measured sensor data is sent to the commercial cloud server and then sent to a web page through Wi-Fi, and messages are sent to smartphone and simultaneously displayed in LCD. However, the sensed ECG information is fed into an android application. Embedded C, Renesas flash programmer and Cube suite programmer were used to feed the values to the pins, so they can be stored in the web page.

11.4.1 Data Sensing and Acquisition

Physiological parameters are measured using various device platforms such as wearable sensors, minor pre-processing hardware and communication software for transmission of measured physiological parameters. The measuring sensors must be light, small, and operate in a wearable package with energy efficiency and also should not block the patient's movements and mobility. Sensors should provide continuous operation without replacing the extended durations. Recently designed biosensors for medical applications are flexible and easy to place on various body parts in contact with the skin. The sensors receive the bio-signals with greater accuracy [46].

11.4.2 Sensor Interface Circuits

The sensor interfacing circuit is used to connect different sensors (e.g., body temperature, pulse, blood pressure, etc.) are then adapted to be prepared for input to the microcontroller. This is accomplished using sensor interfacing hardware that changes over the signals from analogue to digital and further processes the signal to ensure the functionality and compatibility with the microcontroller.

11.5 MICROCONTROLLER

A microcontroller contains memory and a processor, and is embedded with machinery in the computer system including phones, peripherals, and household appliances. Nowadays, most programmable microcontrollers are called 'embedded controllers'. Most of the embedded systems are available with minimum memory and are complex. Input and output devices are connected to the microcontroller, including wearable sensors, relays, displays, and other hardware. Microcontrollers are used to automatically control the connected devices and concentrate the data of the different sensors and communicate such information to the cloud server for additional processing and are used for tracking of the area information.

Our proposed method uses a Reneses microcontroller RL78/G13 single-board device with 20 pins and 128 pins with a flash memory of 16 KB and 512 KB, respectively. The RL78/G13 microcontroller includes the following: on-chip oscillator, real-time clock, power on reset, low voltage detection, watch dog timer, 26 channels of 10 bit, analogue digital converter, 32/32 divider, universal asynchronous receiver/transmitter, LIN, timer array and IEC 60730 hardware. The Reneses microcontroller is a board transmission module that controls the data stream between the distinctive sensors and the microcontroller. The sensor data communicated to the network can be accessed for monitoring the physiological parameters by the doctors. The controller alerts the doctor, nurses and caretaker about the variation output of the parameters. However, the major objective in remote healthcare and monitoring systems is data security and privacy.

11.6 PHYSICAL SENSORS

A sensor is used to detect the change in the parameters or the objects and sends the information to the processor. The sensor is a sub-system which is as basic as light or as complex as a computer. In order to collect patient action and health information, multiple sensors are needed. These sensors should be insignificant to be wearable and used for receiving the data for the proposed system was shown in Figure 11.7.

11.6.1 Temperature Sensor

A temperature sensor is used to measure body temperature. The LM35 series sensor contained an integrated circuit and converts the output voltage in terms of centigrade temperature. Various types of temperature sensors such as thermistors, RTDs, thermocouples and IC sensors are available, however, the proposed system focuses on the LM35 sensor due to its sealed sensor circuitry. The LM35 sensor can

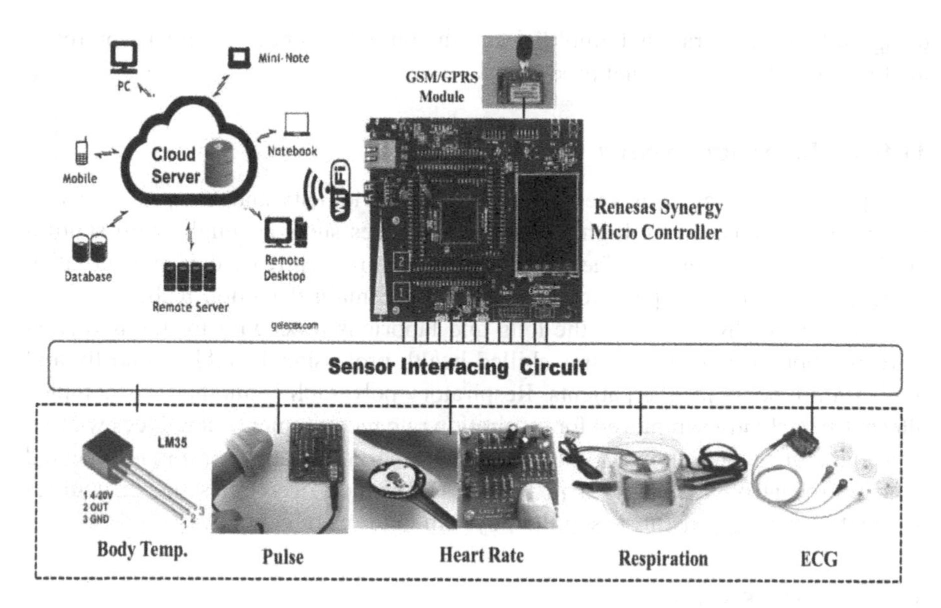

FIGURE 11.7 Proposed IoMT wearable sensor system for patient healthcare monitoring.

measure more accurately than a thermostat. The working temperature range is from −55 ± 0.1°C to 150 ± 0.1°C, with a scale factor of 0.01 V/°C.

11.6.2 Pulse Rate Sensor

The pulse rate is measured using Infra-Red (IR) light absorption of oxygenated and deoxygenated (or reduced) haemoglobin passing though the finger, which shows the difference in light absorption. Hemoglobin with oxygen (or reduced) absorbs more IR and allows more red light to pass through the finger, and hemoglobin without oxygen absorbs more red light and allows more infrared light to pass through the finger. The wavelength of red and infrared are 600–750 nm and 850–1000 nm, respectively. The pulse rate is measured by using a fingertip sensor, which emits light and receives it with a photo detector placed on the other side of the fingertip.

11.6.3 Heart Rate Sensor

The heart rate sensor is comprised of an amplifier circuit and fingertip sensor, as shown in Figure 11.7. The heartbeat sensor comprises an infrared (IR) light emitting diode transmitter and a photodetector receiver. The IR light is transmitted through the fingertip to measure the changes in blood volume in the blood vessel by using a fingertip sensor which is reflected in the light received by photodiode detector. The fluctuation in the blood volume is detected with respect to the pulse and the heartbeat is produced at the output of the photodiode. The signal received from the photodiode is not detected by the Reneses microcontroller directly because of the small amplitude of the signal, which requires amplification. The signal is amplified using an LM324 operational amplifier and the output is sent to the microcontroller analogue pin A_1 for additional processing.

11.6.4 Respiratory Sensor

A respiratory rate sensor is used to monitor the periodicity and non-periodicity of breathing, and controlling highly prevalent diseases such as coughs with sputum production or sleep apnoea. The respiratory rate is frequently used as an alternative to pulse and heart rate parameter variables. The main detection technique used is plethysmography. However, the long and laborious procedure for the detection of respiration of patients requires skilled health professionals, and is a costly and unpleasant process for the patients. Respiratory polygraph is another less complex alternative technique employed for respiration rate measurements; it reduces waiting time and cost, but it is an invasive technique. A smart respiration sensor is connected to a microcontroller, as shown in Figure 11.7, which overcomes the difficulties associated with the existing respiratory system in terms of complexity.

11.6.5 ECG Sensor

An ECG sensor is utilized to monitor the electrical activity of the heart, which includes the measurement of cardiovascular information such as the heartbeat and the basic rhythm of the heart, and prolonged PR and QT intervals. Besides, the ECG sensor gives confirmation of damages that occur in various parts of the heart muscle, and indicates the expanded thickness of the heart muscle. The electrical activity of the heart is measured by using the cost-effective board and the output of biopotential signals from the heart muscles is read as an analogue waveform. IoMT has the potential to monitor the ECG waveforms remotely and can be used to its fullest extent when based on a cloud server [47]. A number of studies have reported on ECG monitoring using IoT [43–53]. ECG signals are measured using three leads AD8232 monitor, which acts as an operation amp to obtain a clear signal from the PR and QT intervals. The AD8232 measures the biopotential signals from the skin surface using an integrated signal conditioner, which is designed to extract, amplify and filter the small biopotential signals in the presence of noisy conditions. The ECG sensor records the biopotential signals of the heart with time, using leads arranged on specific locations of the body. The signals obtained from electrodes placed on the patient are amplified by using an operational amplifier circuit LM324. To begin with, the LM324 has been utilized in the pre- and post-amplifier stages. The output of the amplified signal is specifically sent to the Reneses 2560 pin A_2 for additionally processing. The altium designer and multisim software were used in designing and testing the circuit.

11.6.6 Global System for Mobile Communications (GSM)

A GSM modem is used for connecting the PC to the mobile network, which is operated by using a SIM card to the subscribed mobile operator, just like for a cellular phone. The GSM modem operates over the network to send and receive messages. The GSM communication module acts as a gateway for the receiving module. The receiving module such as GPRS modem, mobile phone or any device sending and receiving SMS is connected with the computer and microcontroller through a USB or serial cable [54]. GSM uses AT (Attention) commands in order to control modems and computers to send, receive, write or delete messages. The IoMT agent will receive the SMS and process the command.

11.6.7 General Packet Radio Service (GPRS)

General packet radio service (GPRS) is a wireless data communication service available as a packet-based that delivers data rates from 56 to 114 Kbps, and is intended to replace the present circuit-exchanged administrations for GSM that associates with cell phone connections and send the Short Message Service (SMS). GPRS disseminates packets of information from several different terminals in the framework over different channels, making a significantly more proficient utilization of the transmission capacity applications such as Internet access. These higher data rates will enable clients to access applications utilizing a portable handset or PC. The cloud server actualizes a wide arrangement of information administrations including sensor and actuator, information handling, information investigation, database stockpiling and information perception.

The received client information is imparted to a cloud server, which is important for the accessibility of information anywhere on the Internet. The cloud server implements a wide set of information management services including sensor and actuator, data processing, data analysis, database storage and data resolution. The data analysis incorporates the stage with the system framework and application, in addition to providing an application program interface (API) and software tools through which the information can be accessed and manipulated. The design and development of the cloud server is shown in Figure 11.8. The cloud server stores the huge database that has enough space to furnish huge amount of data from various sensors for long times and also track the historical backdrop of the user. The database is interfaced to a wide set of data analysis algorithms and APIs for estimation and evaluation. The IoMT client-side computer can easily access the data through the Internet, as shown in Figure 11.8.

11.7 SOFTWARE DESCRIPTION

11.7.1 Embedded C Programming

Embedded C is a language programming expansion of C programming to address the communication between different embedded systems. A large portion of the syntax and semantics of embedded C, for example, main function, variable, data type, loops, functions, arrays and strings, structures and so on resemble standard C programming. In short, embedded C deals with microcontrollers, input/output ports (RAM, ROM), whereas C deals with only memory and operating systems. C is a desktop programming language used for embedding a piece of software code into the hardware to make it function. Here, the program on the wearable sensor operation was written using embedded C.

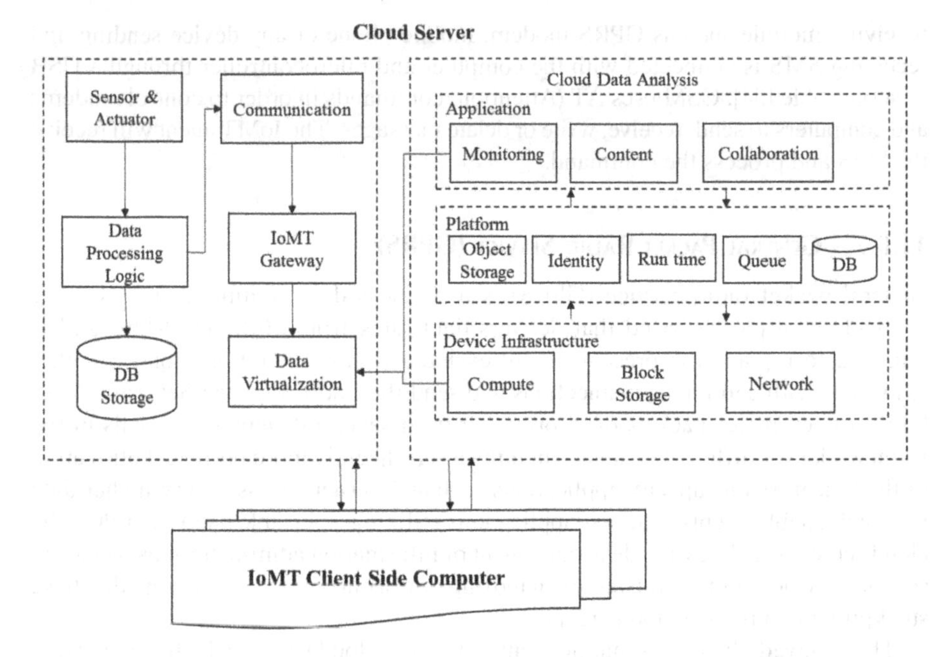

FIGURE 11.8 Cloud server architecture.

11.7.2 Renesas Flash Programmer

Renesas Flash Programmer (RFP) is used to write, erase, and verify the programs with on-chip flash memory mounted onto the target system or program adapter on which a Renesas single-chip microcontroller, E20 emulator, or the on-chip debug emulator with programming function, QB-MINI2 (INICUBE2), or a serial interface is set.

11.7.3 MySQL

It is the most prevalent open source relational SQL database administration framework. It has a perfect convenient database server and is valuable for applications. Also, the SQL supports standard compilation on various stages and has multi-reading capacities on UNIX servers, which enables excellent execution. For non-Unix individuals, MySQL can keep running as an administrator on Windows NT, and as an ordinary procedure in Windows 95/98 machines.

11.8 EXPERIMENTAL EVALUATION

11.8.1 Flow Diagram

A proposed algorithm based on the healthcare of patients shown in Figure 11.9. The parameters can be obtained from the present status of patients. In the case critical condition patients, the system can collect data continuously. The parameters collected

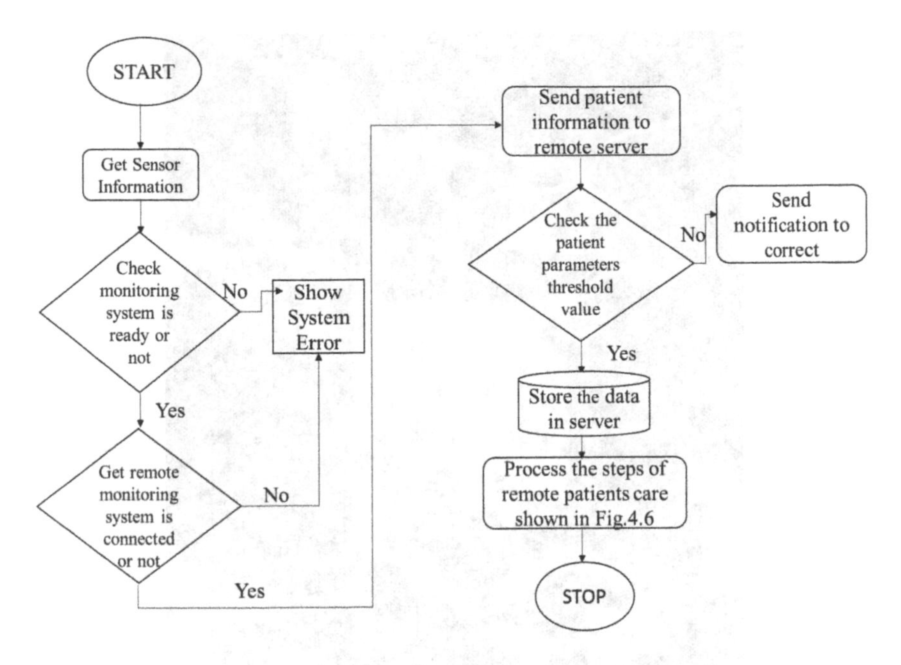

FIGURE 11.9 Flowchart of the remote health monitoring and healthcare system.

are temperature, heat rate, blood pressure, respiration rate, ECG and so on. The proposed system contains an alarm system which generates beeps to notify the concerned doctor/nurse/care taker in case of an emergency. When the data deviates from a preset threshold value, a message is generated to notify the users.

The proposed algorithm is a generalized health-monitoring model that works on the principle of threshold limits, and is customized for individual remote monitoring. It will be designed in a way where the algorithm accepts various threshold limits for each parameter, and estimations of the threshold values can be defined on the basis of the essential parameters of the individuals. The customized algorithm for monitoring aids keep on changing threshold limits throughout the monitoring phase and analyses the health parameters such as heart rate after every 10 minutes and compares it with previous values to find out if any change occurred. A robust scanning mechanism has been applied in alarming module. In this module, the data from heart rate, blood pressure and body temperature are collected through sensors.

11.8.2 Experimental Analysis

In order to validate the proposed idea, a prototype model was developed and evaluated by *assess* of remote patient monitoring using IoMT and assisted living healthcare system as shown in Figure 11.10. The model has been constructed to receive data from all the listed sensors. However, the readings of the other wearable sensors have been validated in the laboratory as well. The step-by-step process of the hardware is connected to the power supply. The proposed system was detected by the SIM and was configured by Wi-Fi, which was indicated on the LCD display as 'OK' message after

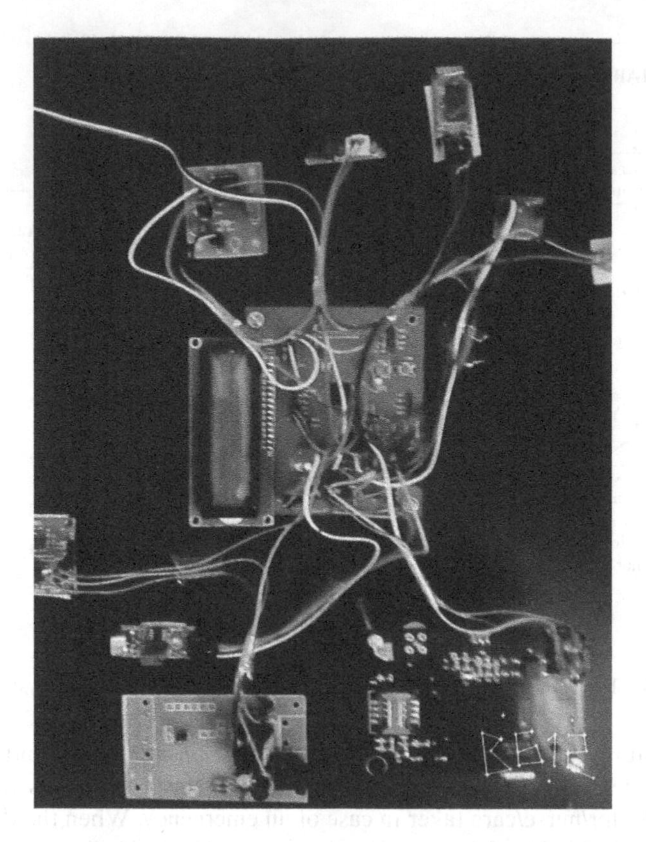

FIGURE 11.10 Prototype of the proposed remote patient monitoring system.

the configuration process was completed and the system was put online. Temperature, pulse rate and heart rate were measured by the sensors and the values were indicated on the LCD display, and simultaneous messages were sent to the doctor, caretaker, and physician through the Wi-Fi cloud. The sent data was viewed on an html webpage with a unique ID and messages were sent to the nurse, caretaker and doctor's mobile phone. The measured ECG waveform was indicated and displayed on the mobile. The wearable sensors with IoMT recorded the physiological parameters of the patients at home and at work. The recorded data was processed and communicated to the doctor, caretaker and nurse or physicians via a mobile phone data connection or Wi-Fi.

Most of the patient-monitoring systems are intended for a particular group or group of individuals suffering from different diseases [55] such as distressing sickness [56], dementia [57], hypertension [58] and diabetes [59], while some other patient monitoring systems are committed to the more established age group patients [60]. Only a few models are displayed for onexclusive portable patient monitoring systems [61,62]. A real-time signal detection algorithm is implemented to measure physical health data such as blood pressure, heart rate, pulse rate, and respiration including ECG signals monitoring in the proposed system. The proposed remote health-monitoring system sends numerous messages during irregular patterns detected in the ECG signal. The Proposed RHM empowers a patient to choose the health data analysis locally or on a centralized server. The local analysis of the received health data from the sensors can alert the clinicians through an alarm, allowing them to cross-examine a patient about his/her health condition. The remote patient monitoring system tends to be secure and safe by sending the monitoring information intermittently to the clinical database, which is typically deployed in mobile base units. The RHM system automatically contacts the healthcare provider or specialist about any irregularity in the measured signals.

As the remote health monitoring and healthcare services are rapidly progressing towards development, current emphasis is on introducing skill into the model for the autonomous decision. Intelligent health monitoring systems can assist doctors in translating the therapeutic medical information, directing and automating the heath monitoring process through intelligent expert systems, and a data management system can successfully encapsulate, extract and interpret real-world context aware information, ensuring doctors get the right information each time [63,64]. The data responds contrastingly relying upon the medicinal information and constant recording of the patients in various conditions. To help these kinds of frameworks, a few methods like data mining (for the presumed present medical condition of a patient), or applying a group algorithm to both current and historic data have been proposed [65,66]. Another approach for simple data mining is introduced as object oriented database system for server and customer [67].

The IoMT is addressing several challenges in sensing, analysing, and visualizing the data before designing the systems for integration into clinical practice. Visualization is integrated with several objects such as sensing, communicating and sharing the information over networks. The data received from the interconnected sensors at regular intervals is processed and the feedback action is provided by a network for analysing, planning and decision-making. The IoMT is a connecting object to the Internet and facilitates remote patient monitoring.

11.8.3 Security and Privacy Concerns of IoMT

The realization of the IoT requires changes in frameworks, architectures and communications that correspond to adaptable, versatile, secure, and unavoidable without being meddlesome [68]. The specialists expect that the technological advancements of the IoMT will present basic improvements in the areas of ethics, information insurance, specialized design, distinguishing proof of organized articles and administration. As more smart systems are connected to the Internet, potential security is required and good connection with feedback administration and information is needed. The security issues (assurance of information protection) may emerge amid data accumulation, transmission and sharing and consequently speaks to a basic segment for empowering across the board appropriation of IoMT technologies advancements and applications. The rapid developments of IoMT healthcare contains risks for security and privacy. Therefore, there is need to develop and integrate viable universal detection systems for healthcare security and privacy, as critical objectives are taken into account. To decrease and understand the risk of security and privacy in healthcare applications, a number of articles have been published [69–75]. The unauthorized use of such private information could present undesirable security risks to the patients [76].

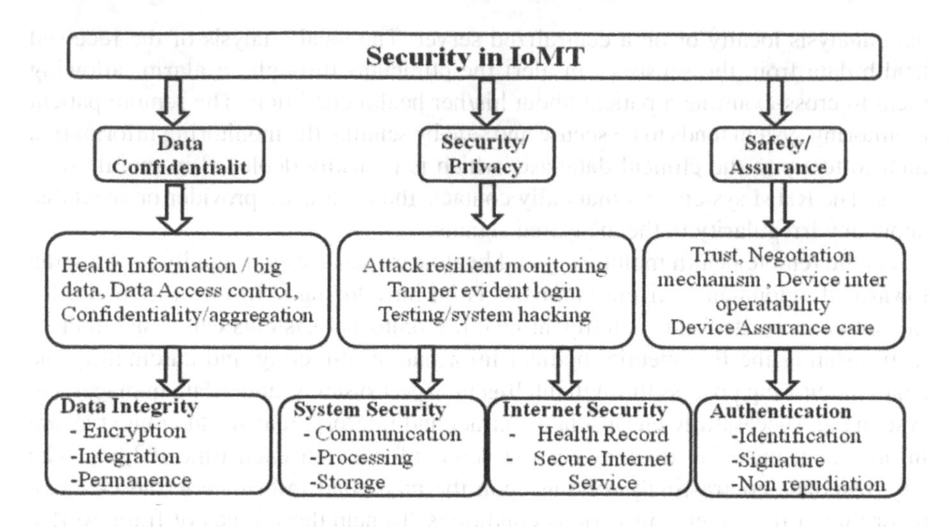

FIGURE 11.11 Security and privacy concerns of IoMT.

Subsequently, compact smart sensors to record, store, and transmit to understand information to a focal server, require secure approaches to transmit recorded information at rapid speed to guarantee that anytime. Ensuring privacy must not be constrained to specialized arrangements such as encryption, ID administration and security upgrading advancements. The prerequisites of IoMT communication systems in healthcare applications could be interoperability is expected to empower diverse things to cooperate in order to give the desired benefit. Bounded idleness and reliability should be conceded when managing emergency circumstances in order for the intercession to be powerful. Authentication, security, privacy and respectability are compulsory when sensitive information is exchanged over the effective (Figure 11.11).

Client verification is imperative in healthcare databases, as it is the initial phase in the whole health data access method. Confirmation systems are expected to check the client's character like: secret key, PIN, unique finger impression, signature, voice design, card, token, and so forth. Using these techniques we can be sure beyond any doubt that the data is sent by the confided in sender. By understanding of information respectability, the transmission of therapeutic reports from the patient to the medical staff can't be changed by any outside or unapproved source. Likewise, the patient's records can't be interpreted by anybody other than the authorized medical staff. Subsequently, confidentiality ensures that the patient's medical documents are protected from uninvolved assaults [77].

The data being produced, accumulated and shared through network devices must be protected with solid and usable verification techniques. As some of the received information might be confidential, and patients would prefer not to share it with others – sometimes even with family doctors – the revealing of such confidential information could lead to prejudice [76]. Subsequently, smart devices can record, store and transmit patient information to a cloud server; therefore, this requires secure approaches to transmit recorded data at rapid rates to ensure individual records can't be compromised at any time. It is imperative that data security, privacy and assurance be deliberately addressed at the planning stage. In summary, security should be fully incorporated into healthcare, from the device to the system, all the way to the server farm.

A solid security system incorporates verification advancements and procedures, confirming patient and supplier identification to guarantee devices are utilized by approved clients. The interchanges channels between the devices inside the IoT should likewise be secure to guarantee the validity of the data going through them. As healthcare frameworks store and process exceptionally delicate information, they ought to have appropriate security and protection structures and systems. Securing privacy must not only be confined to technical solutions such as encryption, ID administration and privacy enhancing measures, but should also include regular activities such as shopper assent, accumulation constraint, transparency, responsibility based self-direction, security accreditation, customer instruction and socio-ethical based customer rights, open mindfulness, divulgence, purchaser support [78]. However, data security, protection and information assurances should be deliberately addressed during the outline stage.

11.9 ADVANTAGE OF REMOTE PATIENT MONITORING

At present, only 10% of the patients are familiar with remote patient health monitoring (RPHM) systems, leaving 90% unaware of this opportunity. Currently, RPHM continues to gain acceptance in hospitals and healthcare centres, and many specialists are expressing interest in the next generation platforms to address the challenges and inefficiencies of the current patient monitoring platforms. Research surveys are stating that the RPHM market is growing rapidly, and in this context the advantage of RPHM for patients, care providers, public health authorities and insurance payers are reported in Table 11.1. IoMT is able to monitor, record and track the changes in

TABLE 11.1
Advantages of Remote Patient Healthcare Monitoring for Patients, Care Providers, Public Authorities and Insurance Payers

	Benefits for Patients	Benefits for Care Providers/ Public Health Authorities	Benefits for Insurance Payers
1	Better outcomes and quality in treatment	Enhance the access of patient health data in real time	Better visibility on patient compliance practices
2	Real-time support and interventions improve the disease management and reduce errors	Continuous monitoring of patient health parameters, regardless of patients location	Better acceptability and accountability from patients and care providers
3	Prevent emergencies and re-admissions	The accuracy, reliability and relevance in reading data enhanced, which would provide better precision in treatment	Reduced costs of care and monitoring
4	Reduced hospital stays	Cost reduction from re-admissions and hospital stays	Real-time processing of patient documents

parameters over a specific time period. Doctors can prescribe the medication and suggest treatment methods based on the recorded changes or the history of the patient. Normal routine check-ups and hospital stays can be minimized. Digital records of patient health parameters on thc clouds arc becoming reliable, and storing data in a computer or memory device has benefits over printing on paper such as minimal chances of data loss.

11.10 APPLICATIONS OF REMOTE PATIENT HEALTH MONITORING AND HEALTHCARE PARADIGM

- The RPHM system mainly focuses on measuring, evaluating and communicating the sensor data of vital parameters, for example, temperature, heart rate, ECG, pulse rate etc;
- RPHM system enables to diagnose and evaluate the obtained data at specific circumstances such as medial practice and facilitate projection for a framework of diseases at an initial stage, prevention, diagnosis, and treatment with overall management;
- RPHM system is able to record long-term clinical data and access the patient's physiological information, which can be sent to the physicians/doctors/caretakers when needed;
- The sensory data of remote patients are easy to analyse and can be presented to physicians in a prescribed format, and are eventually easy to start using their clinical practice.

11.11 LIMITATIONS AND CHALLENGES

The successful implementation of IoMT-based remote patient monitoring systems are facing certain limitations and challenges in terms of technology and networking such as accurate data acquisition, inter-operability between hardware and software, bandwidth, quality of health services, limitation of battery life, sensor biocompatibility and so on. The emerging technology and advances in the field of Internet communication bring many challenges coupled with contribution in the growth of the medical field. The significant challenges in the domain of IoMT are design in hardware and software implementation, design optimization of wearable sensors and real-time processing with low power consumption sensors. Network performance is likely to be constrained in the memory of the system, which influences the performance and management of network-like devices, as well as bandwidth and flexibility of data volume, data security and data privacy. In addition to the network and technological challenges, the most common and general challenges are user attitudes, technological acceptance and barriers, confidentiality, legal issues, ethical and administrative barriers, implementation costs, maintenance of the system and sufficient investments. Overcoming these difficulties in the area of IoMT will enhance the standard of remote healthcare. IoMT provides better and reliable healthcare and health monitoring services, as it bridges the gap between doctors, patient and healthcare services. IoMT enables the doctors and hospital staff to work more precisely and actively with less effort.

11.12 CONCLUSIONS AND FUTURE ENHANCEMENTS

In this chapter, we reviewed the current state of remote patient monitoring using IoMT-related services and technologies in remote healthcare and monitoring. In addition, future perspectives for RPM technologies in clinical practice have also been discussed. A significant number of research challenges have been highlighted, which are expected to be major research areas in the future. In the present work, an IoMT-based smart healthcare system has been designed using a microcontroller. Wearable sensors are used to measure the physiological parameters and the data is transmitted to the embedded controller and sent to the server through a Wi-Fi protocol. The measured physiological parameters are processed and displayed on an LCD so the patient can also monitor his/her health status. The health status of a patient is also sent via an SMS to the doctor/nurse/caretakers' mobile phone through a connected GSM modem where an alarm alerts the caretaker. The real-time data is monitored by the physician/caretakers after logging on to an html web page using a unique IP.

Future undertakings can be designed on the basis of the major challenges in terms of network and technology advances. For instance, there is a need for designing and developing hardware and software modules for better transmission of large and critical data with advanced communication networks. The real-time measurement of physiological parameters of the patient needs to be enhanced. Implementing Li-Fi (Light Fidelity) modules for uploading data may result in faster uploading of data into databases, which could also enhance the functionality of android applications. Future work should perhaps investigate the challenges that are constantly faced by mobile-based systems such as accuracy of critical signals, bandwidth, quality of health services, adaptable wireless technologies, inter-operability between different systems, data encryption and security. However, one shouldn't forget legal and ethical aspects, and to focus on developing user-friendly systems.

REFERENCES

1. Thimbleby H. Technology and Healthcare in future. *Journal of Public Health Research.* 2013;2(3):160–167.
2. White Paper. Internet of Things Strategic Research Roadmap, Antoine de Saint-Exupery, September 15, 2009.
3. Dey N, Hassanien AE, Bhatt C, Ashour A, Satapathy SC. *Book: Internet of Things and Big Data Analytics Toward Next-Generation Intelligence.* Springer, 2018:3–20.
4. Dey N, Ashour AS, Borra S (Eds.). *Classification in BioApps: Automation of Decision Making.* Springer, 2017;26.
5. Tan L, Wang N. Future internet: The internet of things. *3rd International Conference on Advanced Computer Theory and Engineering.* 2010;5:376–380.
6. Wu M, Lu TJ, Ling FY, Sun J, Du HY. Research on the architecture of Internet of Things. *3rd International Conference on Advanced Computer Theory and Engineering.* 2010;5:484–487.
7. Pandikumar S, Vetrivel RS. Internet of things based architecture of web and smart home interface using GSM. *International Journal of Innovative Research in Science, Engineering and Technology.* 2014;3(3):1721–1727.
8. Gómez J, Huete JF, Hoyos O, Perez L, Grigori D. Interaction system based on internet of things as support for education. *Procedia Computer Science.* 2013;21:132–139.

9. Botterman M. Internet of Things: An early reality of the future Internet. In: *Meeting at EU*, Prague, 2009.
10. Sharma M, Siddiqui A. RFID based mobiles: Next generation applications. In: *Inf. Manag. Eng., 2nd IEEE Int. Conf.*, Chengdu, China, 2010;523–526.
11. Ziegler J, Urbas L. Advanced interaction metaphors for RFID-tagged physical artefacts. In: *RFID-Technologies Appl. IEEE Int. Conf.*, Sitges, Spain, 2011;73–80.
12. Pandikumar S, Vetrivel RS. Internet of things based architecture of web and smart home interface using GSM. *International Journal of Advanced Computer Science and Technology*. 2014;3(3):1721–1727.
13. Dimitrov DV. Medical internet of things and big data in healthcare. *Healthcare Informatics Research*. 2016;22(3):156–163.
14. Patel S, Park H, Bonato P, Chan L, Rodgers M. A review of wearable sensors and systems with application in rehabilitation. *Journal of Neuroengineering and Rehabilitation*. 2012;9:21.
15. Yu L, Lu Y, Zhu X. Smart hospital based on internet of things. *Journal of Networks*. 2012;7(10):1654–1667.
16. Yao Q, Tian Y, Li PF, Tian LL, Qian YM, Li JS. Design and development of a medical big data processing system based on Hadoop. *Journal of Medical Systems*. 2015;39(3):23.
17. Lee CH, Yoon HJ. Medical big data: Promise and challenges. *Kidney Research and Clinical Practice*. 2017;36(1):3–11.
18. Kruse CS, Goswamy R, Raval Y, Marawi S. Challenges and opportunities of big data in health care: A systematic review. *JMIR Medical Informatics*. 2016;4(4):e38.
19. Bhatt C, Dey N, Amira A. *Book: Internet of Things and Big Data Technologies for Next Generation Healthcare*. 2016:3–33.
20. Elhayatmy G, Dey N, Ashour AS. Internet of things based wireless body area network in healthcare. In *Internet of Things and Big Data Analytics toward Next-Generation Intelligence*. Springer, Cham, 2018:3–20.
21. Aminian M, Reza Naji H. A hospital healthcare monitoring system using wireless sensor networks. *International Journal of Medical Informatics*. 2013;4:121.
22. Prashob B, Vegnesh N, Sandesh W. Remote health monitoring using IOT. *International Journal of Advance Research, Ideas and Innovations in Technology*. 2017;3(2):23–24.
23. Yadav D, Agrawal M, Bhatnagar U, Gupta S. Real time health monitoring using GPRS technology. *International Journal on Recent and Innovation Trends in Computing and Communication*. 2013;1(4):368–372.
24. Dey N, Ashour AS, Shi F, Fong SJ, Sherratt RS. Developing residential wireless sensor networks for ECG healthcare monitoring. *IEEE Transactions on Consumer Electronics*. 2017;63(4):442–449.
25. Dey N, Ashour AS, Shi F, Sherratt RS. Wireless capsule gastrointestinal endoscopy: Direction-of-arrival estimation based localization survey. *IEEE Reviews in Biomedical Engineering*. 2017;10:2–11.
26. Kakria P, Tripathi NK, Kitipawang P. A real-time health monitoring system for remote cardiac patients using smartphone and wearable sensors. *International Journal of Telemedicine and Applications*. 2015;2015:1–11.
27. Mohammadzadeh N, Safdari R. Patient monitoring in mobile health: Opportunities and challenges. *Medical Archives*. 2014;68(1):57–60.
28. Yin Y, Zeng Y, Chen X, Fan Y. The internet of things in healthcare: An overview. *Journal of Industrial Information Integration*. 2016;1:3–13.
29. Sullivan HT, Sahasrabudhe S. Envisioning inclusive futures: Technology-based assistive sensory and action substitution. *Futures Journal*. 2017;87:140–148.
30. Wang X, Wang JT, Zhang X, Song J. A multiple communication standards compatible IoT system for medical usage. In: *IEEE Faible Tension Faible Consommation*, Paris, France, 2013;1–4.

31. Xu B, Xu LD, Cai H, Xie C, Hu J, Bu F. Ubiquitous data accessing method in IoT-based information system for emergency medical services. *IEEE Transactions on Industrial Informatics*. 2014;10(2):1578–1586.
32. Kolici V, Spaho E, Matsuo K, Caballe S, Barolli L, Xhafa F. Implementation of a medical support system considering P2P and IoT technologies. In: *8th Int. Conf. on Complex, Intelligent and Software Intensive Systems*, Birmingham, UK, 2014;101–106.
33. Sandholm T, Magnusson B, Johnsson BA. An on-demand WebRTC and IoT device tunneling service for hospitals. In: *International Conference on Future Internet of Things and Cloud*, Barcelona, 2014;53–60.
34. Ang LM, Seng KP, Heng TZ. Information communication assistive technologies for visually impaired people. *International Journal of Ambient Computing and Intelligence*. 2016;7(1):45–68.
35. Kamal S, Dey N, Ashour AS, Ripon S, Balas VE, Kaysar MS. Fbmapping: An automated system for monitoring Face book data. *Neural Network World*. 2017;1:27–57.
36. Acharjya D, Anitha A. A comparative study of statistical and rough computing models in predictive data analysis. *International Journal of Ambient Computing and Intelligence*. 2017;8(2):32–51.
37. Chakraborty S, Chatterjee S, Ashour AS, Mali K, Dey N. Intelligent computing in medical imaging: A study. In: *Advancements in Applied Metaheuristic Computing*. 2017;143.
38. Matallah H, Belalem G, Bouamrane K. Towards a new model of storage and access to data in big data and cloud computing. *International Journal of Ambient Computing and Intelligence*. 2017;8(4):31–44.
39. Yilmaz T, Foster R, Hao Y. Detecting vital signs with weable wireless sensors. *Advances in Biosensors*. 2010;10(12):10837–10862.
40. Mane A, Dighe V, Gawali R, Sabale S, Gudadhe S. Location based service and health monitoring system for heart patient using IoT. *International Journal of Innovative Research in Computer and Communication Engineering*. 2017;5(11):11543–116548.
41. Daga N, Prasad S. Smart healthcare system using IoT. *International Journal of Professional Engineering Studies*. 2017;9(4):316–320.
42. Rizal Islam SM, Kwan D, Humaun KM, Mahmud H, Kwacha K-S. The internet of things for health care: A comprehensive survey. *IEEE Access*. 2015;3:678–708.
43. Hu F, Xie D, Shen S. On the application of the internet of things in the field of medical and healthcare. In: *IEEE Int. Conf. on Physical and Social Computing Green Computing and Communications*, Beijing, China, 2013:2053–2058.
44. Soyata T, Muraleedharan R, Funai C, Kwon M, Heinzelman W. Cloud-vision: Real-time face recognition using a mobile-cloudlet cloud acceleration architecture. In: *Proceedings of the IEEE Symposium on Computers and Communications*, Cappadocia, Turkey, 2012:59–66.
45. Hassanalieragh M, Page A, Soyata T, Sharma G, Aktas M, Mateos G, Kantarci B, Andreescu S. Health monitoring and management using internet of things (IoT) sensing with cloud-based processing: Opportunities and challenges. In: *IEEE International Conference on Services Computing*, New York, USA, 2015:285–292.
46. Majumder S, Mondal T, Jamal Deen M. Wearable sensors for remote health monitoring. *Advances in Biosensors*. 2017;17:130.
47. Chavan P, More P, Thorat N, Yewale S, Dhade P. ECG remote patient monitoring using cloud computing. *Imperial Journal of Interdisciplinary Research*. 2016;2(2):368–372.
48. Page A, Kocabas O, Soyata T, Aktas M, Couderc JP. Cloud based privacy-preserving remote ECG monitoring and surveillance. *Annals of Noninvasive Electrocardiology*. 2014;20(4):328–337.
49. Yang G, Xie L, Zheng LR. A health-IoT platform based on the integration of intelligent packaging, unobtrusive bio-sensor, and intelligent medicine box. *IEEE Transactions on Industrial Informatics*. 2014;10(4):2180–2191.

50. Jara AJ, Zamora-Izquierdo MA, Skarmeta AF. Interconnection framework for mHealth and remote monitoring based on the Internet of Things. *IEEE Journal on Selected Areas in Communications*. 2013;31(9):47–65.
51. Rasid MFA, Musa WMW, Kadir NAA. Embedded gateway services for internet of things applications in ubiquitous healthcare. In: *2nd Int. Conf. Inf. Commun. Technol.*, Bandung, Indonesia, 2014:145–148.
52. Yang L, Ge Y, Li W, Rao W, Shen W. A home mobile healthcare system for wheel chair users. In: *IEEE Int. Conf. Comput. Supported Cooperat.*, Hsinchu, Taiwan, 2014:609–614.
53. Castillejo P, Martinez JF, Rodriguez-Molina J, Cuerva A. Integration of wearable devices in a wireless sensor network for an e-health application. *IEEE Wireless Communications*. 2013;20(4):38–49.
54. Pandikumar SA. Model for GSM based intelligence PC monitoring system. *International Journal of Advanced Computer Science and Technology*. 2012;2(2):85–88.
55. Fortier P, Viall B. Development of a mobile cardiac wellness application and integrated wearable sensor suite. In: *SENSORCOMM. 5th IC on Sensor Technologies and Applications*, Nice/Saint Laurent du Var, France, 2011:301–306.
56. Dickerson RF, Gorlin EI, Stankovic JA. Empath: A continuous remote emotional health monitoring system for depressive illness. *Proceedings of the 2nd Conference on Wireless Health*. 2011;5:1–10.
57. Wai A, Fook F, Jayachandran M, Song Z, Biswas J, Nugent C, Mulvenna M, Lee J, Yap L. Smart wireless continence management system for elderly with dementia. In: *IEEE 10th IC on e-Health Networking, Applications and Services, Health Com.*, Singapore, 2008:33–34.
58. Miao F, Miao X, Shangguan W, Li Y. Mobi healthcare system: Body sensor network based m-health system for healthcare application. *e-Health Telecommunication Systems and Networks*. 2012;1(1):12–18.
59. Logan AG, McIsaac WJ, Tisler A, Irvine MJ, Saunders A, Dunai A, Rizo CA et al. Mobile phone based remote patient monitoring system for management of hypertension in diabetic patients. *American Journal of Hypertension*. 2007;20(9):942–948.
60. Bourouis A, Feham M, Bouchachia A. Ubiquitous mobile health monitoring system for elderly (UMHMSE). *International Journal of Computer Science & Information Technology (IJCSIT)*. 2011;3(3), arXiv preprint arXiv: 2011;1107.3695:74–82.
61. Jones V, Gay V, Leijdekkers P. Body sensor networks for mobile health monitoring: Experience in Europe and Australia. In: *4th IC on Digital Society, IEEE Computer Society*, Sint Maarten, Netherlands Antilles, 2010:204–209.
62. Pawar P, Jones V, van Beijnum BJF, Hermens H. A framework for the comparison of mobile patient monitoring systems. *Journal of Biomedical Informatics*. 2012;45(3):544–556.
63. Fotiadis D, Likas A, Protopappas V. *Intelligent Patient Monitoring*. Wiley Encyclopedia of Biomedical Engineering, 2006.
64. Donoghue OJ, Herbert J. Data management system: A context aware architecture for pervasive patient monitoring. In: *Proceedings of the 3rd IC on Smart Homes and Health Telematic*, Nagasaki, Japan, 2005:159–166.
65. Patil D, Andhalkar S, Gund M, Agrawal B, Biyani R, Wadhai V. An adaptive parameter free data mining approach for healthcare application. *International Journal of Advanced Computer Science and Applications*. 2012;3(1):55–59.
66. Kamal MS, Dey N, Ashour AS. Large scale medical data mining for accurate diagnosis: A blueprint. In: *Handbook of Large-Scale Distributed Computing in Smart Healthcare*. Springer, Cham, 2017:157–176.
67. Ranjan R, Varma S. Object-oriented design for wireless sensor network assisted global patient care monitoring system. *International Journal of Computer Applications*. 2012;45(3):8–15.

68. Yan L, Zhang Y, Yang LT, Ning H. *The Internet of Things: From RFID to the Next-Generation Pervasive Networked Systems*. CRC Press, 2008.
69. Shih F, Zhang M. Towards supporting contextual privacy in body sensor networks for health. Monitoring service. In: *W3C Workshop on Privacy and Data Usage Control*, Cambridge, MA, 2010.
70. Appari A, Johnson ME. Information security and privacy in healthcare: Current state of research. *International Journal of Internet and Enterprise Management*. 2010;6(4):279–314.
71. Löhr H, Sadeghi AR, Winandy M. Securing the e-health cloud. In: *Proceedings of the 1st ACM International Health Informatics Symposium*, Arlington, Virginia, USA, 2010:220–229.
72. Tsai FS. Security issues in e-healthcare. *Journal of Medical and Biological Engineering*. 2010;30(4):209–214.
73. Wilkowska W, Ziefle M. Privacy and data security in E-health: Requirements from the user's perspective. *Journal of Health Informatics*. 2012;18(3):191–201.
74. Agrawal V. Security and privacy issues in wireless sensor networks for healthcare. In: *Internet of Things User Centric IoT*, Rome, Italy, 2014:223–228.
75. Kim JT. Privacy and security issues for healthcare system with embedded RFID system on internet of things. *Advanced Science and Technology Letters*. 2014;7(2):109–112.
76. Lake D, Milito R, Morrow M, Vargheese R. Internet of things: Architectural framework for eHealth security. *Journal of ICT Standardization*. 2014, River Publishing, vol 1, 301–328.
77. Divi K, Kanjee MR, Liu H. Secure architecture for healthcare Wireless Sensor Networks. In: *IEEE 6th International Conference on Information Assurance and Security*, Atlanta, GA, USA, 2010:131–136.
78. Hasić H, Vujović V. Civil law protection of the elements comprising the "Internet of Things" from the perspective of the legal owner of the property in question. *Infoteh-Jahorina*. 2014;13:1005–1011.

12 A Comparative Analysis of Classical Cryptography versus Quantum Cryptography for Web of Medical Things (WoMT)

K. Anitha Kumari and G. Sudha Sadasivam

CONTENTS

12.1 INTRODUCTION

In the contemporary world, healthcare-based services undoubtedly play a vital role in ensuring the welfare of people. It is envisioned that by 2020, 40% of Web of Medical Things (WoMT) will be used in healthcare verticals leading to a $117 billion market in patient management [4]. Yet, these services suffer from poor authentication methods, leading to intensive attacks. A security research report presented by the Hewlett-Packard Enterprise (HPE) in 2016 [5], shows that 80% of the devices connected to the cloud fail to provide sufficient password protection. Due to the proliferation of improper security measures in WoMT, the loss incurred due to malicious attacks is increasing rapidly. To address this issue and to enhance the security in WoMT, a basic two-server password authenticated key exchange (PAKE) protocol is proposed using ECC [2,8], and an improved two-server PAKE protocol is proposed using quantum key distribution (QKD). The latter improved model provides a quantum safe holistic E2E security framework to withstand against hazardous attacks occurring suspiciously. The significant contributions of this chapter include comparing the basic model with the improved model and by stating the importance of QKD in WoMT with a formal proof of security.

12.2 WoMT AND SECURITY

The practice of IT in healthcare is considered as the first big step to augment the primary healthcare network across the country [25,26]. Communication protocols of medical IoMT devices are proprietary, heterogeneous and device dependent. Hence, there is a need for a common standard for communication among the devices. The varied nature of healthcare providers necessitates a security framework to provide suitable authentication, and authorization with confidential data transfer between the users and the devices. To enhance trust in the offered services, it is crucial to preserve the privacy and security of patients in the healthcare domain [1]. A new e-Health platform for diabetes care management [19] using WoMT comprises of a physical sensors layer linked to a health Web-portal layer via an existing network infrastructure. The system supports self-management of type 1 diabetes mellitus (TIDM). Medical devices often communicate with the server in their own proprietary protocols [4]. This leads to data islands, undermining the whole idea of the IoT. For this purpose WoMT emerges. WoMT facilitates the integration of all sorts of devices and the applications that interacts using Web standards directly on embedded devices. The conception of

Web of Things (WoT) is to streamline and to standardize the protocols developed by the manifold vendors under one roof by integration without any compatibility issues. The provoking thought behind WoMT is to exclusively leverage the existing tools, techniques and protocols. In general, WoMT focuses primarily on higher layers of the open system interconnection (OSI) model to handle application, services and data under one roof rather than dealing with lower layers. Thus, a holistic framework is achieved with a higher level of abstraction, and security is ensured across all the devices irrespective of protocols/tools used to secure medical big data.

In IoT and in IoMT, the security framework usually adopts Kerberos authentication and role-based authorization. Ticket and assertion-based single sign-on (SSO) [20] combines the advantage of Kerberos and security assertion markup language (SAML) to ensure authentication by adopting mutual authentication. SAML provides authentication, authorization and security in transmission. Context adaptation is much required as the Internet context is significantly different from the IoT context. SSO is considered as a deadly challenge in IoT and in IoMT. Hence, a two key-based scheme is proposed. However, multifactor authentication (MFA) systems are not suitable for the IoMT environment, as most of the devices have limited computational ability. Cost, support, training and maintenance are also considered key challenges in MFA. OAuth protocol uses delegation-based authentication systems in IP-based IoT. The major problem is interoperability and speed that can be solved using Web security standards [21]. Hence, there is a need for integration in the WoMT. Also, intelligence in computing is preferred widely for medical imaging [9,27]. For WoT and WoMT, the security offered by the security agent module is based on the public/private key cryptography infrastructure (PPKI) [22] for authentication and to preserve data integrity. As the WoMT is not an exception against all possible hazardous web attacks, a strong trustworthy authentication protocol is required. Elliptic curve cryptography (ECC) provides better security when compared to the rest of the public/private key cryptography schemes. A phenomenal aspect of ECC is that the small footprint of ECC allows security modules to be implemented on embedded devices. Quadratic residue enhances the security in Diffie-Hellman (DH) algorithm and plays a crucial role in cryptography. It is proven that the quadratic residuosity assumption is stronger than the factoring assumption of the RSA algorithm. DH relies on the assumption that no efficient algorithm exists to ascertain the values of 'a', 'b' from g^{ab}, if 'a', 'b' and 'g' are chosen randomly and independently. Thus, the ECC Diffie-Hellman (ECDH) algorithm is considered as a trustworthy, secure key agreement protocol with high speed. Basically, a two-server model arrangement is preferred to withstand against all possible hazardous attacks, especially with server spoofing attacks. Proofs for two-server PAKE are initially proposed by Szydlo and Kaliski [23]. Yang et al. [24] proposed the first successful two-server PAKE model for authentication and key exchange. Nevertheless, the protocol is liable to active attacks. Anitha et al. [2] proposed a two-server PAKE model to endure all possible invasive attacks using the ECC and Diffie-Hellman mechanism. In addition, applying ECC using wearables in healthcare systems is presented by Kumari et al. [8]. However, if the curve chosen is not a supersingular, it is vulnerable to quantum attacks in ECC. To protect the system against quantum attacks, adopting quantum cryptography techniques is indispensable in the WoMT.

12.3 MATHEMATICAL FOUNDATIONS

Modern cryptographic algorithms treat the information as numbers/elements in a finite space to obey closure property for applying in encryption and decryption procedures [6]. As basic arithmetic operations are not satisfying the closure property, algebraic structures like group, ring and field are used. The characteristics of an algebraic structure 'A' is denoted by char(A). It is said to possess a least one positive integer 'n' such that, $na = 0$ for every $a \in A$.

Some of the commonly used algebraic structures are described below:

12.3.1 GROUP

A group consists of a set of objects with a defined operation (∘), such that, the result of the operation (∘) on any two objects also belongs to the set. It is denoted by $(G,\circ)$, where 'G' is a set with an operation (∘).

A group $(G,\circ)$ satisfies the following axioms:

a. Closure Axiom: $\forall a,b \in G$: $a \circ b \in G$
b. Associativity Axiom: $\forall a, b, c \in G$: $a \circ (b \circ c) = (a \circ b) \circ c$
c. Identity Axiom: $\exists$ unique element $e \in G$: $\forall a \in G$: $a \circ e = e \circ a = a$; the element '$e$' is called the identity element
d. Inverse Axiom: $\forall a \in G$: $\exists\, a^{-1} \in G$: $a \circ a^{-1} = a^{-1} \circ a = i$

Order of a Group: The number of elements in a finite group 'G' is called the order of 'G' and is denoted by $|G|$.

$(Z, +)$ is said to be an infinite additive group, where 'Z' denotes the set of integers. The group $(Z, +)$ satisfies identity and inverse properties with identity element as '0'. Similarly, $(Z^*, \cdot)$, $(\mathbb{Q}^*, \cdot)$, $(\mathbb{R}^*, \cdot)$ and $(\mathbb{C}^*, \cdot)$ denote infinite multiplicative groups that satisfy the identity property with identity element as '1'.

Group 'G' is said to be cyclic, if there exists an element $a \in G$, such that for any $b \in G$, there exists an integer $i \geq 0$ satisfying $b = a^i$. The element 'a' is called a generator of 'G' and is denoted as $G = \langle a \rangle$. A generator of a cyclic group is also called the primitive root of the group's identity element. For instance, let's say that there exists a multiplicative group Z_p^* with an order $p - 1$, where 'p' is a prime number with a generator 'a'. Discrete logarithms are defined for these types of cyclic multiplicative groups and, therefore, they play a significant role in modern cryptography.

12.3.2 RING

A ring 'R' can be defined as a set with two operations, addition (+) and multiplication (·), by satisfying the following properties:

a. 'R' is an abelian group with an additive identity '0' under addition (+). An abelian group is a set of satisfying closure, associativity, identity, inverse and commutative properties
b. 'R' satisfies the closure, associativity and identity property under multiplication (·)

c. Commutativity axiom: $\forall a, b \in R$: $a \cdot b = b \cdot a$
d. Distribution axiom: $\forall a, b, c \in R$: $a \cdot (b + c) = a \cdot b + a \cdot c$

The number of elements in a ring '*R*' is called the order of '*R*' and is denoted by $|R|$. A ring is said to be an abelian group under addition when it satisfies the associative property under multiplication. For instance, for any $p > 0$, Z_p is a ring under addition and multiplication modulo '*p*'. Similarly, Q, R and C are all rings under addition and multiplication operations.

12.3.3 Field

A field '*F*' is a ring that can form a group of non-zero elements under multiplication. For any '*a*', '*b*' $\in F$, $ab = 0$, implies that either $a = 0$ or $b = 0$ to denote that '*F*' is not a field. Set of integers '*Z*', is said to be a ring, but not a field, as not every integer has a multiplicative inverse. $Z_p\{0, 1, 2, 3, \ldots, p - 1\}$ is a field under addition and multiplication modulo '*p*', such that '*p*' is a prime number.

Some of the concepts connected with the field are as follows:

- The number of elements in a finite field is called the order of '*F*' and such order exists only if the order is a prime power p^r where '*p*' is a prime number and '*r*' is a positive integer;
- The characteristic of any field can be either '0' or a prime number. '*p*' is called the characteristic of a field when adding an element '*r*' times results in zero. A field is called a finite characteristic field or field of positive characteristic if it has a non-zero characteristic; and
- A multiplicative group is said to be a finite field consisting of non-zero elements. When all non-zero elements are expressed as powers of a single element, called a primitive element of the field, it is said to be cyclic. In field theory, a primitive element of a finite field F_p is a generator of the multiplicative group of the field. In other words, $a \in F_p$ is called a primitive element, when all the non-zero elements of F_p can be written as a^i for some positive integer.

Let's say there exists an integer $a \in Z_p^*$ then '*a*' is called as a quadratic residue (QR) modulo '*p*', if $x^2 \equiv a(\bmod p)$ for $x \in Z_p^*$. Otherwise, '*a*' is called as a quadratic non-residue (QNR) modulo '*p*'. QR_p denotes a set of quadratic residues modulo '*p*' and QNR_p denotes a set of quadratic non-residues modulo '*p*'. For example, QR_7 denotes the set of all quadratic residues modulo 7. Then $QR_7 = \{1^2, 2^2, 3^2, 4^2, 5^2, 6^2\}$ (mod 7) $= \{1, 2, 4\}$. The obtained quadratic residues are most widely used in modern cryptographic algorithms to enhance security.

12.4 CRYPTOGRAPHY

Cryptography is the science of protecting secret information. It deals with secrecy and protecting the information from prying eyes [7]. The primary objective of cryptography is to protect the user details from attacks by offering confidentiality, integrity and availability.

12.4.1 Classical Cryptography

Classical cryptography is categorized as symmetric key/secret key cryptography, asymmetric key public key cryptography and hash functions.

12.4.1.1 Symmetric Key/Secret Key Cryptography

Symmetric key cryptography uses a single key for both encryption and decryption. Therefore, it is more pliable to attack. This technique is well-suited to encrypt huge volumes of data. Data encryption standard (DES), advanced encryption standard (AES) and blowfish belong to this type of cryptography.

12.4.1.2 Asymmetric Key/Public Key Cryptography

The asymmetric key technique uses two keys, namely, private and public. The public key and the private key are used in the encryption and decryption process. Albeit different, the keys are mathematically connected and ascertaining the private key from the public key is infeasible. Most of the public key algorithms are based on discrete logarithm/integer factorization/elliptic curve relationships for which no efficient solution exists till now. Due to this, it is commonly applied in authentication protocols, key exchange processes and in sharing secrets. Diffie–Hellman (DH) key exchange protocol, digital signature standard (DSS), digital signature algorithm (DSA), ElGamal cryptosystem, ECC schemes, Rivest–Shamir–Adleman (RSA) scheme and Cramer–Shoup cryptosystem are some examples of public key cryptography. To perform the key exchange process, asymmetric key cryptography is highly preferred as it is robust against attacks. To transmit data effectively, the symmetric key algorithm is commonly used. Combining the advantages of a public key cryptosystem for key encapsulation and symmetric key cryptosystem for data encapsulation is referred to as a hybrid cryptosystem. Pretty good privacy (PGP) and secure sockets layer (SSL)/transport layer security (TLS) schemes utilize this procedure.

Syntactically, any public key cryptosystem uses the following terminologies:

A plaintext message ‘*m*’
A ciphertext message ‘*c*’
An encryption key ‘*e*’
A decryption key ‘*d*’
A competent key generation algorithm: $k \rightarrow e\, e \times d\, e$
A strong encryption algorithm A: $m \times e \rightarrow c$
A strong decryption algorithm B: $c \times d \rightarrow m$
A cryptosystem is called symmetric, if $d = e$ and is called asymmetric/public, if $d \neq e$. It is computationally infeasible in practice to compute ‘*d*’ from ‘*e*’.

12.4.1.3 Hash Functions

A function $H(x)$ is said to be a hash function, if it possesses the quality of mapping a set of symbols from one group to another in an easy manner. Usually, the hash functions are irreversible and the benefits obtained by this approach include collision resistance, pre-image resistance and efficiency. Further, hashing provides high randomness, hassle free secure computation with greater speed.

An authentication protocol provides high protection to credentials against threats and attacks. Essential properties of authentication protocols are:

- Credentials must be transmitted in encrypted form
- Credentials must be transmitted through a secure link
- Credentials must be stored after being encrypted/one-way hashed
- Keys established for the session should be one-way hashed using acceptable hashing algorithms
- Limited login attempts
- No traces of previously used session keys
- Protecting credentials from commonly occurring attacks

12.4.2 Quantum Cryptography

Quantum cryptography is a cutting edge technique based on the properties of quantum mechanics to assure the secure exchange of secret keys [3,11]. Transmission between two parties happens via the elementary particles called photons. In quantum, a bit denotes either zero or one. For brevity, quantum component bits are referred as qubits. In contrast, qubits can be a zero and one simultaneously, and thus the speed of quantum computing originates from exploiting this parallelism. Quantum states are denoted by vertical bars followed by an angle bracket. For instance, bit 0 and 1 are represented as $|0>$ and $|1>$ corresponding to two orthogonal states in a quantum system. The establishment of quantum cryptography is done through embracing the Heisenberg uncertainty principle, which says, while measuring the polarisation of a photon, it is not possible by an onlooker to measure the other. In particular, a basis can be horizontal/vertical orthogonal polarisation states. Furthermore, a couple of bases is said to be connected; the first basis can completely randomize the second basis that utilizes a filter in the 45°/135° basis to measure the photon. Presume an onlooker is measuring a qubit $|\Psi\rangle$ to recognize it as $|0\rangle$ and $|1\rangle$. Once the measurement is over, qubit will turn out to be either $|\Psi\rangle \rightarrow |\Psi'\rangle = |0\rangle$ or $|\Psi\rangle \rightarrow |\Psi'\rangle = |1\rangle$, which is exclusively relying upon the measurement result; hence, by drawing a conclusion the values before and after measurement differs considerably. Thus quantum entanglement acts as one of the strongest pillars of quantum mechanics.

12.4.2.1 Secret-Key Distillation

In information theoretical sense, to preserve the secrecy of the established quantum key and to minimize the transmission errors, secret-key distillation must be used, since the interceptor has a probability 1/4 of introducing an error between receiver and sender bits [10]. It comprises a reconciliation phase and a privacy amplification phase. Reconciliation is used to detect and mitigate the unusual set of large number of transmission errors 'e' occurring between the key elements. Privacy amplification is an essential procedure to distill a secret key from a common random variable in order to protect the secret key from the intruders, such that it is infeasible to derive any useful information from the key.

12.5 PROPOSED MODELS

The standard notations used in the chapter are listed below:

Ø – Empty set
N – Set of natural numbers
ℂ – Set of complex numbers
ℝ – Set of real numbers
ℚ – Set of rationals
Z – Set of integers
p – A large prime number
q – A large prime number of satisfying the condition $1 < q < p - 1$, such that 'q' and 'p' are relatively prime
$Z_p{}^*$ – Integer Group 'G' under multiplication modulo 'p'
QR_p – Set of quadratic residues modulo 'p'
QNR_p – Set of quadratic non-residues modulo 'p'
F_p – Finite field of 'p' elements
$E_p(a,b)$ – Elliptic curve over a finite field F_p
G_1, G_2, G_3, G_4 – Generators of field F_p
x_1, x_2 – Private keys $\in E_p$
y_1, y_2 – Public keys
$b_1, b_2, a_1, a_2, r, r_1, r_2 \in F_p$
P – Password
Hash() – Secure one-way hash function
$b_2 = b_1 \oplus \text{Hash}()(P)$
K/K' – Secret key
Q_K – Quantum key
∀ – for all
∃ – there exists

In the basic model, the two-server PAKE methodology is proposed using ECC and the Diffie–Hellman key exchange mechanism. PAKE is a mutual authentication scheme that establishes a secret key between the communicating parties based upon the prior knowledge of the confidential information with a low entropy password [12]. The secret key established guarantees secure communication among the parties. ECC provides better security when compared to ElGamal with smaller keys, and it is ideal for environments that require high security with less key length [14]. The base key length of ECC is 160-bits and the most extreme length can reached up to 512-bits. ECC with 256-bit key length provides equivalent security to RSA with 3072-bits (minimum) and ElGamal with 1024-bits (minimum) [13]. As an effect, developed countries adopt ECC in their frameworks. It is known to be secure under the DDH assumption, as it is NP-Hard. ECC offers equivalent security as that of RSA and ElGamal cryptosystems with less key length to enhance speed [13,15]. Generally, elliptic curves are defined over a finite prime field F_p [16]. It is a probabilistic encryption scheme, as encrypting the same message several times yields different ciphertexts. The proposed framework executes in three stages, namely initialization, registration and authentication and key exchange.

The following algorithm illustrates the ECC cryptosystem:

Algorithm: Elliptic Curve Cryptosystem

Step 1: Choose a prime field F_p of characteristic greater than 3.

Step 2: Let $E_p(a,b)$ be the elliptic curve over F_p of 'p' elements in the form $y^2 = x^3 + ax + b \bmod p$, satisfying the condition $4a^3 + 27b^2 \bmod p \neq 0$. To form a group, an extra point at infinity denoted by 'O' is included. It is formulated as $O = (x, \infty)$.

Step 3: Algorithm: QR selection and point generation

```
if y² = f(x) = x³ + ax + b = 0
    return (0, 0)
  else
    assign s←0, y←1, j←1,x ←0
    for j←1 to p
     do
          s← (s+j) mod p
          if y² = s
               x = y²
               return (x,y)
          else y←y+1, j←j+2
  if j – p
      return (-1,-1)
```

From the returned point, 'x' is referred to as the quadratic residue element. Points on the given curve is $\left(x, \sqrt{f(x)}\right)$ and $\left(x, -\sqrt{f(x)}\right)$.

Step 4: Choose a generator/base point 'G', such that 'G' is a QR of F_p. The following algorithm illustrates the selection of generator.

Algorithm: Generator Selection

```
for each x ∈ Fp
     check whether x is a QR of Fp
       if satisfied
            generate the points on the curve
       else
            continue for loop;
from the generated points, randomly select a point G
     check whether G = hG by satisfying the condition y² = x³ + ax + b
```

where 'h' is the co-factor of order of the base point 'n' and equal to the ration $\#E(F_p)/n$ such that $\#E(F_p)$ is the number of points in the curve (curve order).

```
       if satisfied
            G is the generator/base point
       else
            continue with next point
```

Step 5: Let $x \in F_p$ be the ECC private key.

Step 6: Compute the ECC public key as $y = G \times x$.

Step 7: Ciphertext (A, B) is computed as,

$$A = C_1 = G \times a \bmod p$$

$$B = C_2 = M + y \times a \bmod p$$

where 'a' is a random number in F_p, 'm' is message and 'M' is the value obtained after mapping 'm' as a point in 'E'.

Finally, $(A, B) = (C_1, C_2) = (G \times a \bmod p, M + y \times a \bmod p)$.

Step 8: M is obtained by,

$$M = C_2 - x * C_1$$

where x is the decryption key, C_1 and C_2 are the ciphertexts.

Obtaining '*m*' from '*M*' is discrete logarithm problem and is NP-Hard.

For calculating $P \times a$, either point addition or point doubling is applied. Assume that P (x, y) is a point on the curve 'E' and 'a' is an element of F_p. Obtaining 'a' from $P \times a$ is the highest challenge in ECC, when the order is of large prime. The difficulty involved in breaking $P \times a$ to find 'a' is computationally infeasible in polynomial time. This is referred to as an elliptic-curve discrete logarithm problem, or ECDLP for brevity. This hard-core predicate makes the proposed 3D ECC DH PAKE protocol highly secure against attacks.

a. *Point Addition*: If the points $P = (x_p, y_p)$, $Q = (x_q, y_q)$ are similar, $R\,(x_r, y_r)$ is obtained as follows:

$$x_r = \lambda^2 - x_p - x_q$$

$$y_r = \lambda(x_p - x_r) - y_p$$

$$\lambda = \frac{y_q - y_p}{x_q - x_p}$$

b. *Point Doubling*: If the points $P = (x_p, y_p)$, $Q = (x_q, y_q)$ are different, R (x_r, y_r) is obtained as follows:

$$x_r = \lambda^2 - x_p - x_q$$

$$y_r = \lambda(x_p - x_r) - y_p$$

$$\lambda = \frac{3x_p^2 + a}{2y_p}$$

12.5.1 Overview of Basic Protocol

The basic protocol runs in three phases, namely, initialization phase, registration phase and authentication and key exchange phase.

12.5.1.1 Initialization Phase

During initialization, public parameters $\{F_p, p, E_p(a,b), G_1, G_3, \text{Hash}\}$ are agreed and published by the entities *C*, *S*1 and *S*2. $E_p(a, b)$ is the elliptic curve over a finite field F_p of '*p*' elements in the form $y^2 = x^3 + ax + b \bmod p$, satisfying the condition $4a^3 + 27b^2 \bmod p \neq 0$. G_1 and G_3 are the generators of field F_p and Hash is a highly secure one-way hash function. G_2 and G_4 are private values known only to *S*1 and *S*2 to avoid man-in-the-middle and client impersonation attacks.

12.5.1.2 Registration Phase

The registration process is illustrated as follows:

Input: Public parameters $\{F_p, p, E_p(a,b), G_1, G_3, \text{Hash}\}$.
Process:
C:

1. Choose username and password.
2. Compute $G_3 X P$.
3. Select the decryption key $x_i \in F_{p i=1,2}$ and compute the encryption key as, $y_i = G_1 \times x_{i\ i=1,2}$ for servers $S_{i\ i=1,2}$.
4. Randomly choose $b_1 \in F_p$ and calculate b_2 as given in Equation 12.1.

$$b_2 = b_1 \oplus \text{Hash}(P) \tag{12.1}$$

5. Forward the authentication information $\{\text{Username}, G_3XP, x_i, a_i, a_j, b_i, b_i, y_j\}_{i=1,j=2}$ and $\{\text{Username}, G_3XP, x_j, a_i, a_j, b_j, y_i\}_{i=1,j=2}$ to the servers S1 and S2 respectively.

S1:

1. Calculate M_1 from $G_4\, X\ (G_3 X\, P)$.
2. Encrypt the point M_1 with the encryption key of the server S2 as given in Equation 12.2.

$$(A_2, B_2) = (G_1 \times a_2 \bmod p,\quad M_1 + y_2 \times a_2 \bmod p) \tag{12.2}$$

3. Store $\{A_2, B_2, x_1, a_1, b_1\}$ in database.

S2:

1. Calculate M_2 from $G_4\, X\ (G_3 X\, P)$.

2. Encrypt the point M_2 with the encryption key of the server S1 as given in Equation 12.3.

$$(A_1, B_1) = (G_1 \times a_1 \bmod p, \quad M_2 + y_1 \times a_1 \bmod p) \tag{12.3}$$

3. Store $\{A_1, B_1, x_2, a_2, b_2\}$ in database.

Output:

Successful registration of C with S1 and S2.

12.5.1.3 Authentication and Key Exchange Phase

The mutual authentication process is explicated as follows:

Input: Username, password $\rightarrow G_3\, X\, P$ and $G_1\, X\, r$.
Process:
S1:

1. Calculate M_2 from $G_4\, X\, (G_3 X\, P)$.
2. Verify M_2 against the values A_1', B_1', V_2 received from S2 to authenticate client and S2.
3. Upon successful verification, compute R, K and H from the values received from S2.
4. Transmit R and H to client.
5. Generate the secret key K.

S2:

1. Calculate M_1 from $G_4\, X\, (G_3 X\, P)$.
2. Verify M_1 against the values A_2', B_2', V_1 received from S1 to authenticate client and S1.
3. On successful verification, transmit the key generation parameter to S1.

C:

1. On receipt of R and H from S1, validate the servers S1 and S2.
2. Generate the secret key K'.

Output:

1. Mutual authentication of entities C, S1 and S2.

A complete step-by-step process entailed in authentication is shown below.

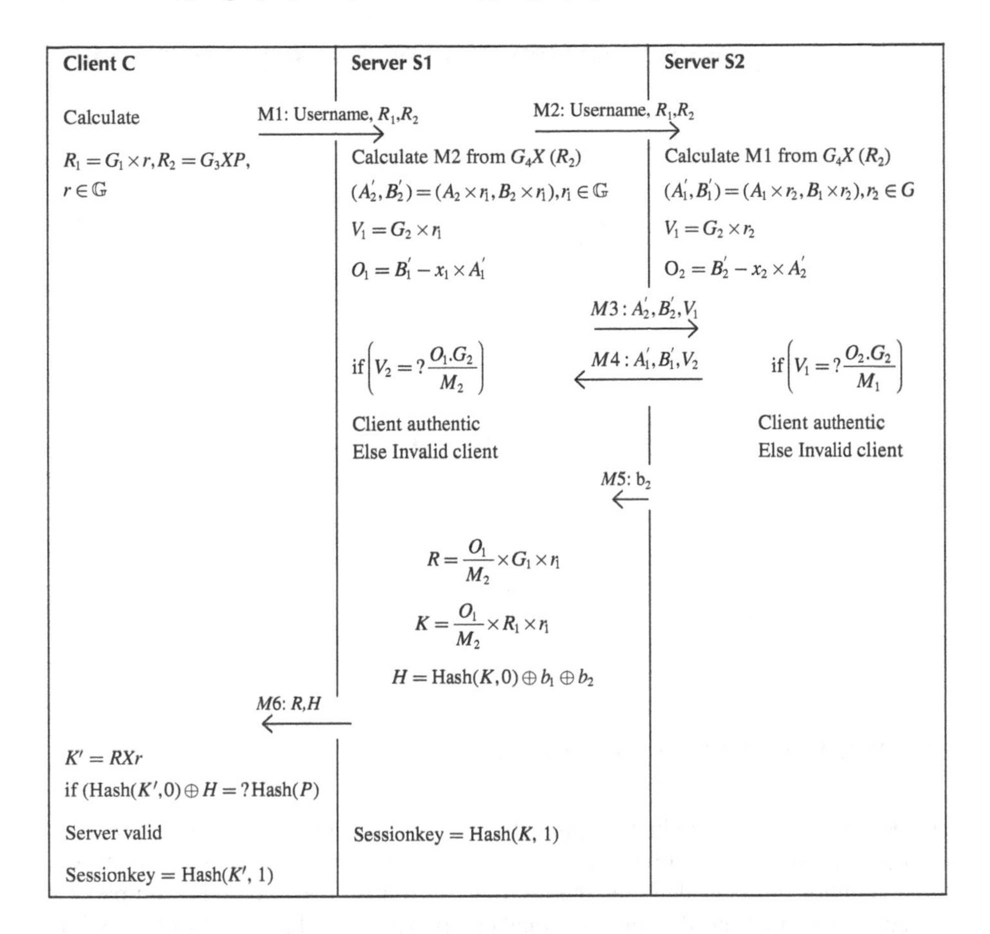

12.5.1.4 Design and Analysis of Basic Protocol

The basic protocol is tested for a healthcare-based web application. The proposed protocol is assessed using the automated validation of Internet security protocols and applications (AVISPA) to check the design strength of the protocol [17]. It is an automated validation tool for security protocols. AVISPA ensures that it is impossible to launch the commonly occurring attacks in the basic protocol by representing the summary as 'safe', as shown in Figure 12.1. ECC DH PAKE key length of 384-bits provides equivalent security as ElGamal DH PAKE of 2048-bits and DH PAKE of 3072-bits. Thus, the speed of the protocol is increased without violating the security.

12.5.1.5 Communication and Computation Complexity Analysis of Basic Protocol

Communication and computation complexity of the proposed basic and improved protocol is examined and tabulated, as shown in Table 12.1. The number of group elements in communication are measured in terms of 'L' and the number of hash values in communication are measured in terms of 'l'. The total runtime of the basic

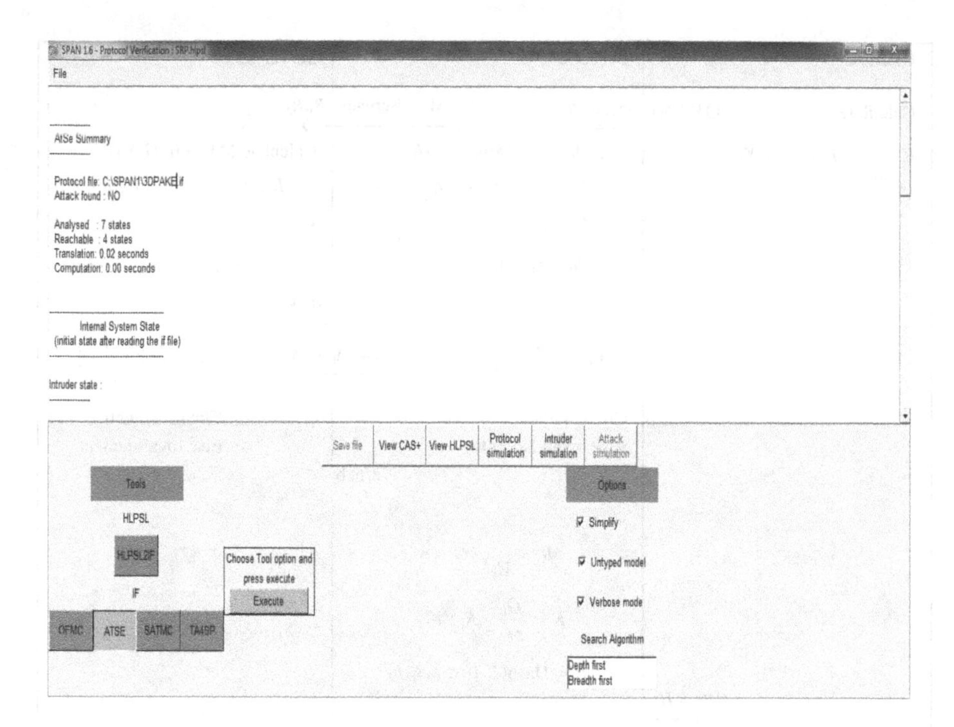

FIGURE 12.1 ECC DH Protocol design analysis using AVISPA.

protocol is 11L + 21. Modular exponentiations, qubit computation, modular/scalar multiplications, hash functions and bitwise exclusive (XOR) operations, with respect to each entity, are considered as the metrics to evaluate computational complexity. Based on these metrics, the computational complexity of the basic protocol is 24. It is to be noticed that the computational cost and the number of rounds has been reduced visibly, as shown in Figure 12.2.

12.5.1.6 Attack Resistance Capability of Basic Protocol

Table 12.2 outlines the direct/indirect attacks occurring in security protocols. The attack resistance capacity conveys the resistance capacity of the proposed protocols out of the total dangerous attacks listed. The basic protocol is highly resistant against side-channel, whereas the classical schemes are prone to these attacks. However, it is liable to quantum attacks.

12.5.1.7 Security Property Analysis of Basic Protocol

Table 12.3 summarizes the primary security functionalities to be satisfied by the security protocols. Security properties satisfied by the basic protocol shows a moderate improvement over the existing protocols.

12.5.1.7.1 Remark

Albeit the basic protocol shows improvement in terms of computation and communication, it is prone to quantum attacks when the curve chosen is a supersingular

TABLE 12.1
Communication and Computational Complexity Analysis of Basic Protocol

Participants	Yang et al. Protocol	Yi et al. Protocol	Basic Protocol
Client: Communication (bits)	2L + 2l	3L + 4l	3L + 1l
Client: Communication (rounds)	4	4	2
Client: Computation	7	21	7
Server S1: Communication (bits)	6L + 3l	6L + 3l	11L + 2l
Server S1: Communication (rounds)	8	5	6
Server S1: Computation	15	12	11
Server S2: Communication (bits)	4L + 1l	6L + 3l	8L + 1l
Server S2: Communication (rounds)	4	5	4
Server S2: Computation	6	12	6
Total Running Time (Client–Server Side)	Comm: 9(6L + 3l) Client – S1 – S2	Comm: 11 (7L + 4l) Client – S1 Client – S2	Comm: 13(11L + 2l) Client – S1 – S2
	Comp: 28 Client – S1 – S2	Comp: Worst case: 45 Best case: 33 Client – S1 Client – S2	Comp: 24 Client – S1 – S2
	Rounds: 8 Client – S1 – S2	Rounds: 6 Client – S1 Client – S2	Rounds: 6 Client – S1 – S2

curve. Hence, it paves the way for developing an improved protocol under the hood of quantum cryptography.

12.5.2 Overview of Improved Protocol

Phases involved in the improved protocol remain similar to those of the basic protocol. The phenomenal aspect is that layered security is provided in the improved protocol to withstand quantum attacks by satisfying and ensuring long-term secrecy. Also, the obtained secret key is infeasible to copy by any eavesdropper with the aid of privacy amplification. Albeit, quantum indistinguishability is attained through entanglement, thought behind privacy amplification is to exploit the interceptor/eavesdropper to know nothing about the secret key [10,11]. Sender and receiver determine the secret with the help of a function '*f*' of their key elements so as to spread interceptor's partial ignorance over the entire result. In classical cryptography, such a function is termed a hash function. Hence, in the improved model to distill the secret key, privacy is amplified with the help of a hash function, SHA-512.

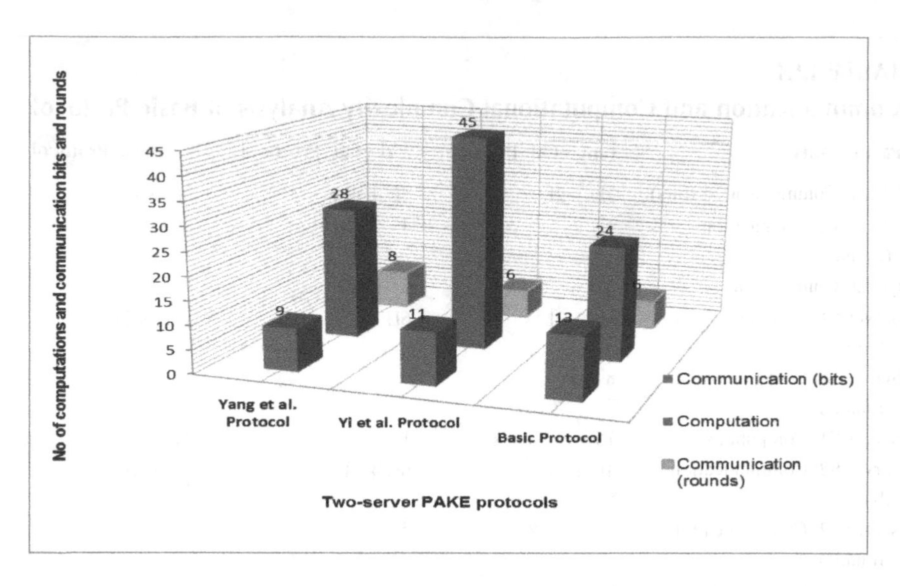

FIGURE 12.2 Complexity analysis of the basic protocol.

TABLE 12.2
Attack Resistance Capability of Basic Protocol

S. No	Attacks	Yang et al. Protocol	Yi et al. Protocol	Basic Protocol
1.	Impersonation attack by inside adversary	Possible	Not Possible	Not Possible
2.	Offline dictionary attacks	Possible	Possible	Not Possible
3.	Online dictionary attacks	[a]	[a]	Restricted
4.	Low-encryption-exponent attack	Possible	Possible	Not Possible
5.	Known and chosen ciphertext attack	Possible	Not Possible	Not Possible
6.	Known and chosen plaintext attack	Possible	Not Possible	Not Possible
7.	Known-key distinguishing attack	Not Possible	Not Possible	Not Possible
8.	Chosen-key distinguishing attack	Not Possible	Not Possible	Not Possible
9.	Sniffer attack	Possible	Not Possible	Not Possible
10.	Replay attack	[a]	[a]	Restricted
11.	Man in the middle attack	Not Possible	Not Possible	Not Possible
12.	Interleaving attacks	Not Possible	Not Possible	Not Possible
13.	Lowe's attack	Not Possible	Not Possible	Not Possible
14.	Cross-site scripting attacks	[a]	[a]	Restricted
15.	SQL injection attack	[a]	[a]	Restricted
16.	Side channel attacks	[a]	[a]	Restricted but not limited
17.	Rainbow table attack	[a]	[a]	Restricted
18.	Quantum attacks	Possible	Possible	Possible

[a] Not discussed in literature.

TABLE 12.3
Security Property Analysis of Basic Protocol

S. No	Security Functionality	Yang et al. Protocol	Yi et al. Protocol	Basic Protocol
1.	Known key security	Yes	Yes	Yes
2.	Forward secrecy	Yes	Yes	Yes
3.	Key control	Yes	Yes	Yes
4.	Key confirmation	Yes	Yes	Yes
5.	Speed of the protocol based on key size (As per NIST & OWASP recommendation)	Low [3072-bit]	Medium [2048– bit]	High [384 – bit]
6.	Zero-knowledge proof	Yes	Yes	Yes
7.	Implicit key authentication	Yes	Yes	Yes
8.	Explicit key authentication	Yes	Yes	Yes
9.	Anonymity	Not Possible	Not Possible	Not Possible
10.	Key freshness	Yes	Yes	Yes
11.	Reciprocity	Yes	Yes	Yes
12.	Impersonation resilience	Yes	No	No
13.	Disclosing the password by spoofing the servers	Possible	Possible	Not Possible
14.	Client side complexity	Low	High	Low
15.	Dependency on server S1	High [Computation high in S1]	Low [Computation equal in S1 and in S2]	Low [Computation of S2 close to S1]
16.	Security of password-based variant in the database	Very low [No encryption]	High	High
17.	Efficiency based upon computation and communication	High	Low	Medium
18.	Security based upon authentication parameters	Low	Low	High
19.	Quantum Safe	No	No	No

12.5.2.1 System Architecture

Detailed system architecture of the improved protocol is presented in Figure 12.3. Most of the network functions are allowed to run in the cloud environment. A distributed edge cloud is deployed close to the accessing devices. These edge servers are connected to the centralized cloud and the security interfaces are allowed to run in the edge cloud. The solutions that are proposed in this paper hold for communication between edge servers and centralized cloud environments, as shown in the Figure 12.3.

The communication from devices to edge cloud uses simple random number generators and encryptors for confidentiality. Between the edge servers and central cloud, QKD is used to generate the secret key that can be hashed and used for message authentication using HMAC. Beforehand, the communicating parties should be authenticated for generating the secret key using QKD approach. Hence, a flexible and robust 3D ECCDH PAKE protocol is applied.

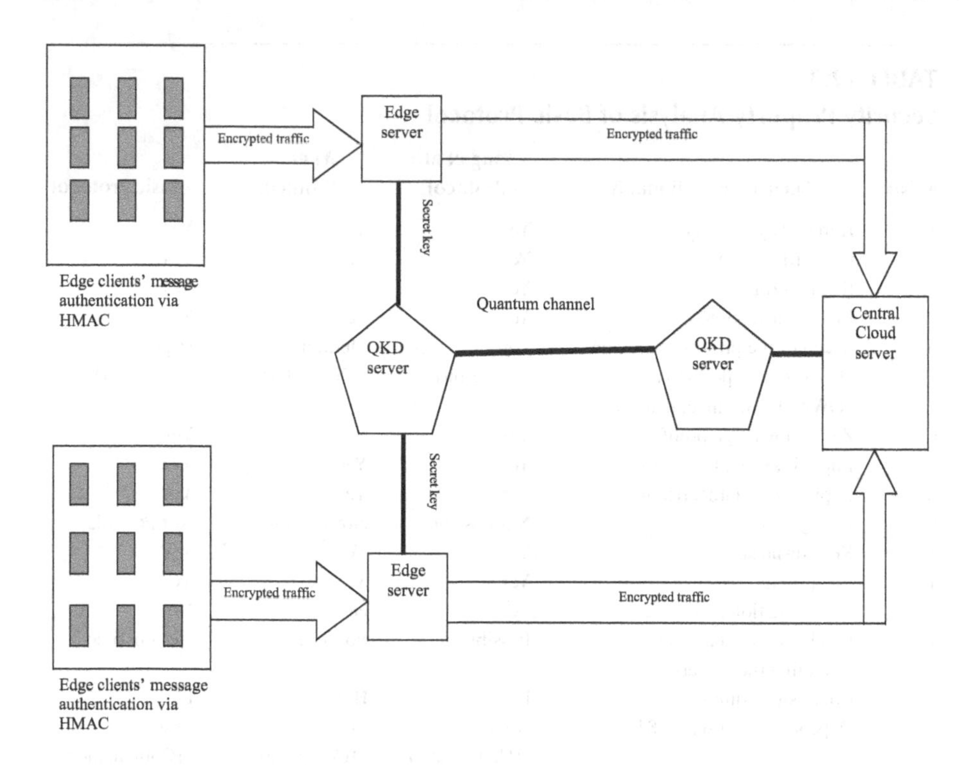

FIGURE 12.3 Overall system architecture of the improved protocol.

12.5.2.2 Quantum Key Generation and Distribution

QKD process is described below:

1. The keys K/K' generated at the end of authentication phase are used as random input (α) for generating the quantum key
2. The QKD transmitter (edge server) polarises α into quantum bits (qubits) by randomly choosing the bases for each bit, either rectilinear or diagonal basis. $\beta = [+ + \times + \times \times \times +]$ that results in polarised key (qubits), $p = [\,|\,-\backslash\,|\,\backslash\,/\,/\,-]$
3. The qubits p are transmitted through the quantum channel to the receiver
4. The receiver generates a random basis, β' $[+ \times \times \times + \times + +]$ to measure the incoming photons
5. On passing through β', p transforms to p'. Polarisation is retained if the basis selected in transmitter and receiver is the same, otherwise it is changed. Since β' is randomly chosen, the polarised photons of p' are estimated to be 50% same with p. Therefore $p' = [\,|\,/\,\backslash\,/\,-\,/\,-]$. Thus, $\alpha' = [0\,0\,1\,0\,1\,0\,1\,1]$
6. Match between β' and β is identified and the result r is obtained where $r = [\surd \times \surd \times \times \surd \times \surd]$. ($\surd$ = matching, $\times$ = not matching)
7. Only the bits with the matching bases are retained, the bits with different bases are discarded. A shifted key Q_K, is produced as $Q_K = [0\,\_\,1\,\_\,\_\,0\,\_\,1]$
8. To satisfy privacy amplification, $Q_K = [0\ 1\ 0\ 1]$ is applied with a hash function results in the generation of the hash-based shared secret key

The protocol ensures equal contribution from servers S1 and S2 for authenticating the client. Using ECC encryption and a quantum key ensures high speed and strong security. The fundamental laws of quantum physics that support the security properties of QKD include:

1. Heisenberg uncertainty principle that states that measuring an unknown quantum-mechanical state changes it physically. Hence, an eavesdropper can only physically change the qubits in the QKD channel;
2. The 'No Cloning Theorem' makes it impossible to replicate an unknown quantum state. Hence, an adversary cannot replicate qubits to hide eavesdropping; and
3. The quantum entanglement property imposes a threshold on the information leaked to an unauthorized user.

These laws ensure that an eavesdropper will not be able to obtain the secret key in real time or offline. A complete step-by-step process entailed in authentication and key exchange is shown below.

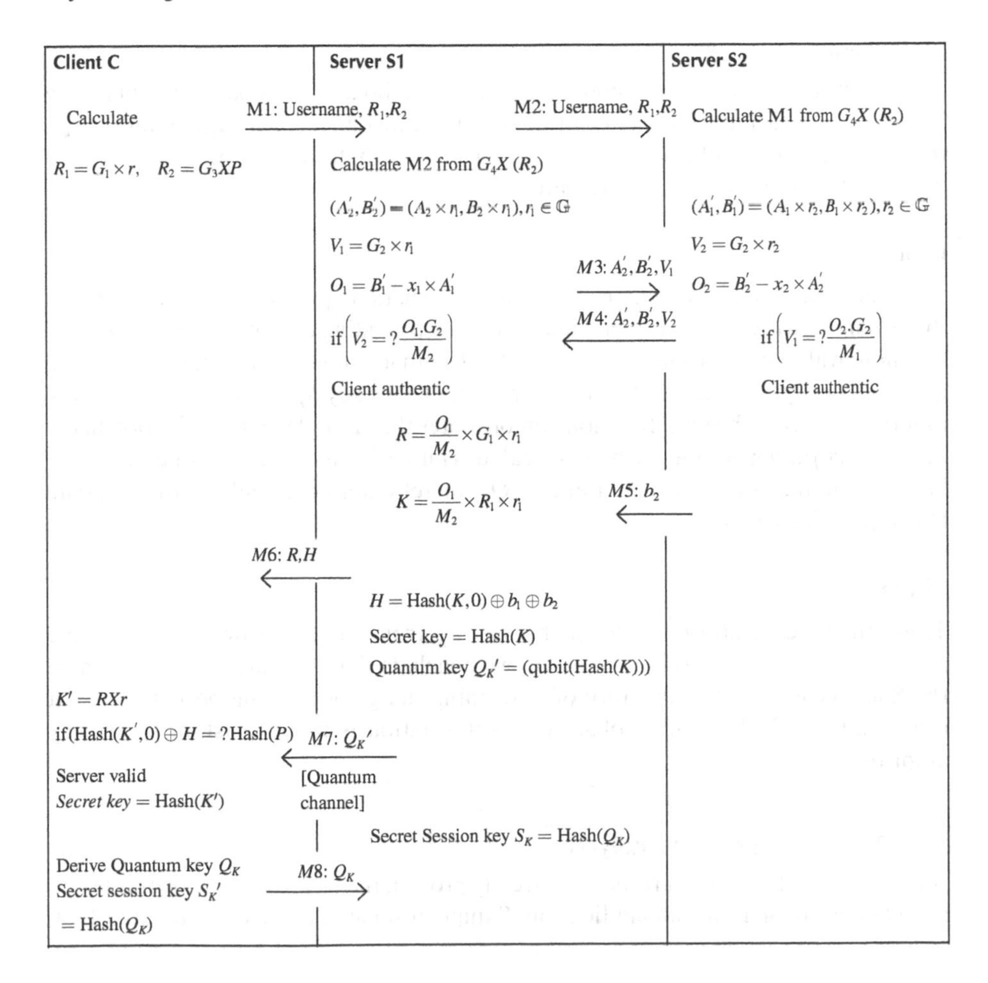

To restate, thus improved protocol is strenuous against quantum attacks. And both the protocols satisfy the perfect secrecy property stated by Shannon in 1949 [18] as $H(K) \geq H(P)$, where K is the random variable modelling the key and P is the plain text [18].

12.5.2.3 Security Analysis of Improved Protocol Against Interception Attacks via Privacy Amplification

THEOREM 1

The proposed improved protocol $\beta(t_1, \varepsilon_1)$ is secure against interception attacks by amplifying the privacy under the random oracle model with a collision-resistant hash function represented as 'Hash'. The Hash function is universal, for any distinct Q_K and Q_{K^1}, the probability that $\text{Hash}(Q_K) = \text{Hash}(Q_{K^1})$ and obtaining the secret key S_K from Q_K is at most 1/2. Here 't_1' denotes the running time and 'ε_1' denotes the probability that β succeeds.

Statement

Assume that $\gamma(t, \varepsilon)$ is an interception algorithm that breaks the protocol to obtain the secret key S_K with the knowledge of the k deterministic bits of the quantum key Q_K transmitted between client and server S1. Here, 't' denotes the running time and 'ε' denotes the probability that γ succeeds.

Proof

The statement is proved using proof by a contradiction approach. The fact remains that *Hash* is a universally acceptable function by providing complete randomness for the input with notable security. Based on the fact that although γ has the traces of the quantum key Q_K, it is infeasible to generate the secret key S_K, which is a completely random function. The hash function demonstrates that a single bit modification in the input results in generating a new hash value. Thus, γ is unable to generate the secret key S_K with the traces of quantum key Q_K, thereby not being able to differentiate $\text{Hash}(Q_K)$ with $\text{Hash}(Q_{K^1})$.

Effect

Thus, the case of attaining the secret key S_K from the quantum key Q_K by the adversary γ is unsuccessful. Therefore, we conclude that $\Pr_\beta[\text{Succ}] = \varepsilon_1 > \varepsilon$ where $\Pr_\beta[\text{Succ}]$ denotes the probability of β winning the game and the probability is at most ½. Conceivably, this probabilistic information is more resistant to privacy amplification.

12.5.2.4 Experimental Analysis

As the design of the ECC protocol is already proved, the improved protocol is tested for a healthcare-based web application. Sample test cases are shown in Table 12.4.

TABLE 12.4
Test Cases of Improved Protocol

S. No	Username	Password	Run Time (ns)	Secret Session Key (bits)
1.	Mary23	yaguacire95	5.15569569E8	b5645f97c20e6f71801334ef3b47276 f240a2bb8282fe961e72ca17b7f7380 458bb1f7830baffbb93f2aa6d778544 4741ecd92aaafe7fa196f1aa13ded93ef2
2.	David	re1ns+@ll	5.33543845E8	453bbc3ef019f059d858e5efdde60 474afa343d92061c4f6a9d6138cfb1 f2db080232e8139851cced55fb5850 d64c75692ddc15d72b6c3ce26a2eb6 ae4b6cc2b
3.	Dev	tenant-atwill	4.61211023E8	1d79dfba0e0720e866271f20686083 be8b8a4d31ce153a36f3cab18117ad 8637b30a2626de9a288b3db77040f6 ce78639688e1fa293eb5c9d4b1ec61 a51ab83f
4.	antony3	Rebecca	4.64196035E8	0c6c34a3496b65690400cb5c5589235 f1b10cbeadee17dc9991b463fed4bbb 8c63d96024bd32f8fce0d1e937a5d8f 8959c573ed44412eab7d7dc61fc71aa 2c85
5.	Joshua	un!ver$a1!+y	4.42021694E8	797cecb8ab3c9c61166704a45735205 060f612c4c2aad9e77ee5b11a7ab84b b169e7b461b85b9a47aaeeb04b8cb61 8bf76b5594

12.5.2.5 Communication and Computation Complexity Analysis of Improved Protocol

Communication and computation complexity of the proposed improved protocol is examined and tabulated, as shown in Table 12.5. Modular exponentiations, qubit computation, modular/scalar multiplications, hash functions and Bitwise Exclusive (XOR) operations, with respect to each entity, are considered as the metrics to evaluate computational complexity. Based on these metrics, the computational complexity of basic protocol is 28. The total runtime of the improved protocol is 13L + 21. Albeit there is a slight increase in communication and computation cost for the improved protocol, the basic protocol is found to be liable to quantum attacks. On the other hand, the improved protocol preserves and guarantees long-term secrecy. For better understanding, the computational and communicational complexities are shown in Figure 12.4.

12.5.2.6 Attack Resistance Capability of Improved Protocol

Table 12.6 outlines the invasive attacks occurring in security protocols. The improved protocol is highly resistant against side-channel and quantum attacks, whereas, the

TABLE 12.5
Communication and Computational Complexity Analysis of Improved Protocol

Participants	Yang et al. Protocol	Yi et al. Protocol	Basic Protocol	Improved Protocol
Client: Communication (bits)	2L + 2l	3L + 4l	3L + 1l	5L + 1l
Client: Communication (rounds)	4	4	2	4
Client: Computation	7	21	7	9
Server S1: Communication (bits)	6L + 3l	6L + 3l	11L + 2l	13L + 2l
Server S1: Communication (rounds)	8	5	6	8
Server S1: Computation	15	12	11	13
Server S2: Communication (bits)	4L + 1l	6L + 3l	8L + 1l	8L + 1l
Server S2: Communication (rounds)	4	5	4	4
Server S2: Computation	6	12	6	6
Total Running Time (Client–Server Side)	Comm: 9 (6L + 3l) Client – S1 – S2	Comm: 11 (7L + 4l) Client – S1 Client – S2	Comm: 13 (11L + 2l) Client – S1 – S2	Comm: 15 (13L + 2l) Client – S1 – S2
	Comp: 28 Client – S1 – S2	Comp: Worst case: 45 Best case: 33 Client – S1 Client – S2	Comp: 24 Client – S1 – S2	Comp: 28 Client – S1 – S2
	Rounds: 8 Client – S1 – S2	Rounds: 6 Client – S1 Client – S2	Rounds: 6 Client – S1 – S2	Rounds: 8 Client – S1 – S2

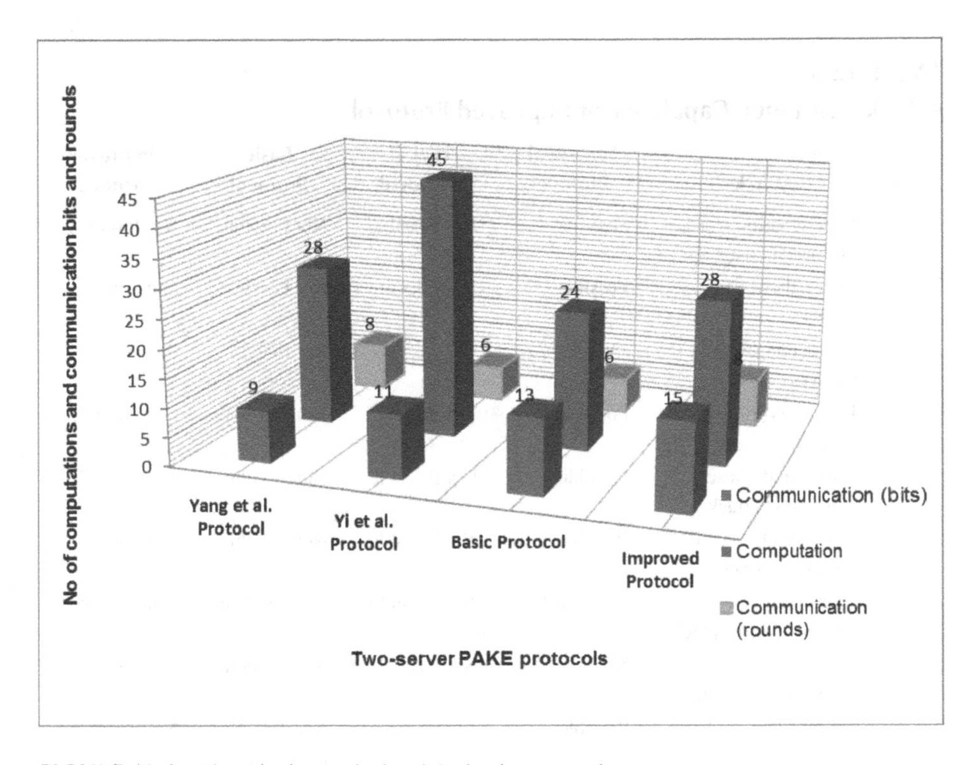

FIGURE 12.4 Complexity analysis of the basic protocol.

classical schemes and basic protocol are prone to these attacks. Thus, the attack intensity is reduced visibly in a quantum key-based improved protocol.

12.5.2.7 Security Property Analysis of Improved Protocol

The most prominent property to be satisfied by any security protocols is listed in Table 12.7. It is inferred from the table that the improved protocol is quantum-safe when compared to existing and basic protocols.

From Figure 12.5, it is concluded that the improved protocol is quantum safe by assuring that the filing of attacks is infeasible in the proposed protocol.

12.6 CONCLUDING REMARKS

This chapter presents a detailed comparison of classical cryptography with quantum cryptography on a two-server PAKE protocol for WoMT. It indeed assures the prominence of quantum cryptography by distributing the quantum keys to the client machines/devices to withstand against quantum attacks. The adversary model demonstrated in Section 12.4.2 assures the privacy of the quantum-based secret key. It is observed, that the improved protocol performs reasonably well against all possible contemporary attacks including quantum attacks, as discussed in Tables 12.6

TABLE 12.6
Attack Resistance Capability of Improved Protocol

S. No	Attacks	Yang et al. Protocol	Yi et al. Protocol	Basic Protocol	Improved Protocol
1.	Impersonation attack by inside adversary	Possible	Not Possible	Not Possible	Not Possible
2.	Offline dictionary attacks	Possible	Possible	Not Possible	Not Possible
3.	Online dictionary attacks	[a]	[a]	Restricted	Restricted
4.	Low-encryption-exponent attack	Possible	Possible	Not Possible	Not Possible
5.	Known and chosen ciphertext attack	Possible	Not Possible	Not Possible	Not Possible
6.	Known and chosen plaintext attack	Possible	Not Possible	Not Possible	Not Possible
7.	Known-key distinguishing attack	Not Possible	Not Possible	Not Possible	Not Possible
8.	Chosen-key distinguishing attack	Not Possible	Not Possible	Not Possible	Not Possible
9.	Sniffer attack	Possible	Not Possible	Not Possible	Not Possible
10.	Replay attack	[a]	[a]	Restricted	Restricted
11.	Man in the middle attack	Not Possible	Not Possible	Not Possible	Not Possible
12.	Interleaving attacks	Not Possible	Not Possible	Not Possible	Not Possible
13.	Lowe's attack	Not Possible	Not Possible	Not Possible	Not Possible
14.	Cross-site scripting attacks	[a]	[a]	Restricted	Restricted
15.	SQL injection attack	[a]	[a]	Restricted	Restricted
16.	Side channel attacks	[a]	[a]	Restricted but not limited	Restricted
17.	Rainbow table attack	[a]	[a]	Restricted	Restricted
18.	Quantum attacks	Possible	Possible	Possible	Not possible

[a] Not discussed in literature.

and 12.7 of Section 12.4.2. Though building QKD network and deciding the power requirement is considered as the limitation of quantum cryptography, it is concluded that the end-to-end security is guaranteed in the WoMT. Future work will shed light on producing a cost-effective solution without affecting the performance and security rendered.

TABLE 12.7
Security Property Analysis of Improved Protocol

S. No	Security Functionality	Yang et al. Protocol	Yi et al. Protocol	Basic Protocol	Improved Protocol
1.	Known key security	Yes	Yes	Yes	Yes
2.	Forward secrecy	Yes	Yes	Yes	Yes
3.	Key control	Yes	Yes	Yes	Yes
4.	Key confirmation	Yes	Yes	Yes	Yes
5.	Speed of the protocol– based on key size (As per NIST & OWASP recommendation)	Low [3072-bit]	Medium [2048– bit]	High [384 – bit]	High [384 – bit]
6.	Zero-knowledge proof	Yes	Yes	Yes	Yes
7.	Implicit key authentication	Ycs	Yes	Yes	Yes
8.	Explicit key authentication	Yes	Yes	Yes	Yes
9.	Anonymity	Not Possible	Not Possible	Not Possible	Not Possible
10.	Key freshness	Yes	Yes	Yes	Yes
11.	Reciprocity	Yes	Yes	Yes	Ycs
12.	Impcrsonation resilience	Yes	No	No	No
13.	Disclosing the password by spoofing the servers	Possible	Possible	Not Possible	Not Possible
14.	Client side complexity	Low	High	Low	Low
15.	Dependency on server S1	High [Computation high in S1]	Low [Computation equal in S1 and in S2]	Low [Computation of S2 close to S1]	Low [Computation of S2 close to S1]
16.	Security of password based variant in the database	Very low [No encryption]	High	High	High
17.	Efficiency based upon computation and communication	High	Low	Medium	Medium
18.	Security based upon authentication parameters	Low	Low	High	High
19.	Quantum Safe	No	No	No	Yes

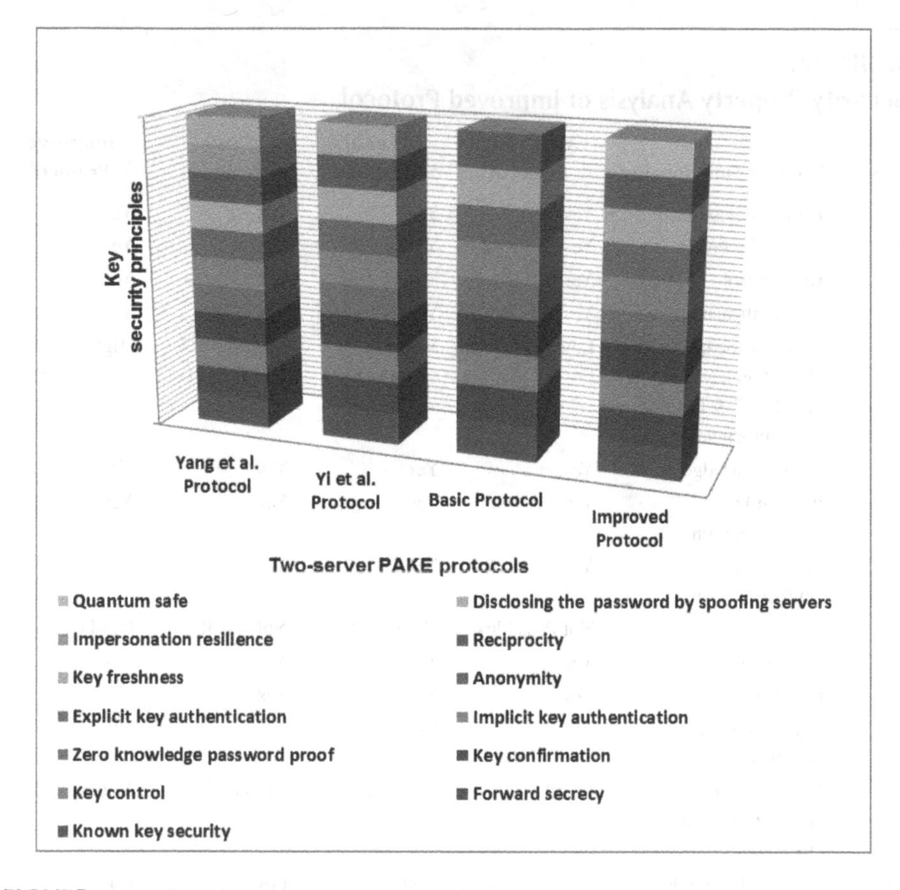

FIGURE 12.5 Security property analysis of the improved protocol.

REFERENCES

1. Lucy L. Thomson, Healthcare Data Breaches and Information Security Addressing Threats and Risks to Patient Data, in *Data Breach and Encryption Handbook*, ABA, Chicago, pp. 57–85, 2012.
2. Anitha Kumari K, Sudha Sadasivam G, Rohini L, An Efficient 3D Elliptic Curve Diffie-Hellman (ECDH) based Two-Server Password-only Authenticated Key Exchange Protocol with Provable Security, *IETE Journal of Research, T & F*, vol. 62, no. 6, pp. 762–773, 2016.
3. 'Quantum safe crypto', Available at: http://www.idquantique.com/quantum-safe-crypto/.
4. Bauer H, Patel M, Veira J, The Internet of Things: sizing up the opportunity, 2016, Available from: http://www.mckinsey.com/industries/high-tech/our-insights/the-internet-of-things-sizing-up-the-opportunity.
5. Securing the Internet of Things The Conversation Every CIO Needs to Have with the CEO, 2016, Available from: https://www.eiuperspectives.economist.com/sites/default/files/images/EIU_HPE%20SecuringIoTArticle_PDF.pdf.
6. Koblitz N, *Algebraic Aspects of Cryptography*, Springer, New York, 1998.
7. Delfs H, Knebl H, *Introduction to Cryptography*, Springer, New York, 2007.

8. Anitha Kumari K, Sudha Sadasivam G, Akash SA, A Secure Android Application with Integration of Wearables for Healthcare Monitoring System using 3D ECCDH PAKE Protocol, *Journal Of Medical Imaging And Health Informatics (JMIHI)*, vol. 6, no. 6, pp. 1548–1551, 2016.
9. Chakraborty S, Chatterjee S, Ashour AS, Mali K, Dey N, Intelligent Computing in Medical Imaging: A Study, in *Advancements in Applied Metaheuristic Computing*, IGI Global, USA, 2017.
10. Van Assche G, *Quantum Cryptography and Secret-key Distillation*, Cambridge University, UK, 2006.
11. Bouwmeester D, Ekert A, Zeilinger A, *The Physics of Quantum Information*, Springer, New York, 2000.
12. Bellovin SM, Merritt M, Encrypted key Exchange: Password-Based Protocols Secure Against Dictionary Attacks, in *Proceedings of the IEEE Symposium on Research in Security and Privacy*, pp. 72–84, 1992.
13. Ahmed MH, Alam SW, Qureshi N, Baig I, Security for WSN Based on Elliptic Curve Cryptography, in *Proceedings of the International Conference on Computer Networks and Information Technology*, pp. 75–79, 2011.
14. Koblitz N, Menezes A, Vanstone S, The State of Elliptic Curve Cryptography, *Designs Codes And Cryptography*, vol. 193, no. 2, pp. 173–193, 2000.
15. Pateriya RK, Vasudevan S, Elliptic Curve Cryptography in Constrained Environments: A Review, in *Proceedings of the International Conference on Communication Systems and Network Technologies*, pp. 120–124, 2011.
16. Menezes AJ, Vanstone SA, Elliptic Curve Cryptosystems and Their Implementations, *Journal Of Cryptology*, vol. 6, no. 4, pp. 209–224, 1993.
17. Vigano L, Automated Security Protocol Analysis with the AVISPA Tool, *Electronic Notes In Theoretical Computer Science*, vol. 155, pp. 61–86, 2006.
18. Shannon CE, Communication Theory of Secrecy Systems,, *Bell System Technical Journal*, vol. 28, pp. 656–715, 1949.
19. A. Al-Taee M, Sungoor AH, Abood SN, Philip NY, Web-of-Things Inspired e-Health Platform for Integrated Diabetes Care Management, in *IEEE Applied Electrical Engineering and Computing Technologies (AEECT)*, Amman, Jordan, pp. 1–6, 2013.
20. Adriano W, An IdM and Key-based Authentication Method for Providing Single Sign-on in IoT, in *IEEE Global Communications Conference (GLOBECOM)*, San Diego, USA, pp. 1–6, 2015.
21. Chen Y, Xia B, Wu B, Shi L, Design of Web Service Single Sign-on Based on Ticket and Assertion, in *IEEE International Conference on Artificial Intelligence, Management Science and Electronic Commerce (AIMSEC)*, Dengleng, China, pp. 297–300, 2011.
22. Web of Things Security – http://github.com/w3c/web-of-things-framework.
23. Szydlo M, Kaliski B, Proofs for Two-Server Password Authentication, *Lecture Notes In Computer Science*, vol. 3376, pp. 227–244, 2005.
24. Yang Y, Deng RH, Bao F, A Practical Password-Based Two-Server Authentication and Key Exchange System, *IEEE Transactions on Dependable And Secure Computing*, vol. 3, no. 2, pp. 105–114, 2006.
25. Bhatt C, Dey N, Ashour A, *Internet of Things and Big Data Technologies in Next Generation Healthcare*, Springer, 2016.
26. Dey N, Hassanien AE, Bhatt C, Ashour A, Satapathy SC, *Internet of Things and Big Data Analytics towards Next Generation Intelligence*, Springer, 2018.
27. Dey N, Ashour AS, Computing in Medical Image Analysis, in *Soft Computing based Medical Image Analysis*, pp. 3–11, Elsevier, USA, 2018.

Index

C

D

G

H

J

K

L

M

S

T

U

V

W

X

Y

Z